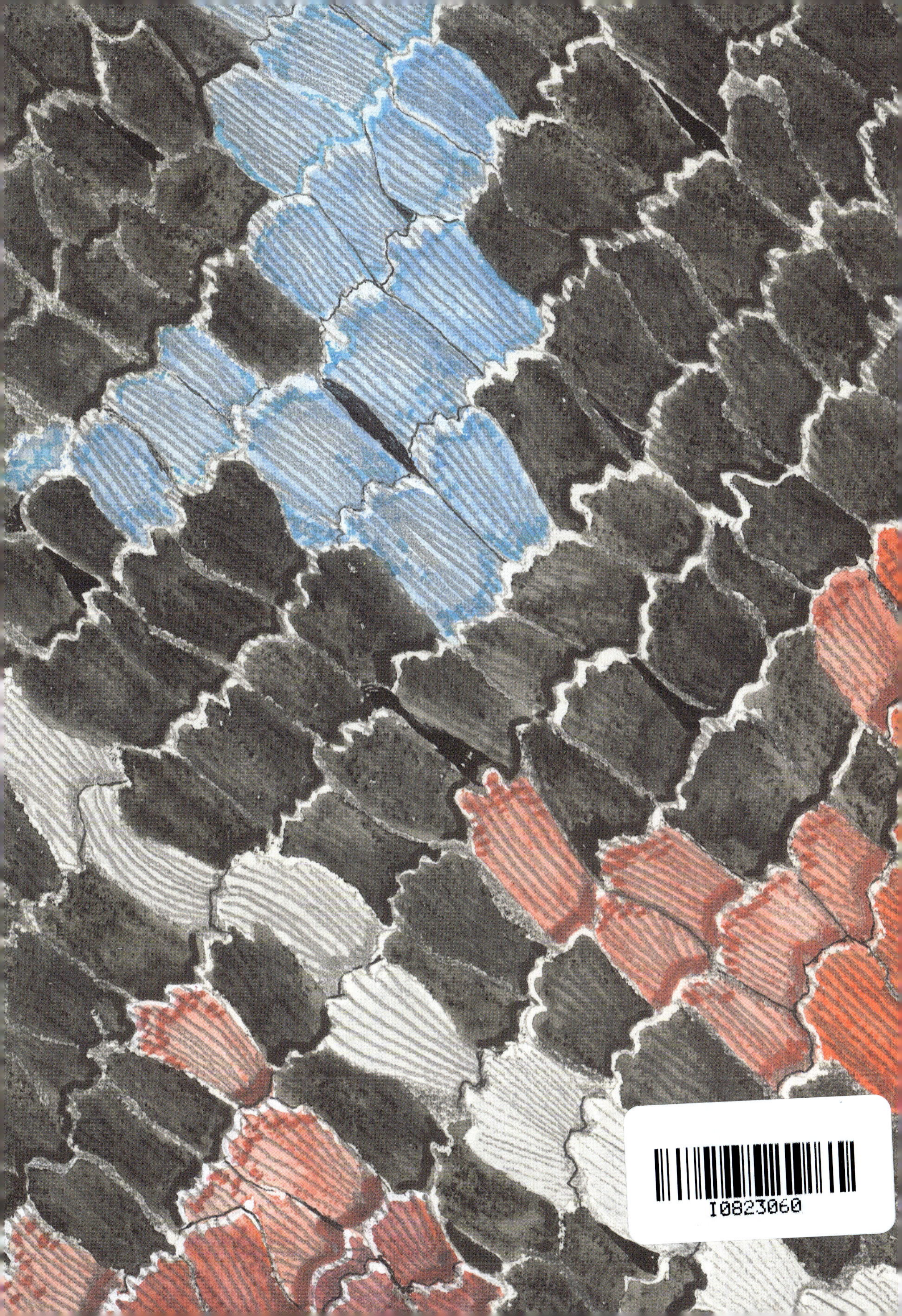
I0823060

BUTTERFLIES OF THE BAY AREA

and (Slightly) Beyond

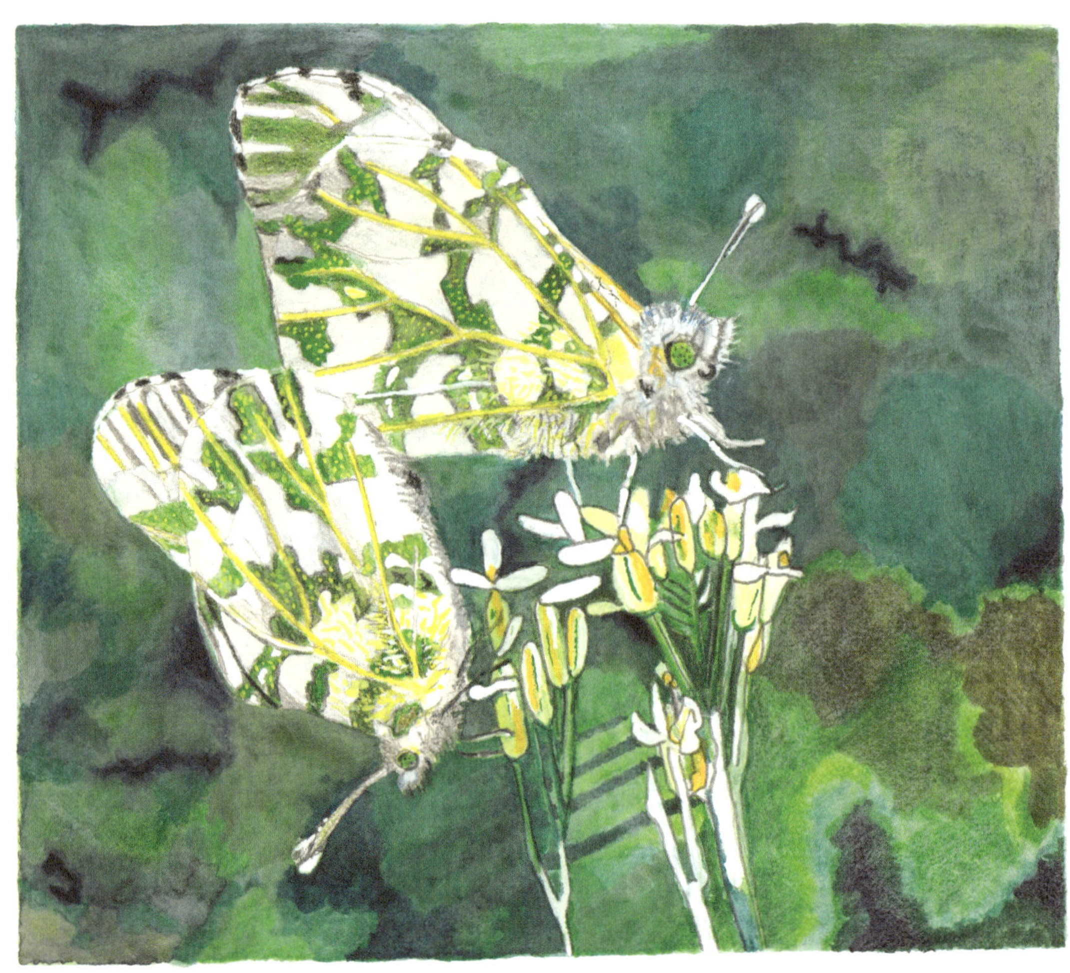

Large Marbles on tower mustard

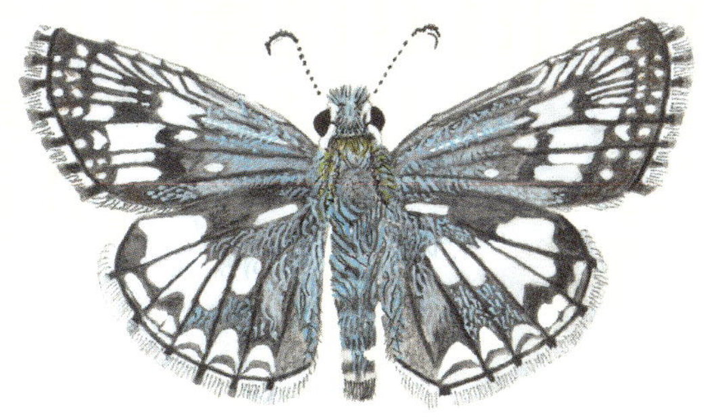

BUTTERFLIES OF THE BAY AREA

and (Slightly) Beyond

AN ILLUSTRATED GUIDE

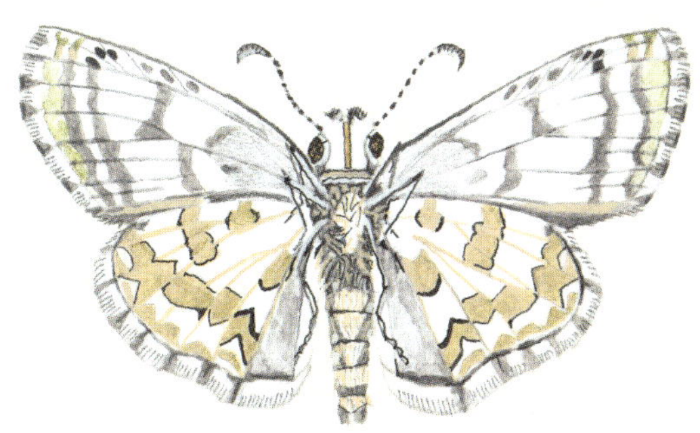

Written and Illustrated by

LIAM O'BRIEN

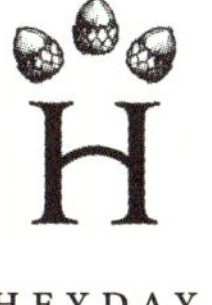

HEYDAY
BERKELEY, CALIFORNIA

Library of Congress Cataloging-in-Publication Data is available.

Cover Art: Liam O'Brien
Cover Design: Archie Ferguson
Interior Design/Typesetting: Victor Mingovits
Endpapers: Liam O'Brien, painting of butterfly scales

Published by Heyday
P.O. Box 9145, Berkeley, California 94709
(510) 549-3564
heydaybooks.com

Printed in East Peoria, Illinois, by Versa Press, Inc.

10 9 8 7 6 5 4 3 2 1

to Dr. Jerry Powell,
who with a small piece of paper
changed the trajectory of my life

Anything you do
Let it come from you
Then it will be new
Give us more to see

STEPHEN SONDHEIM
"Move On," *Sunday in the*
Park with George

Contents

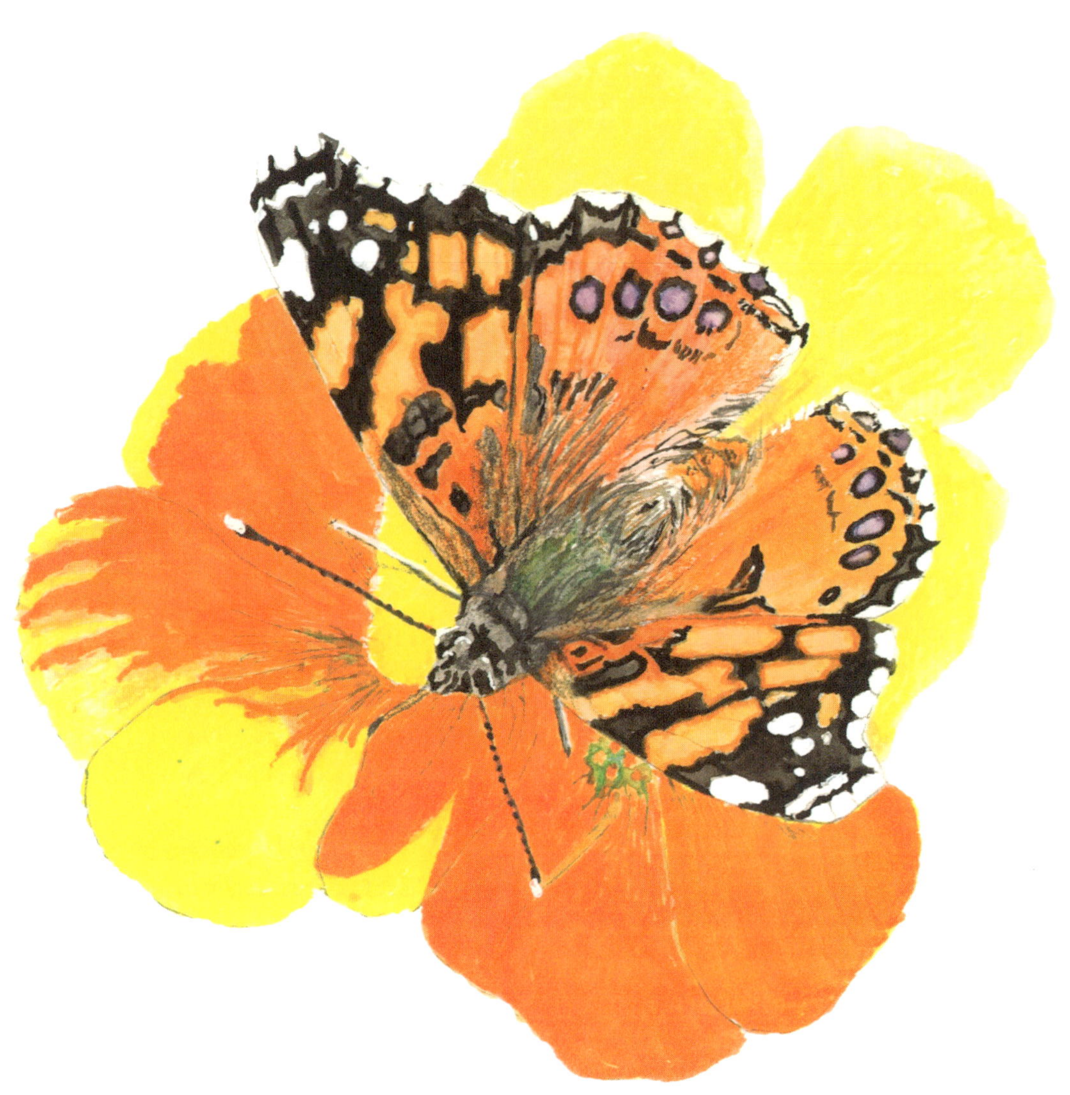

West Coast Painted Lady

Introduction

Twenty-five years ago I wrote in a journal:

> Everything changed the other day. A doctor came back into the exam room and stared at me. "I have to do something that I've never had to do with a patient. Your test came back positive. You are HIV positive."

I'm so codependent I immediately started to comfort him for having had to deliver the news. Then, I drove to my parents' home. To this day I cannot shake the look of terror in my mother's eyes when I told her. "We'll get through this, Liam," she said.

I drove back to my apartment in Benicia, and I remember picking up the *Audubon Society Field Guide to North American Butterflies* by Robert Michael Pyle, perhaps as a way to handle this pyroclastic flow I was now traveling on. Why this life preserver? I can't really explain why I picked it up—butterflies had come into my life as early as 1996 but nothing with any hyper focus. I think something about the guide's order, the taxonomy, and their beauty somehow seemed . . . larger and more important than the horror movie I'd just been handed. I also wrote, "I'm not sure where all of this is going. As the virus progresses, I hope it . . . can't get my spirit. I hope."

* * *

It was just fate that I acquired the condition right when it was changing from a death sentence to a chronic illness. I am not starting my butterfly book with this story to garner your pity. Please do not feel sorry for me; I'm doing great since starting the cocktail of medication soon after the diagnosis. A friend of mine said at the time, "Liam, the disease is set up for the poor and the wealthy, and unless you are planning to win the lottery tomorrow, I'd advise you to go make yourself poor." A great deal has changed in the management of HIV since he said that. It's still a rather complicated bureaucracy to negotiate. I bring all of this up because what I am about to say next might be

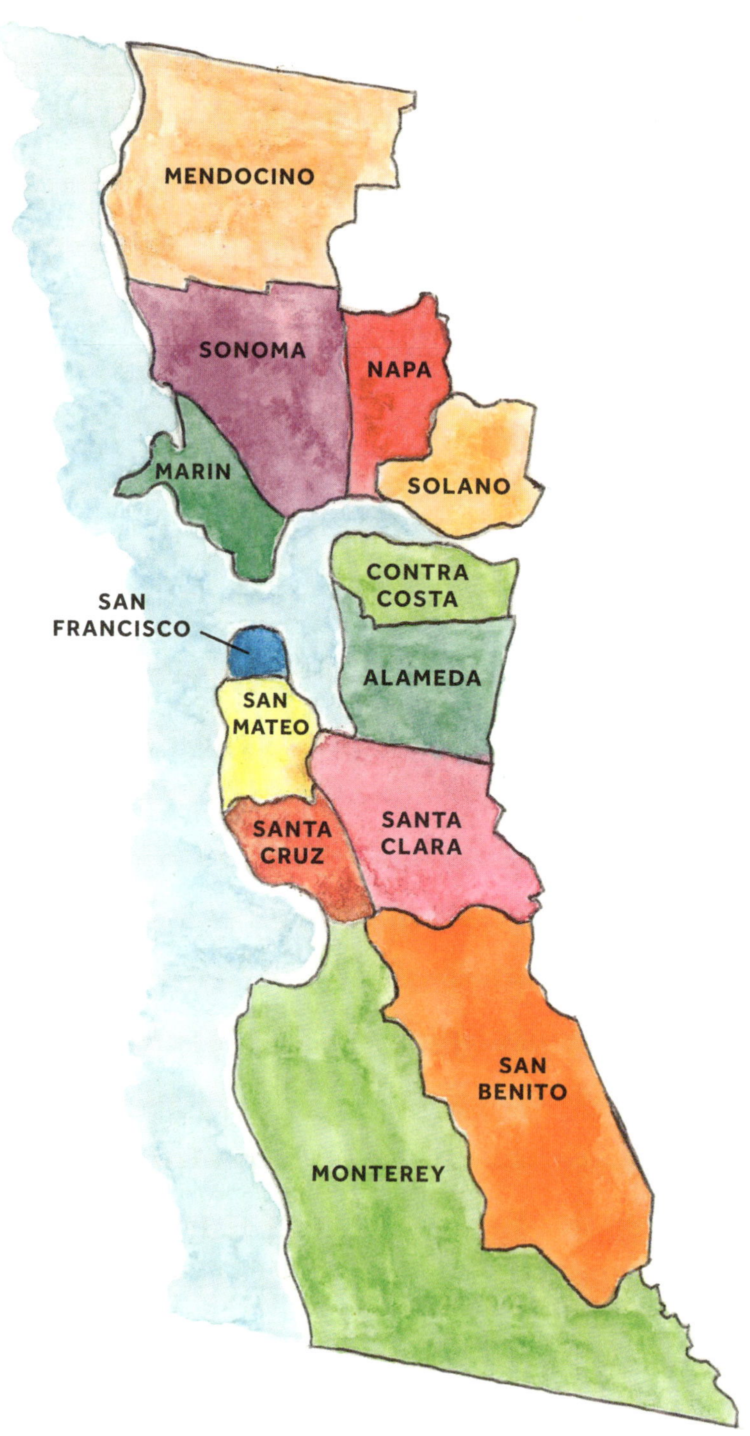

The geographic scope of this book

difficult to understand: the virus gave me butterflies. And I don't regret a second of hosting it in my body.

I used to be a professional stage actor. I studied, garnered skills, and entered the Actors' Equity union as a twenty-two-year-old. My secret weapon: a double barrel shotgun singing voice. Auditions came by the hundreds, but I learned early that people want to work with confident people. I also had luck. Lots of luck and a career in repertory theater that really took off. I got a job on Broadway in *Les Misérables* eight days after landing on the island of Manhattan, a dream many actors never fulfill. I stayed with the show for three years. Something, however, was always missing—drive. Try living in New York City without drive. Not possible. By then theater was something I just did, with almost no fulfillment. Performing in beautiful, cavernous places to audiences packed to the rafters, then home alone on a bus.

And then a butterfly flew into the yard . . .

* * *

I'd moved back to San Francisco and now was playing the role of Prior Walter in *Angels in America* at the American Conservatory Theater. Living off the Duboce Triangle section of the city, I was fortunate enough to have a beautiful backyard, and one day I noticed a butterfly wafting down among the flowers. I hurried down to the garden to watch it. While gorging itself on pink cosmos flowers, this large, lemon-yellow insect let me approach quite closely. I ran back upstairs and got some paper and started sketching, something I hadn't done since my college days. I can only now look at that moment and see that my entire life changed thanks to a visit from one Western Tiger Swallowtail (*Papilio rutulus*). I went down a rabbit hole that day in the yard, and I think I'm still down there. I joined the Lepidopterists' Society soon thereafter. (*Lepidoptera* is a fancy word that means "scaled wings" in Latin, and once you get the pronunciation down—Lep-pi-DOPT-er-rist—you'll soon be impressing all your friends.) I started keeping a graphic journal, painting the places where the butterflies and moths were, then inserting them in the painted scenes, sort of a field guide in reverse.

I heard retired professor Jerry Powell, a titan in the moth world and the man this book is dedicated to, say once in a lecture: "Learn where you live," and I thought at the time, *God, San Francisco, what a crap hole for butterflies.* Interestingly enough, no one had inventoried the city's species in decades, so, having quit the acting business, I decided to do that.

While combing the city, I started to consider an idea for something that we just might do to help a small green butterfly still hanging on in the Upper Sunset. I had absolutely no background in conservation, but it soon became a river I found myself a guide on, becoming the go-to person for all things butterflies in the city of San Francisco. One trains as an actor to have a dispassionate objectivity in observing and re-creating human behavior. It was an easy enough shift to apply that to the insect world. Not only do I find butterflies enthralling (who doesn't, right? Except perhaps those few who suffer from lepidopterophobia, the fear of butterflies and moths), but I'm equally interested in our relationship to these creatures as well.

In this book I'm aiming for something a little different from the classic field guide.

Here is a book that celebrates realistic paintings, pithy anecdotes, conservation, and the downright joy these bugs have given me. Give all of us. A book for the dead of winter that gives you the ability to dream up spring adventures where you might see some of the more obscure butterflies. But you can also stay home with this book, no matter the weather. If you see a butterfly in your yard or out your window, you'll probably find it here.

Born in Redwood City, I grew up in one of the first tract developments in south San Jose, called Almaden, in the Santa Clara Valley. It was nothing but plum orchards in any direction you looked back then. When not riding our bikes or building forts, my brothers and I were exploring Ross Creek, a tributary of the Guadalupe River. I remember how cool the water of Ross Creek felt on my bare feet, turning my toes instantly pink. The creek ran with clear water and was full of frogs and tadpoles for kids to catch.

Since I moved back to the Bay Area as an adult, I've spent more than 30 years living in San Francisco, and the city has been a strategic center to radiate out from as I've explored our great variety of ecosystems, and the butterflies that live in each. We have oak woodland, riparian, even xeric communities represented in the Bay Area. Heck, I'm even going to get you excited about what you can find in vacant lots here. To the core counties of the Bay Area, I've added three more: Mendocino, shrouded in fog and home to some of our more obscure skippers; San Benito, long dismissed as not having any major natural interest (hello, the Pinnacles?); and Monterey with its tough-to-reach butterfly jewels. I want you all to get to all of these places and hopefully see the butterflies I've seen.

It's hard to discuss butterflies with other humans sometimes, people who don't want to go any deeper than a Hallmark card. Butterflies are so weighed down in myth and misinformation that I'm somewhat stunned they can fly at all. In these pages I crowbar back a lot of emotion we've projected onto them and show them from new angles. I like the metaphor that I'll be restoring these frescoes, painting them as true as I can possibly get. We have a great deal of buildup to strip away. Butterflies have been symbols for so long, but they are in truth living, wild, opportunistic insects, not pretty objects, nor as Jeffrey Glassberg, president of the North American Butterfly Association, puts it, "party favors for the human circus" (Glassberg.1999.1).

* * *

Now for the hypocritical part. I LOVE the beauty of them too. Hopefully you'll come to appreciate the actual way they look and not some fantastical representations our society is saturated with. Though I'm not so hard-hearted to pull a crayon out of a child's hand, I believe the realism I'm going for here makes objectifying them more difficult in the long run. By the time this book is published, I will have traversed every intimate dot, stripe, and false eye on over 130 species. Some may be wondering, *Why painting, since photography has come so far?*

The basic answer is that I am a romantic. I want to pay homage to all the great butterfly illustrators of the past: W. H. Howe, W. J. Holland, Titian Peale, and even John James Audubon (who painted them in reverse on glass plates—no small feat). I wanted to travel down every scale on every wing they traveled like a slow, aerial drone feasting on color and form, to bring forward some of the more subtle prejudices. For instance, why is the female of the species relegated in these kinds of books to being below and to the right in a smaller picture than the male? The female butterfly has her day in my book.

A painting will never obtain perfection, but it's important to set a high bar if your desire is to help people. Peter Brastow, founder of Nature in the City in San Francisco said, "Learning the name of the thing in front of you is the first moment in conservation" (Brastow.2007.pers.comm). I took my obligations to the reader seriously, and as they say, the portrait of an artist is revealed in their mistakes anyway.

* * *

The number one thing folks with HIV are supposed to stay away from is stress. It eats the immune systems T cells (the good guys) like Ms. Pac-Man eats those dots. So, what could be less stressful, I ask you, then a single human being watching a butterfly? The pause, the gaze, the concentration, the delight. How ironic is all of that?

The virus has connected me to nature in a profound and surreal way. I house a deadly, parasite-like thing within me just as many butterfly caterpillars do. (Butterflies co-evolved with parasitic wasps and flies. That's not a bad thing, folks. It just is.) The difference is that infected butterflies don't make it to the flying, pretty phase we all love, and I got a bonus twenty-five years. I'm thinking now of all the people that didn't make it, people I shared dressing rooms with, who just missed the lifesaving cocktail, and them being robbed of their new chapters. I have a lot of survivor's remorse about that. Why me? Can you imagine all the butterflies they would have seen?

The world of butterflies was handed to me by the fate of the disease. What a blessing. What a gift. The virus has been a good tenant through the years—asymptomatic, low key, responds to a regime. Pays its rent on time. No drama.

And no, I don't think it's touched my spirit.

LIAM O'BRIEN
October 2024

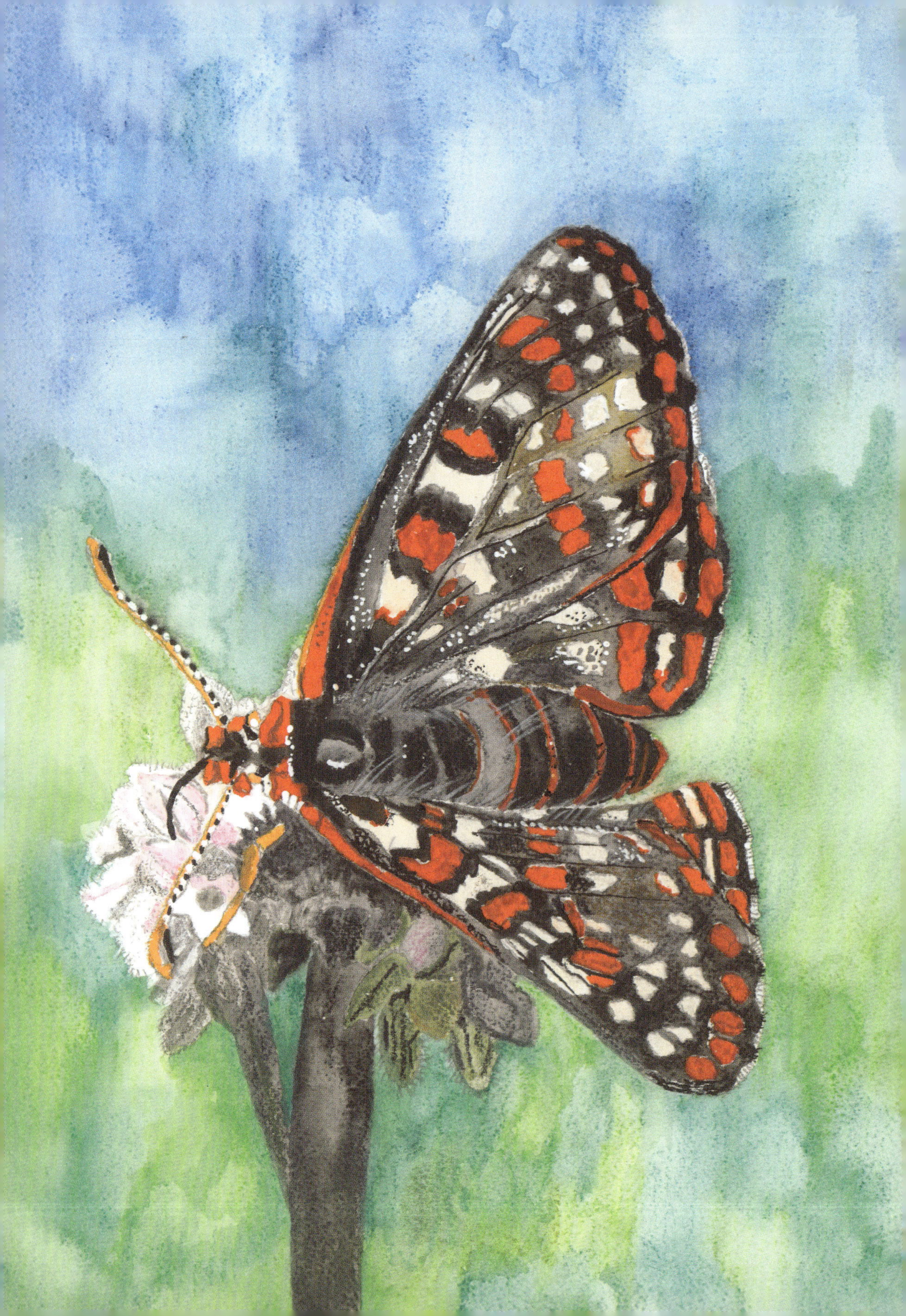

The federally threatened Bay Checkerspot

What Is a Butterfly?

Well, let's start with science. We are in the Kingdom Animalia, the Subkingdom Invertebrata, the Phylum Arthropoda, the Class Insecta, and the Order Lepidoptera. Lepidoptera covers both butterflies and moths and literally translates to "scaled wings."

If you were into beetles, you'd be a coleopterist, which means "hard wings." Skippers used to be in the Subfamily Hesperioidea when they were considered distinct enough to earn this distinct classification, but the category has been removed, and they are lumped back now into the lone Superfamily Papilionoidea, the butterflies. Moths are a whole different line in Lepidoptera.

There are six families of butterflies:

- **Hesperiidae**—the skippers
- **Papilionidae**—the swallowtails and parnassians (We used to have two parnassians in the Greater Bay Area, one in Santa Cruz that is now extinct and one in Marin that is now extirpated from the county.)
- **Pieridae**—the whites, the marbles, the orangetips and sulphurs
- **Nymphalidae**—also known as the brushfoots (the group with the most representatives)
- **Lycaenidae**—the blues, coppers, and hairstreaks
- **Riodinidae**—the metalmarks

What is a butterfly? It's a flying insect that, for the most part, stops most people while they are walking or seated thinking for just a few seconds. Its size (from smaller than a dime to something around the size of a saucer plate), its shimmering beauty (a fantastic array of words used to describe their color: *scintilla*, *fulvous*, and *aurora* to name a few), and its ability to fly have all seduced us from the beginning. Hell, I even still get a dopamine rush to this day seeing a Cabbage White. Their capricious flight and wandering we find enthralling.

This beauty combined with the butterfly's ability to transform itself through metamorphosis has given rise to a poetic array of names in a great deal of different languages. It's called "God's fire" in Scottish Gaelic (*dealan-dè*) (Dupré.1974.6). In my research of the meaning of the word *butterfly* I found there were enough different stories of its origins to boggle the mind. "Flies to butter" and other folk etymologies make it difficult to name any as authentic. I welcome you to enter that morass on your own.

For me a butterfly is a thing of memory—one of my earliest, somewhere between walking and kindergarten—lying on the cool, green lawn in our backyard in San Jose watching Fiery Skippers (*Hylephila phyleus*) zip about. I was little, so my eyes were parallel with their world. I was then rather indifferent to them for the next three decades. Vladimir Nabokov, the famous Russian author and lepidopterist, said in *Speak, Memory*, "It is astounding how little the ordinary person notices butterflies" (Nabokov.1989). Though many appreciation societies have risen since he said that, I find that his words for the most part are still true to this day.

The Parts of a Butterfly

One of the most intimidating areas of learning these things are all the big, new words you are going to encounter with butterflies. But, let me tell you, they all help tremendously once you get them down. (Though despite years of practice, trust me, I'm butchering the names of body parts in the field to this day.)

Getting specific with terminology helps especially when going over complicated patterns like the Duskywings (*Erynnis* species) discussed here. We all know the basic insect head, thorax, and abdomen we learned in elementary school, but there is a plethora of words, particularly regarding the wings of butterflies, that create a handy shortcut. Lots of things happen in the area of the discal cell of the forewing and hind wing, so pay attention to its important location.

It may take a few seasons for these to settle in, but once they do, they'll come into play when identifying butterflies on the wing. I first read it in Kenn Kaufman's *Field Guide to Butterflies of North America* many years ago, and I thought, *No way*. But if you really throw yourself into this, you will come out with the ability to not only identify a butterfly on the wing, but sex it as well. You won't even need them to land. It's not an idle promise. Most advanced lepidopterists I know can do it, and you can learn to do it too. But it does take practice.

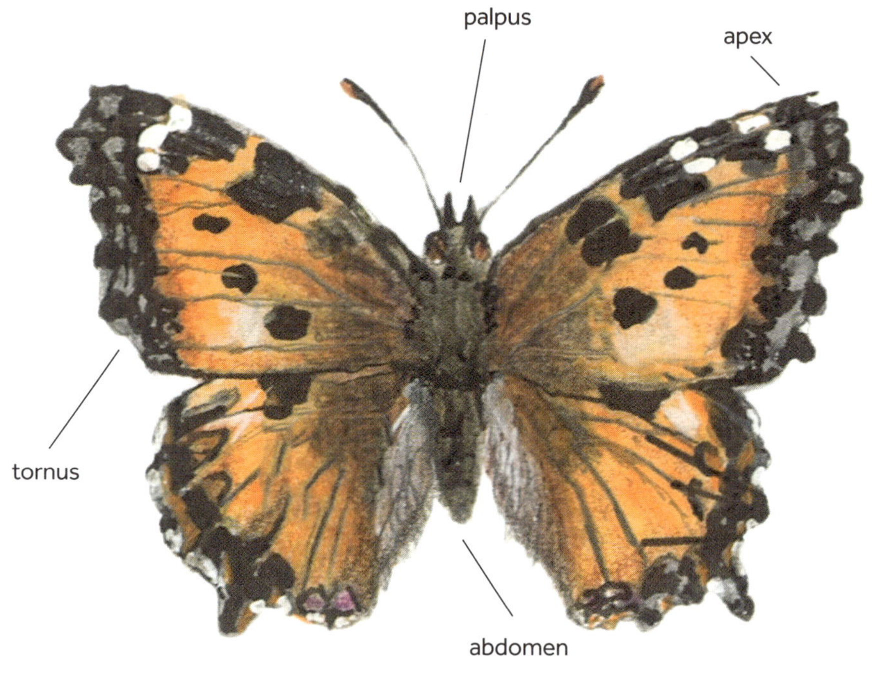
palpus
apex
tornus
abdomen

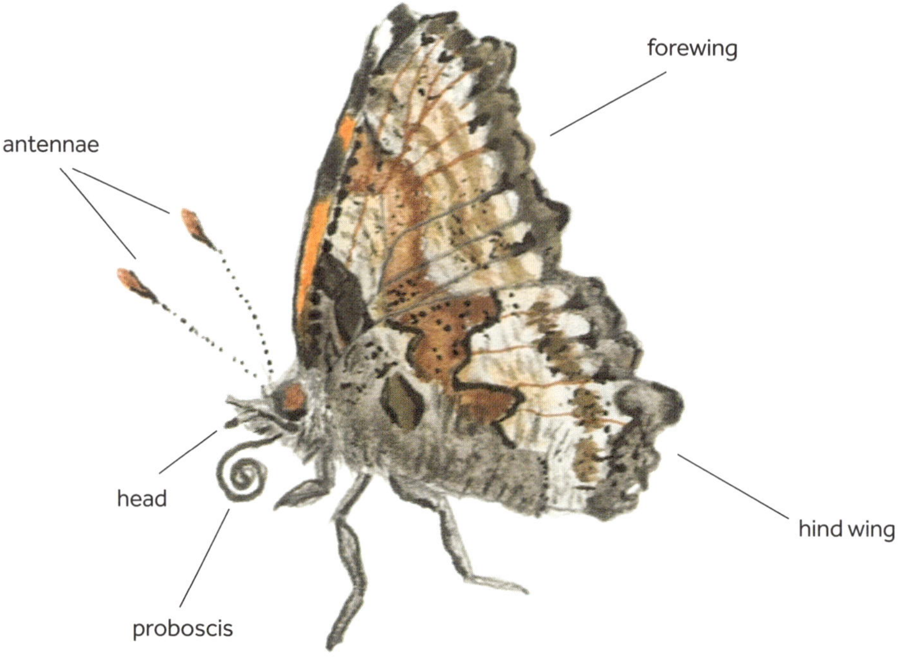
forewing
antennae
head
proboscis
hind wing

PARTS OF A SKIPPER

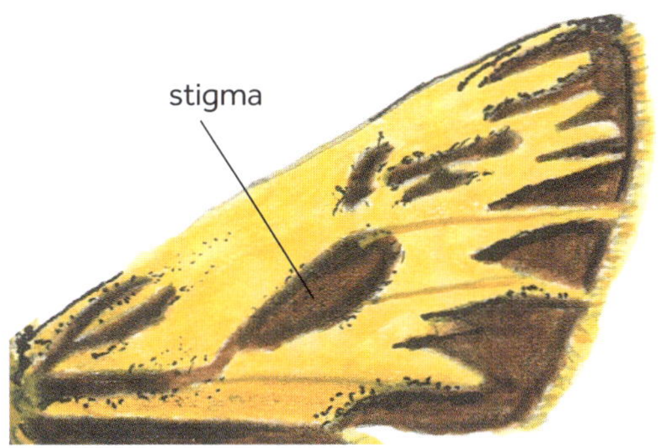

Fiery Skipper

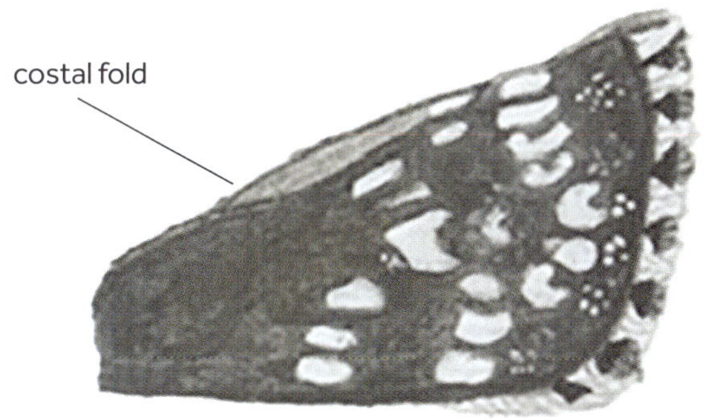

Two-Banded Checkered Skipper

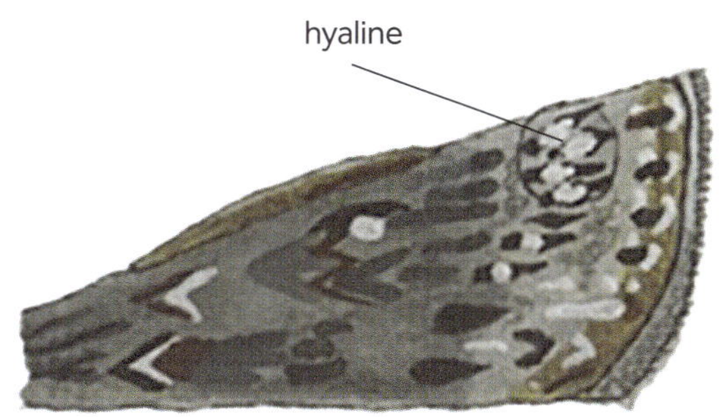

Propertius Duskywing

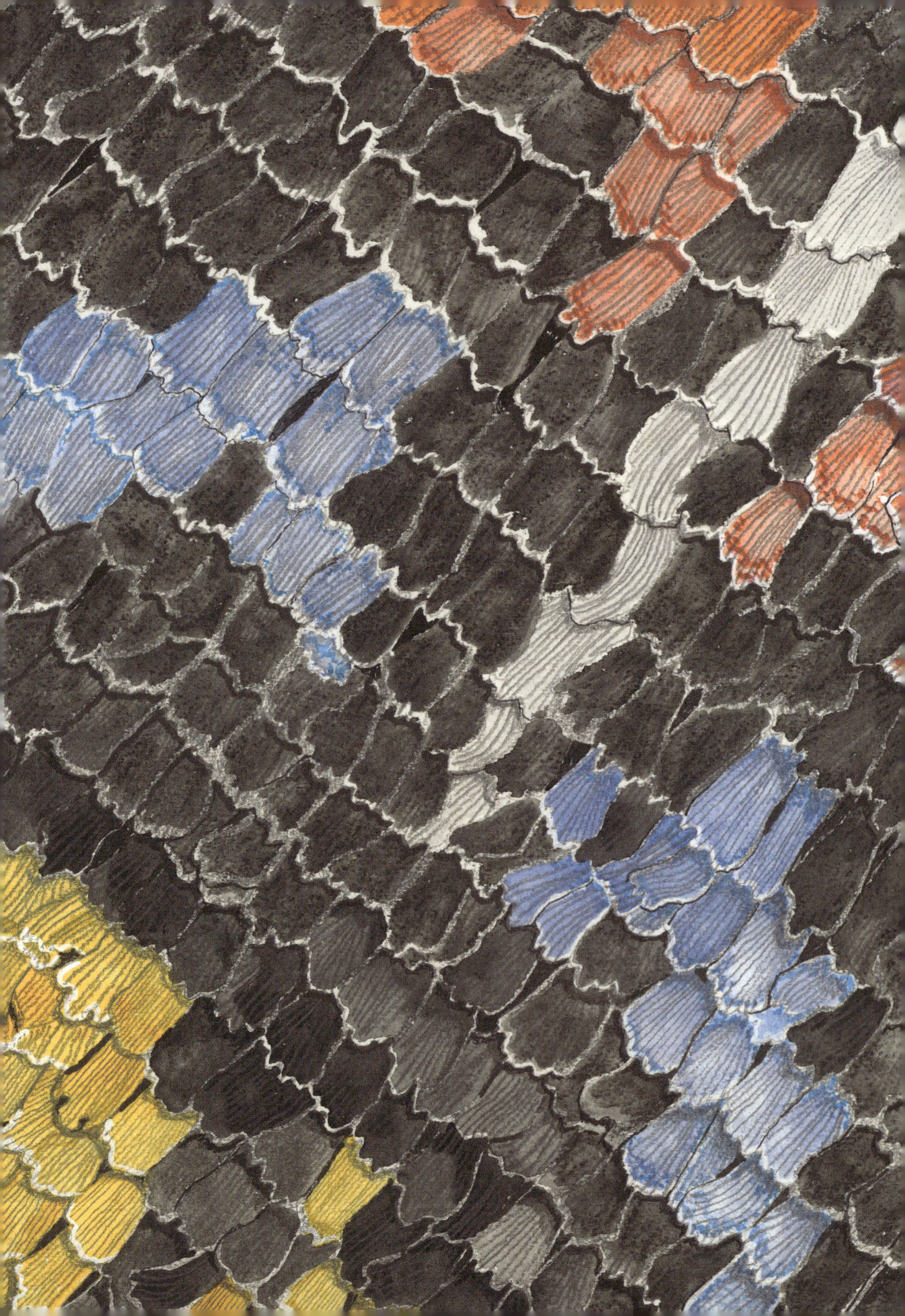

The Scales Have It

I find this part of a butterfly slightly confusing and complex, and anything that sounds like physics I'm going to try and take very slow. The wings are covered in minute scales laid like overlapping shingles on a roof. The wings of a Green Hairstreak (*Callophrys viridis*) are not actually green like paint on a wall. The color of this butterfly is made up of small structures within the scales that bend and redirect the wavelength of light as it passes through those structures—removing some colors or bolstering others, ultimately making it seem like greenness to our eye.

Within the scales one would find a multitude of shapes, the most important called a gyroid. All these shapes work together to either bend the light, enhance the wavelengths, or block them out. This blocking out helps our eyes perceive a male Mission Blue (*Icaricia icarioides missionensis*) as a uniform, pigmented color and helps us see iridescent hues, like the scintilla spots on an Acmon Blue (*Icaricia acmon*), that enhance the blue.

Take a look at the underside of the American Painted Lady (*Vanessa virginiensis*—considered by many to be the Most Beautiful Underside of any butterfly in the USA); the owl spots stand out in iridescence. All the silver spots on a Callippe Fritillary (*Argynnis callippe*) shine like buffed Victorian silverware. Or marvel at the male Pipevine Swallowtail (*Battus philenor*), electric in its gunmetal blue hind wings. These are all examples of saturated or refracted light bouncing off these shingled scales. It's how we get the blue sky.

As I was beginning the three years of painting for this book, I thought glitter would be the medium I could use to bring to life these iridescent hues, so in some sample paintings, I laid each piece of glitter in these scintilla spots like mosaic tiles. "Yeah," said the art director one fateful day. "That's not gonna really translate when we scan the originals. It'll come out muddy. Glitter doesn't scan." My hopes were dashed. No problem—making them glitter without the glitter is the real challenge.

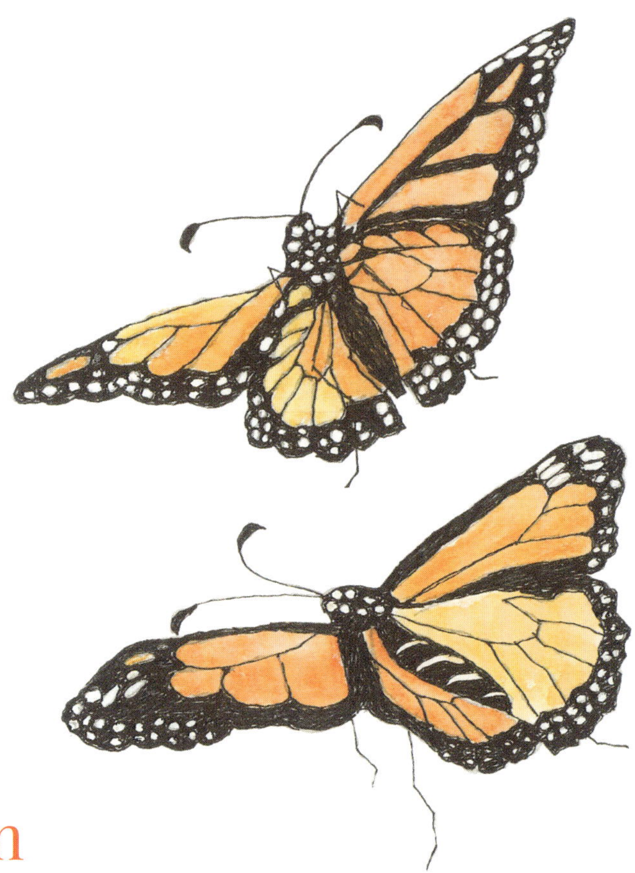

On Migration

I remember living in the Richmond District of San Francisco along Golden Gate Park for a while, and there was a particular day when I was walking and noticed a butterfly flying above me. As I watched it, I saw another, and another, all blasting down Fulton Street. They seem to be coming out of the park, then heading due west up the road. Another and another, all just going over the tops of cars and my own head. I looked around madly to connect with another human and to verify what I was seeing. It was a California Tortoiseshell (*Nymphalis californica*) "movement"—an explosion of them popping out of their chrysalises after a few days of extended heat, and I was caught up in it, awestruck.

Migration of butterflies doesn't seem to be as clear cut as those who do and those who don't. For instance, a vast majority of the blue butterflies don't fly much beyond where they were born. Many, like Gray Buckeyes, (*Junonia grisea*) definitely move about within their range, but they aren't true migrants like the Painted Ladies (*Vanessa cardui*) and the most famous traveler, the Monarch (*Danaus plexippus*).

So why do butterflies migrate? California Tortoiseshells are following the new flowers on plants that are edible in the early spring and then proceed upslope to higher elevations into summer. Then there is a slight reverse migration in the fall. Monarchs travel in search of warmer weather to overwinter in. Butterflies migrate for different reasons. One rarely sees the full picture of what's going on because it might take a few generations to complete the journey.

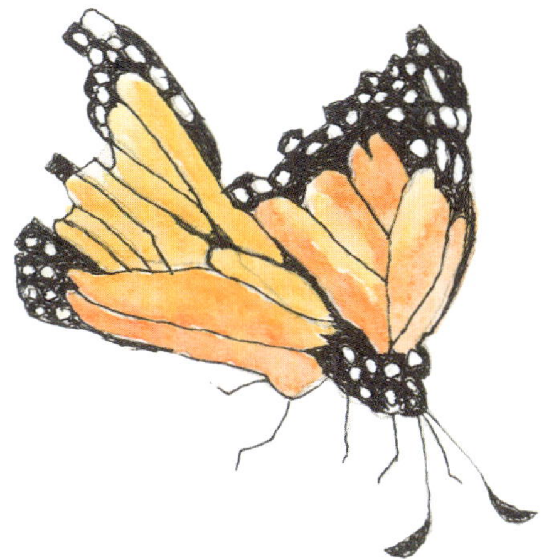

Painted Ladies emerge by the bazillions out of the southern realms of California. A great place to witness this is at Anza-Borrego Desert State Park. In the early spring, depending on the year, they can be seen flying in a north–northwesterly trajectory. This butterfly has different migratory patterns throughout the world.

King of all the butterfly migrations is that of the Monarch. It is the supreme example of long-distance displacement. Four full generations make the complete journey with the first three having individuals that live up to three months and the final, overwintering generation living four to five. Those living on the western side of America move from basins in Eastern Washington and Idaho in a south–southwesterly direction down to approximately the San Diego area. The ones in the east famously move south along the Eastern Seaboard, round the gulf, and settle in the state of Michoacán in Mexico. Numbers fell to a historic low a few years back throughout its western migration, then rallied like thunder the next year, leading many to conclude that we still have so much to learn with this butterfly. A large group of people are dedicated to this most famous of phenomena.

When butterflies come into an area, they are referred to as immigrants, and when they leave the area they are called emigrants. Those butterflies I saw on the street that day were emigrating to the foothills of the High Sierra. It's a complicated behavior that starts to become more clear after many years of watching.

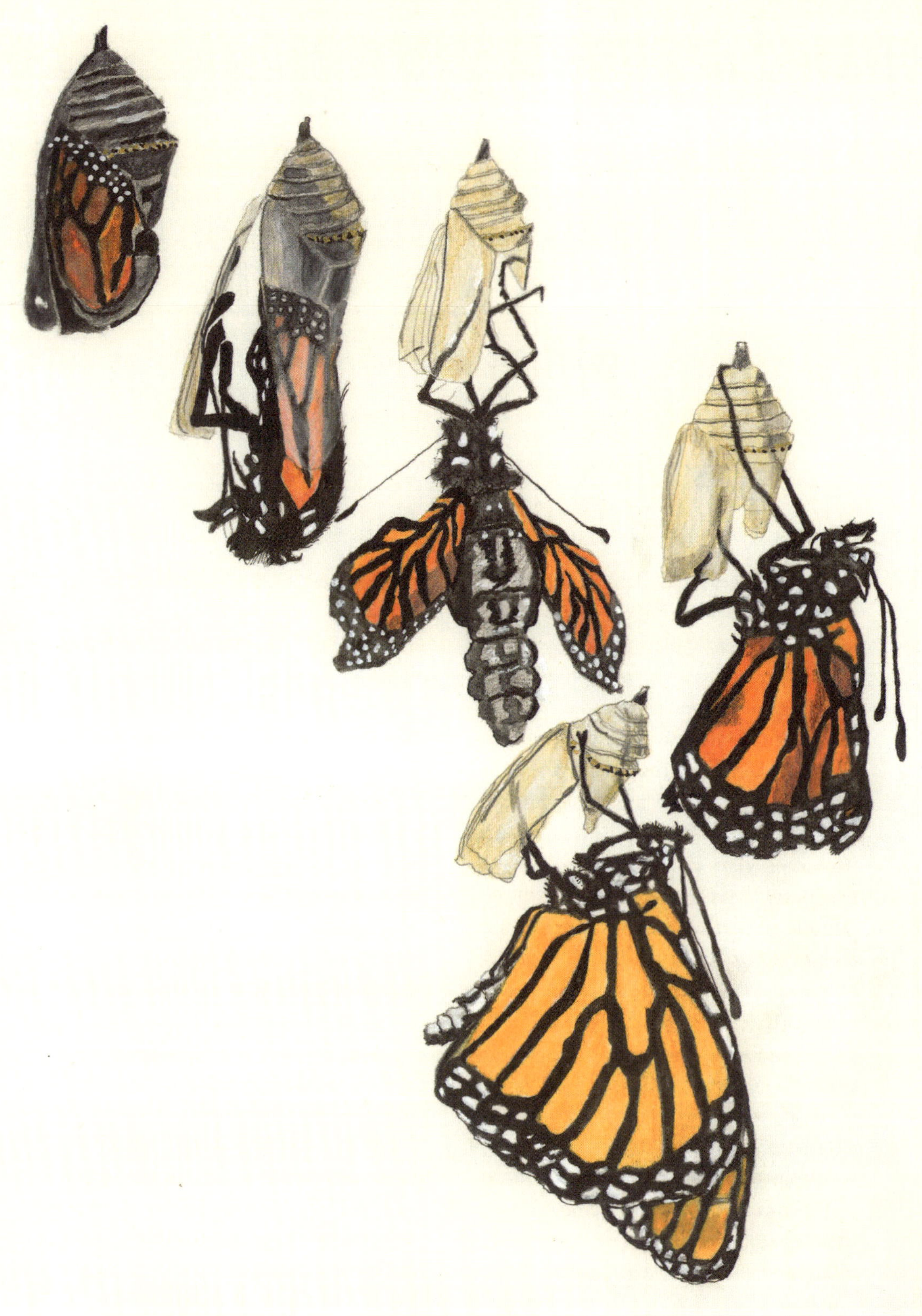

On Metamorphosis

Butterfly caterpillars, when done feeding, set out for their new phase of transformation. They leave the host plant (although some transform on it) and find a solitary place to change into a cocoon or chrysalis. Many moths do their transformation as a pupa within the safety of a cocoon that they weave with silk, but butterflies have a naked pupa called a chrysalis. At the back end of the larva, the butterfly creates with silk an anchor, or cremaster, that acts as a holdfast to the plant or other surface. Some hang straight down and some point upward at an angle with a strand of silk that acts as a girdle. Pupae (also referred to as chrysalides) are, in general, either brown like the ground or green like a plant. There are some colorful exceptions.

Inside a cocoon or chrysalis, the caterpillar dissolves, digesting its body with enzymes. If you cut open a cocoon at a certain time, caterpillar soup would pour out. Sometimes this is because the native wasps and flies the butterfly co-evolved with have penetrated the chrysalis and adopted it as their host. For the life of me I can't understand why this upsets people. We have to take care not to be so fixated on the adult, pretty phase of a butterfly and understand that at all four life phases: egg, larva, pupa, and adult (really cool word we seem to have dropped for the adult: the *imago* and its plural *imagines*) are basically food for everyone else. Yes, the noble butterfly pollinator story is taught every day in every elementary school, but how come we aren't teaching kids that it's equally noble to be eaten?

The contents of that poured-out liquid are not quite the slurry they appear to be. Certain cells known as imaginal discs are present. There is a disc for every new body part the adult will need. Once this reconfiguration is achieved, the final phase of the butterfly is finished. Once it ecloses (comes out of the pupa) it begins its find-a-mate-lay-eggs-and-die phase, the phase on the wing most of us are enthralled by.

ONE GENERATION

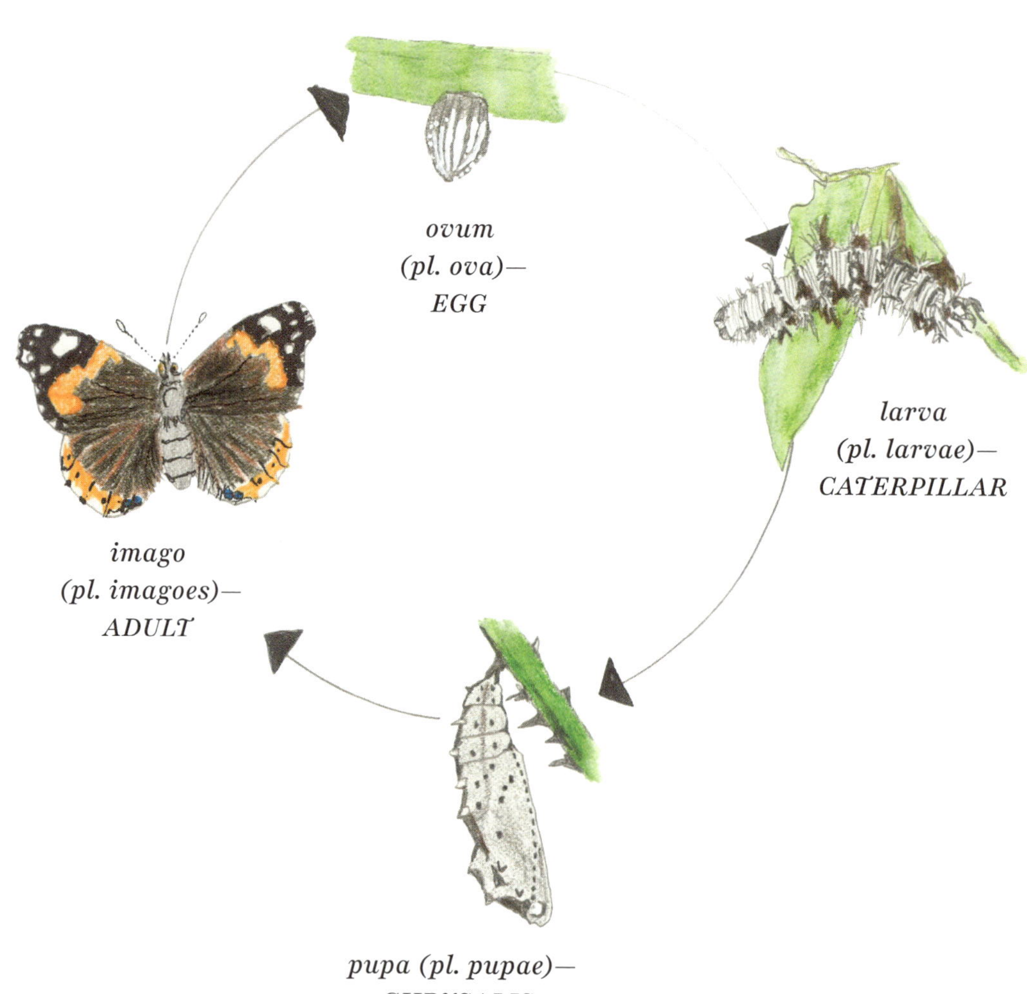

CHRYSALIDES

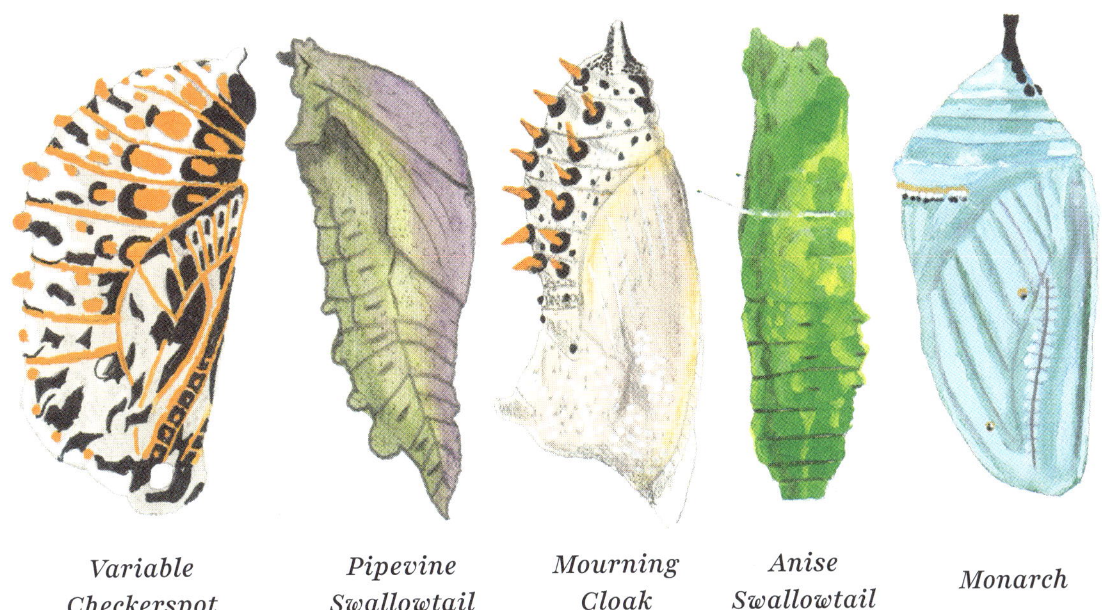

Variable Checkerspot · *Pipevine Swallowtail* · *Mourning Cloak* · *Anise Swallowtail* · *Monarch*

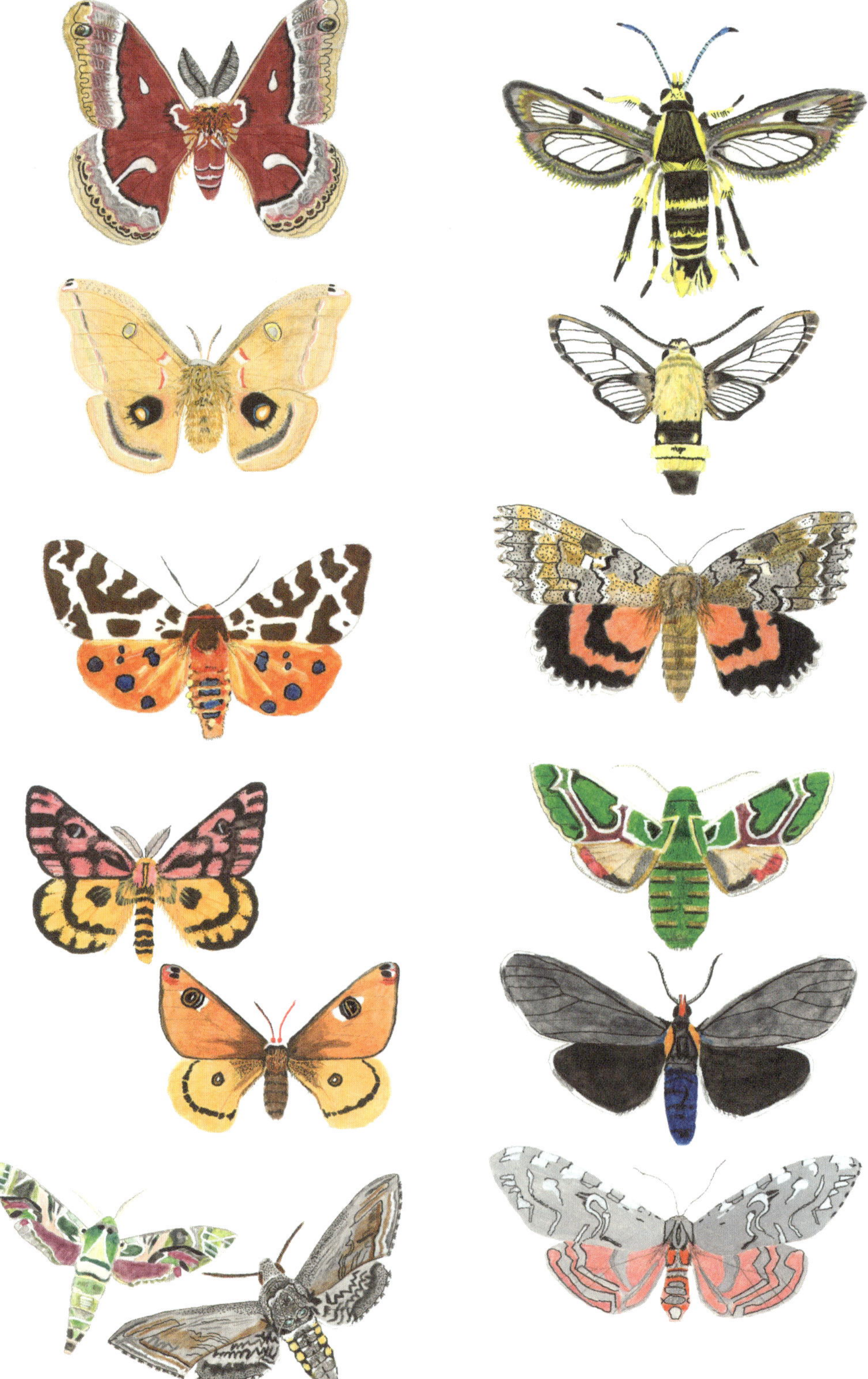

A Shout-out for the Moths

Though not specifically addressed in this book, moths are impossible to ignore. There are over sixteen thousand species of Lepidoptera in North America, 94 percent of which are moths. If one thinks about it, butterflies are just a branch of Lepidoptera that found the daylight to thrive in; the bigger story has always been moths. When I first started out in the deep end of the pool with entomologists from UC Berkeley, no one took you seriously if butterflies were your only focus; you had to step up to moths. I'm glad this happened, as it instilled in me a profound respect for the often-overlooked moths. I actually read once that the difference between butterflies and moths is this: "butterflies are pretty, moths are ugly." They have a bad rap. Behold a few moths on these pages—they should reassess your definition of ugly. If you are intimidated by just how many there are, then start with the bigger-in-size families: the Saturniidae (Silk Moths) and the Sphingidae (Hawkmoths). It was one of the prouder moments of my life when I could stand at a moth light with Professor Powell and pick out one of the Tortricid moths, a family on which he was a world authority, on a large bedsheet. "Is that one?" I asked, pointing to something smaller than a pencil eraser. "Yes," he replied, immediately collecting it.

And promise me just one thing. That you will find yourself someday near Nogales, AZ, along the Mexican border (I suggest Peña Blanca Lake) at a moth light and sheet. You'll see creatures that look like they followed Alice back out of Wonderland landing before you, and your mind will be blown apart from all the color, shape, and sizes you'll see in the beauty of moths.

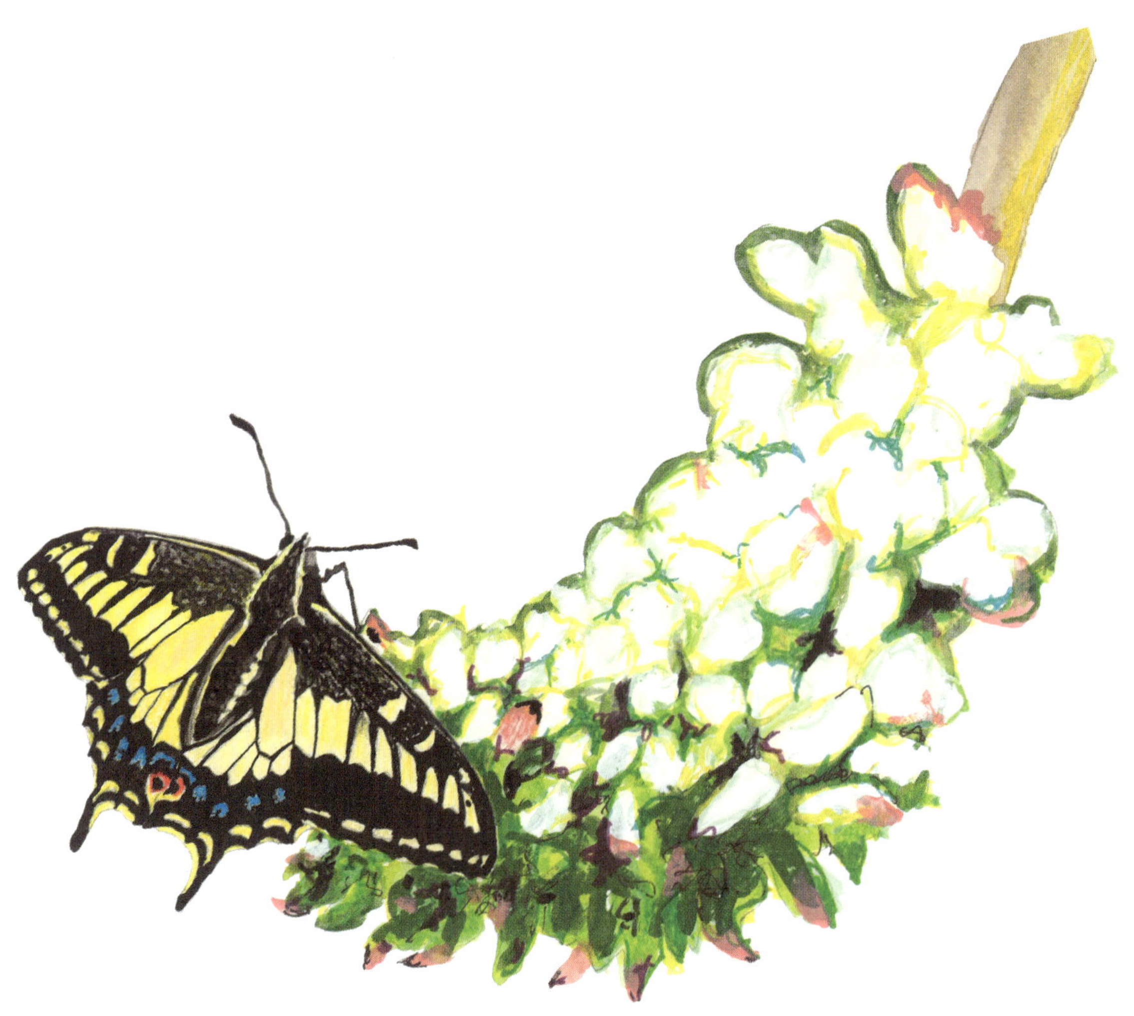

Anise Swallowtail on Buckeye Blossom

On Nets and Collecting

> The lure of the chase has always motivated collectors; rare and exciting examples of swallowtails were traded for a great deal of money and, sadly, some still are. The passion to kill and collect runs very deep in lepidopterists—a curious phenomenon and justified by some, even in high places but the vast majority of lepidopterists abhor such collecting.
>
> —Feltwell.1993.25

In full disclosure, I began as a collector. Right about when my interest in butterflies was growing, I met a grad student from UC Berkeley by the name of James Kruse. Jim was new to the Bay Area from Wisconsin and eager to see and collect as many moths in the genus *Hemileuca* as he could. Most important, he had a car, and I would split the gas with him as we went exploring isolated regions of the California, Nevada, and Utah tristate area. Jim's very pregnant wife would allow him forty-eight hours on these excursions, and we'd drive madly through the night.

One time we were in the Big Bear region of the San Bernardino Mountains in California searching in the narrow, endemic range of the Baird's Swallowtail (*Papilio machaon bairdii*)—a butterfly known to have low numbers even when it flies. After a full day of seeing them on the wing, Jim actually netted one. The only one netted on the trip. He looked in the net, turned it inside out, and I watched it fly off. I was shocked. "A female," he said. "Far more important than the boys."

It was an important ethical moment for me. One must be aware of the bigger picture of a species and collect responsibly.

Truth be told, now I recognize we probably shouldn't have even been attempting to collect the boys in this small, stressed population. We drove home empty-handed. I drove home wiser.

I may have been living in my own spin zone, but because I generally took only one male and one female when I was collecting so that I could paint them later, I felt I was having a low impact. The adult butterfly, which on average only lives ten days to two weeks on the wing, would serve a greater purpose for the painting, I told myself.

Though I no longer collect like I did in my early days, I still carry a butterfly net. Awareness has gotten better, and now I am questioned incessantly on a trail—as if I am carrying a loaded automatic assault weapon. It's a multipurpose tool, not always an instrument of death. Comes in awful handy over your head if you enter a mosquito zone. I carry it out of habit for sweeping purposes over a field to stir up things, use it for catch-and-release when I'm not sure about something with the smaller guys and when a specimen seems necessary for the taking.

The naturalist like myself who earnestly believes that carrying a net is all right might be surprised to discover how few places one can actually carry it. One cannot wield it in a national, state, or county park. It is allowed in a national forest but not in a national monument. Among the handful of places one is allowed to collect is BLM (Bureau of Land Management) land. And be forewarned: carrying it but professing you're not collecting or "I didn't know" isn't going to get you out of a serious ticket or even worse a violation of the Lacey Act, which states one cannot "harass the wildlife," especially if it's in the area of an endangered species. It almost doesn't seem worth it, does it? I've never understood why a permitting system, like an annual hunting/fishing license, which limits one's catch and the fees go to conservation, has not been set up for the butterfly collector. Heck, we don't even eat them (but I guess we are "consuming" them in a way).

I think it has to do with the thought that killing a pretty invertebrate is more profound than killing a slimy fish. The pious disdain for the occasional collection seems hypocritical compared to the daily slaughtering of butterflies by cars. Don't believe me? Go out and look at the grill of your car after a long road trip. I stopped collecting because my interest waned, conservation became more challenging to me, and photography jumped to the fore when I discovered iNaturalist, an app that helps you learn about the natural world around you.

Now there is a whole new set of givens since I began: global warming, insect apocalypse, and the rising sense of one's place in an environment. One's impact and legacy on the earth.

Did you help our planet or hasten its demise? I fight so hard to steer clear of butterfly clichés, but it's looking more and more like each one truly does matter, especially those butterflies beholden to a single host plant. Photography, video, field sketching—all of these other ways of enjoying butterflies have far less impact than collecting. iNaturalist lets you immediately add valuable information to known ranges of species with the click of your camera. I write this in 2025 when the West Coast experienced cataclysmic rains in the earlier part of the year. Many butterflies at the larval stage were drowned. (The coup de grâce of any butterfly population is weather, not the lone person with a net.) Though I may sound like a hypocrite, now that I am older and wiser, I think we should all hold off on our collecting for a while until butterfly populations hopefully rebound.

There is an exception to this though—the child that must. You'll be able to tell quickly as a parent if this is not a fanciful whim but an internal, hyper-focused desire that is no doubt a window into their adulthood. It takes a special parent to guide this child. If they just have to make a collection, they will just transcend the obstacles. Please do not crush this exception to the rule. They may be the next Jane Goodall or E. O. Wilson. Let them fly with it.

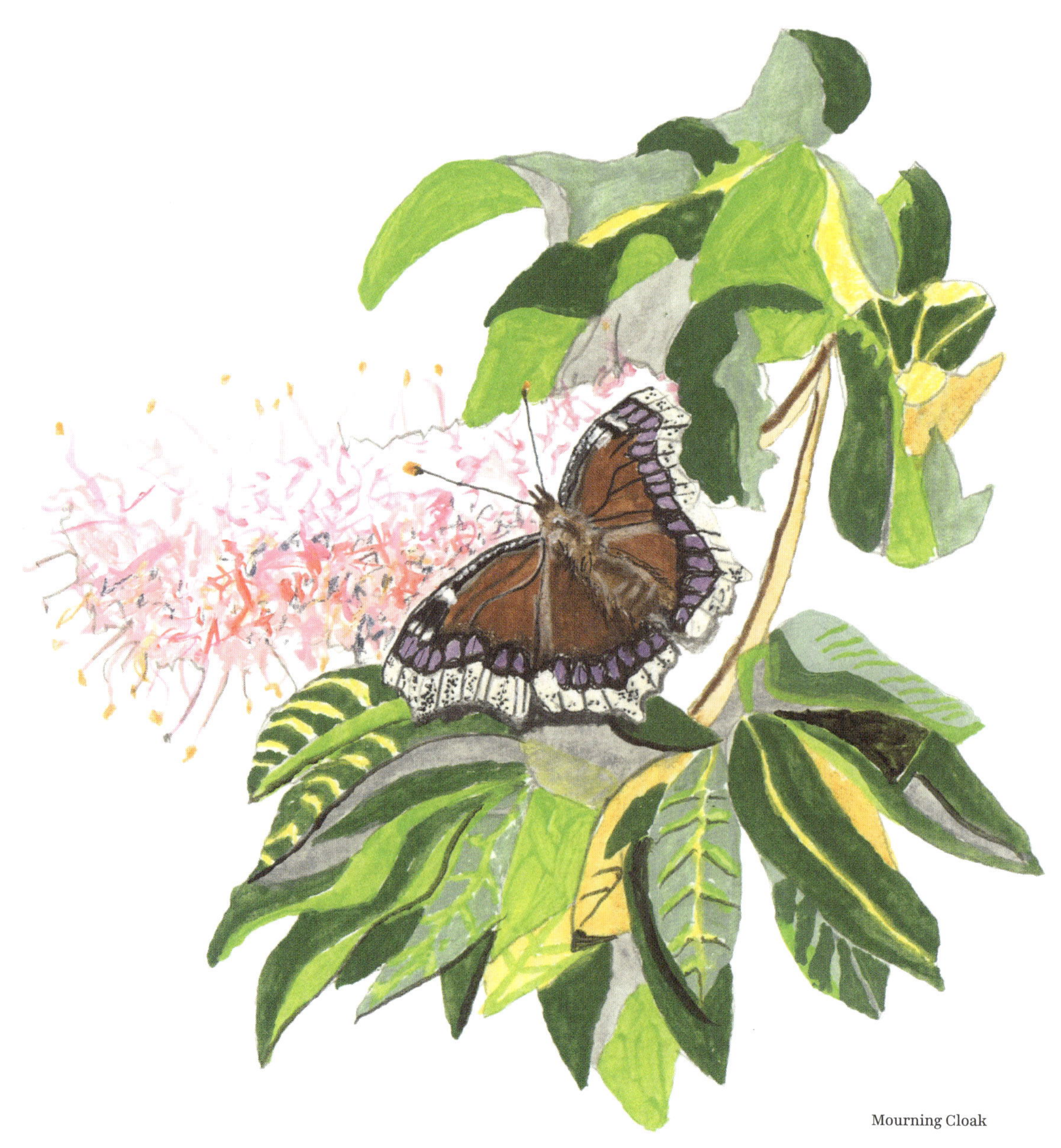

Mourning Cloak

On Common Names and Latin Names and the Big Bag of Crazy Those Are

Linnaeus did not create the binomial system to make you feel bad. He's not laughing from the grave every time you slaughter the Latin. It's actually there to unify us through varying, disparate languages and variations of common names even within a single language. Allow yourself to get it wrong. I would tell Jerry Powell I saw a bunch of larva. "How many did you see?" he'd shoot back, knowing quite well he was giving me a semantics class. "Five larva? Larvi? Five larvae?" The last guess was the correct one. Not all will be so lucky to have such a drill sergeant barking at them like I was. Be wrong really *big* for a good, long while and *do not* give up. It'll click.

I was on a hike early on in my career here on San Bruno Mountain. I said, "Look! A frit*ill*ary! (Accent on the second syllable.) A guy leaned over to me in a very condescending manner and said, "Actually, it's *frit*illary." (Accent on the first syllable.) Then a third guy with a British accent leaned over to us both. "Actually," he said, "it's both."

I tell that story now because when I tried to learn the names, especially the common ones, they were always shifting. I guess I'm just a little more used to it now because it continues to this day. They used to be Anglewings; now they are Commas. They used to be Bramble Hairstreaks, now they are Lotus Hairstreaks. When I told Powell that the "Coastal" Green Hairstreak had reverted from *Callophrys dumetorum* back to *Callophrys viridis*, the eighty-year-old curmudgeon said, "Oh good. That's what it was when I was in college." What I've learned is that names are always in a state of flux. Even while writing this book, I've had to incorporate new genera and species changes.

The North American Butterfly Association (NABA) tried to codify all the common names floating out there a decade or so back by picking just one for each butterfly. I guess they were somewhat successful at this, but my worry is where does all the regional lore go when you kill off all those other names? How did they get there? Personally I'm not going to jump off the roof if I hear someone talking about a Spring Azure (*Celastrina ladon echo*), or calling it by its even older common name, Echo Blue; they both led to the now acceptable Echo Azure (*Celastrina echo*). A lot of the confusion arises when subspecies are bumped up to species status. In parts of the country, the Common Wood Nymph (*Cercyonis pegala boopis*) is called the Ox-eyed Satyr; in other parts it's called the Blue-eyed Grayling. Fantastic names dumped for unification.
I don't want to lose any of those. And don't we, when you come to think about it, have the capacity to handle a bunch of names for a single creature? Our mind has an extraordinary ability to stretch. Go ahead and learn them all. Who cares if you get it wrong. Those in charge of "right" get it wrong all the time.

On Butterfly Counts

The Greater Bay Area has the highest density of annual butterfly counts of anywhere else in the nation. About ten counts are held here each year, with another ten counts throughout Central and Northern California. What is a butterfly count? Started by the Xerces Society and now run by the North American Butterfly Association (NABA), it is a day set aside for people to inventory the butterflies of a certain area. People (and anyone is welcome) spread out from a fixed point within a fifteen-mile-across circle and (1) count all the butterfly species and (2) count all the individuals. I cannot encourage you enough to go on a butterfly count if this passion is taking hold for you. It's really how I got started. I went on my first one in 1996 and have run the San Francisco Count for the last twenty-five years. The positive energy and comradery on these days is palpable.

Now you think to yourself, *I don't really know my butterflies so I can't really help.* First of all, everyone's set of eyes plays into the success of the day. If I'm looking one way, and while you're looking the opposite way you say, "I see one!" Then together we can ID it. Even if you don't know the species, you'll be paired with someone who will. The second count I ever did, the Berkeley Count, is one of the oldest in the nation. They handed me the clipboard with all the names of things we could possibly see, and Latin started to be shouted out on the trail. There was something quite thrilling about being in the deep end of the pool and flipping pages back and forth. Count days are not the best way to learn your butterflies, but through osmosis you can glean a great deal from all the folks gathered and their shared knowledge. Things move at a fast pace, and a novice can become overwhelmed. You are, however,

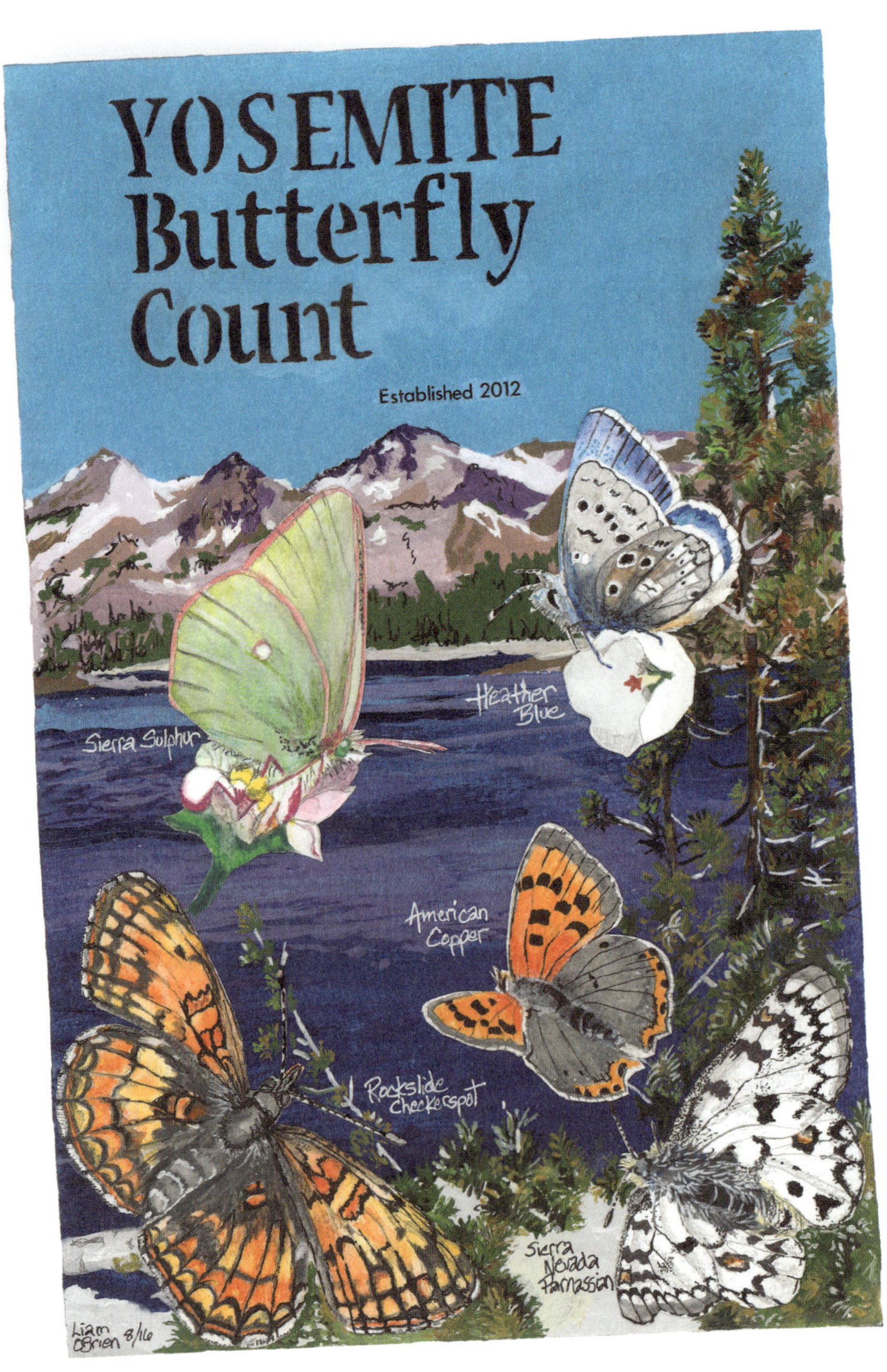
YOSEMITE
Butterfly
Count
Established 2012
Heather Blue
Sierra Sulphur
American Copper
Rockslide Checkerspot
Sierra Nevada Parnassian
Liam O'Brien 8/16

among people that don't care if you get them all right. It's a very supportive community.

Aside from the two aforementioned counts, in the Greater Bay Area you have the option of those held in Marin, Point Reyes National Seashore, Mount Diablo, Benicia, Pinnacles National Park, Hastings Ecological Reserve, and Monterey. Further to the north are the counts of Mount Shasta, Big Creek/Chico, and the Warner Mountains, and you'll find the Yosemite Count to the east. (These are just a few examples.) Although counts are also run in the early spring and late fall, the ones I've mentioned are concentrated between June 1 and July 31 and are referred to as the Fourth of July Butterfly Counts. Check with NABA for butterfly count days at naba.org. Leaders usually pick days for their counts and post them by March or April.

Ultimately, and probably not unlike the Annual Christmas Bird Counts (which the butterfly count is based on), you get to know the fellow enthusiasts and you get to be among nature for the day. Each season adds to the knowledge. I've made lifelong friends from counts, and I look forward to the count season tremendously each year. You will too.

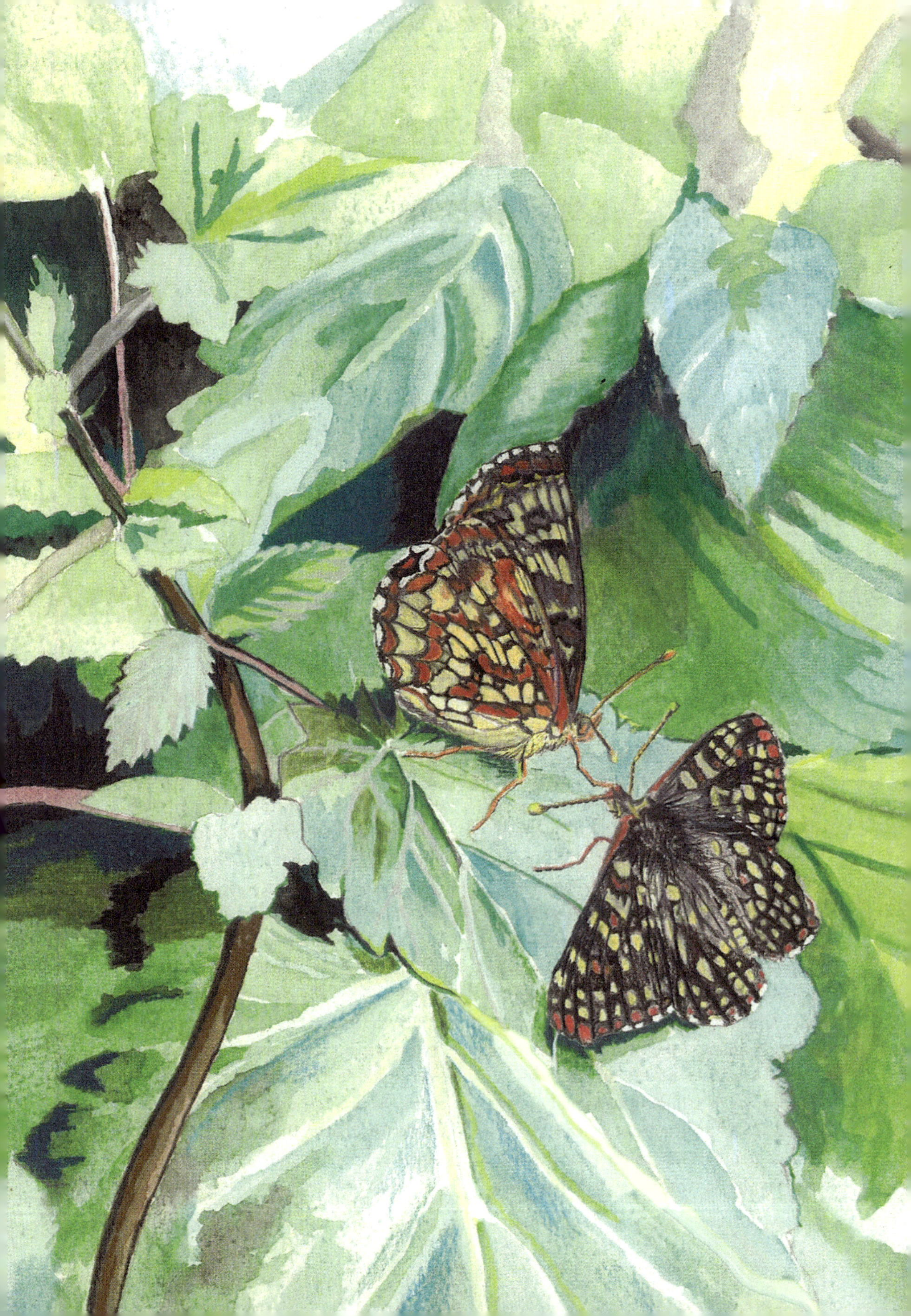

A pair of courting Variable Checkerspots

On Butterfly Watching

If you have sunshine and flowers (and even better the two simultaneously), the odds are with you that you'll see some butterflies. They are little solar panels and they need to be in the high fifties in temperature to really get going. I guess I would have to add a third thing for the ideal recipe—no wind. Butterflies seem to abhor the wind, which makes sense with all their patrolling and following scent trails of pheromones. When looking for butterflies on any windy day, check the leeward side of any hill and you'll most assuredly find them sitting out of the wind. You may even find them in what we refer to as dappled sunlight—where the shade and sun create glades in a forest or in and out of rows of mature trees on a street. In these areas one might see Tiger Swallowtails, Margined Whites, and all of the Commas. A whole handful of skippers: Common Roadsides, Arctics, and Silver-spotteds seem to prefer this habitat as well.

If you get a little more advanced in your species identification, you develop the ability to "Rolodex" (for you youngsters that was a card index fixed on a wheel people usually had on their desks for phone numbers. Back in the olden days called the seventies.) each environment as you enter it. Say the hike starts in sun at a trailhead, so you think about those species that live only in the sun, then the trail enters full shade. A perfect example of this: One usually only sees a Cabbage White (*Pieris rapae*) in full sun and almost never in the shade of a forest. Its counterpart, the Margined White (*Pieris napi*), is a denizen of the forest and isn't seen in the open sun. Coastal scrub, riparian corridors, rocky outcrops, and grassland all have for the most part their own community of butterflies.

It is sad how much healthy, native habitat our species has altered, but do you want to know an interesting fact? We've actually created butterfly habitat with roadcuts, agricultural field borders, trails, and disturbed areas. Disturbance is great for the generalists (butterflies that lay their eggs on many host plants) but not so good for the specifests, the species obliged to a single plant, which is usually a native. Fill your gardens with native plants, folks. The generalists will do fine on their own. Remember to look up. It may seem obvious to turn your head to the sky to see a butterfly, but in general people walk looking forward and at the ground before them. There is a whole slew of species up in the canopy of the trees you are either walking below or by. California Sisters, Mournful Duskywings, and Golden Hairstreaks skip and glide between the oak leaves on a hot afternoon. Keep your eyes moving low, mid-shoulder, and high.

Two other known behaviors of butterflies work toward your advantage in finding them.

The first is the phenomenon of mud-puddling. Males (and sometimes females) need minerals, salts, nectar, and protein to sustain them while in flight. They also play into the male's need to create a spermatophore packet to pass to the female while mating. Salts are primarily found in the mud. You'll mostly see blues and swallowtails there, but other species as well in the early afternoon sitting on the banks of creeks or simply using a puddle to extract compounds from the mud. One can also see them on animal scat extracting minerals and proteins.

The second behavior is hill-topping. Many butterflies, but not all, are internally programmed to go to the nearest hilltop to find a mate. Without this it would be extremely rough to find one another. Why a hilltop? Why do humans go to a single's bar? It's an excellent way to guarantee procreation for the next generation. It usually gets going about 3:00 p.m., but some species like the Anise Swallowtail (*Papilio zelicaon*) are known to start earlier. There are many more boys at a hilltop waiting for girls to fly by. Once you witness this activity on your own it will be difficult to unsee it on every hike with a hill you take from then on. It's wonderful to see all the aerial maneuvering going on at a mountain summit while trying to tell them apart.

Many butterfly enthusiasts carry binoculars, and this is something you might want to consider purchasing. They've made quite a few advances in recent years in close-range binocs, and these most assuredly will help you zoom in on those crucial details. Personally I never got into the habit of carrying them, and they would be rendered useless if I tried now on account of my freakish right eyeball.

To successfully photograph a butterfly, start by remembering that an adult butterfly has the equivalent of ten thousand eyes in those compound optical structures. They are programmed just after emerging to dodge anything coming too close—they've got mates to find and babies to make. So what is my advice? Don't act like a bird. No sudden movements. You will be shocked how close you can get to a butterfly if you move incredibly slow. Slower than what you think is slow. Take your record shot furthest away, then slowly move in for more. If you see the wings go up as you approach and you want that pretty dorsal shot, take a slow step back. (Think about it—wings up means the next movement for the butterfly is a downward exit thrust.) Normally when you take that step back, the butterfly relaxes and drops its wings. If it leaves, don't chase it. If you stay still, nine times out of ten the butterfly returns right to where it was. As they age on the wing each day, they become easier to approach.

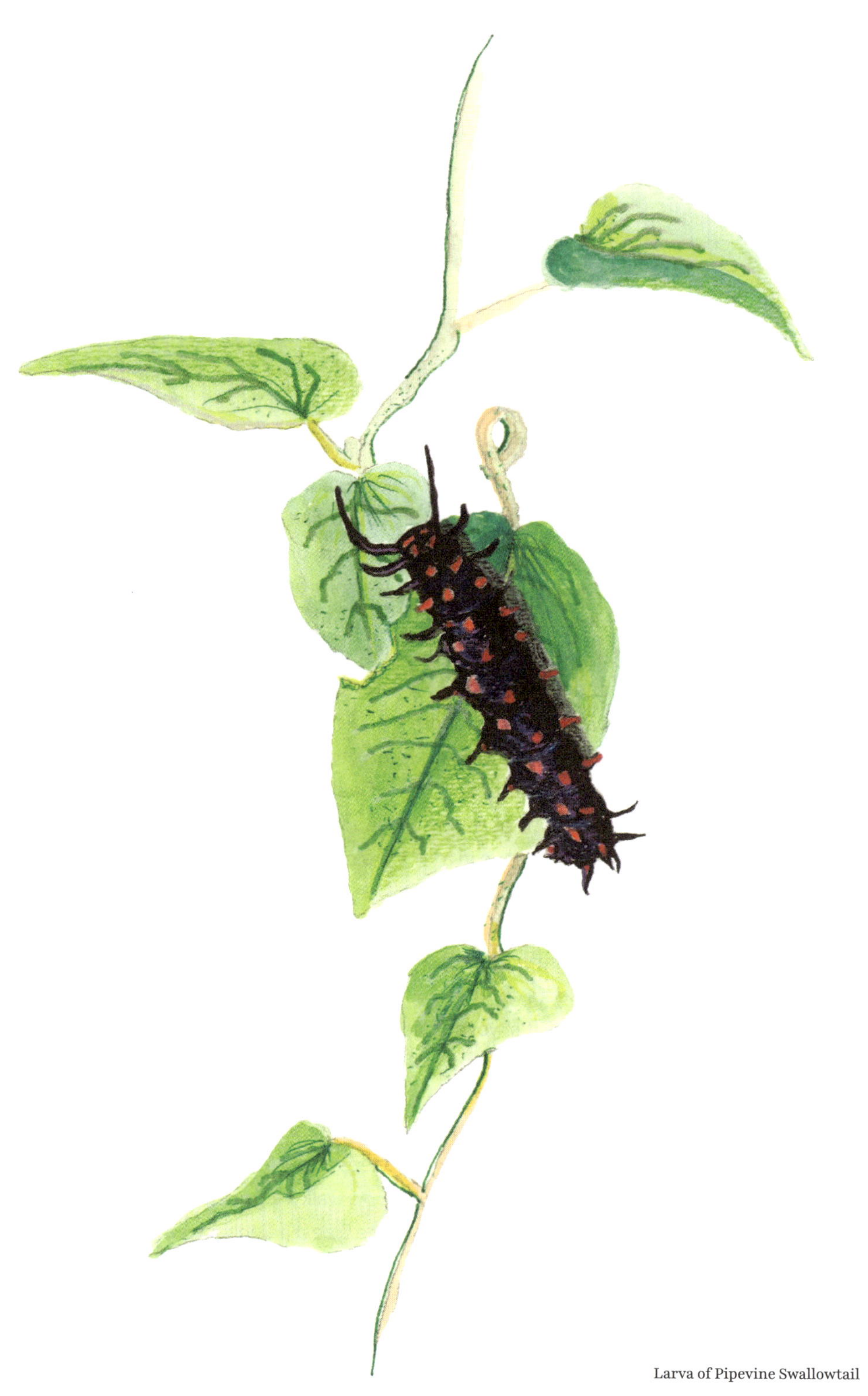

Larva of Pipevine Swallowtail

The Best Butterfly Walks in the Greater Bay Area

ALUM ROCK PARK

Santa Clara County
When: Early February–May

From the 680 freeway in San Jose, get yourself to the Alum Rock Avenue exit going east. Stay on this till Alum Rock Road; make a left and this should lead you to the park. Once inside, proceed in your car till the last possible parking lot). Pay your parking fee (automated kiosks).

The grassy area under the trees against the parking lot once held the most Mourning Cloaks (*Nymphalis antiopa*) I'd ever seen in a day. There is a known migration or movement of these creatures, and it almost seemed like they were using it like a lek—there were that many. A lek is a place where insects congregate usually for mating purposes. It's worth checking out. Cross through the gate from the parking lot and proceed on to the main trail. There is usually a large puddle from the winter rains near this gate, and I've photographed Silvery Blues (*Glaucopsyche lygdamus*) and Echo Blues (*Celastrina echo*). There are lots of smaller trails that go down to the creek that spur off the main, paved one. Always work your way back to the main trail heading east.

So what is this somewhat bizarre place called Alum Rock Park? In the Victorian and Edwardian eras people came here to "take the waters" of the mineral springs. The water was warm, smelled of rotten eggs (primarily sulfuric), and was once believed to be therapeutic. (My grandparents went to Calistoga, CA, for the same reason.)

The park is a wonderland of statues and grottoes, all paying homage to various gods of water. It's ornate, detailed, and somewhat gaudy. The place was so popular it had its own train from downtown San Jose. One can still see the train trestles while driving in. When the vogue ended, the statuary and faux ruins fell into real ruin. The reason butterfly enthusiasts come here is for one species that is difficult to find anywhere else in our area. This is the most accessible population.

The trail meanders along the Upper Penitencia Creek through towering sycamore trees. I saw four species of white butterflies on a walk here once in the meadows below the trees: a Cabbage White (*Pieris rapae*), a Margined White (*Pieris napi*), a Large Marble (*Euchloe ausonides*), and the last one is kind of a tricky, advanced one—the female form "alba" of the Orange Sulphur (*Colias eurytheme*).

You'll cross a bridge or two, and you'll start to notice that the trail begins to hug the slope on your left. Lotus Hairstreaks (*Callophrys dumetorum*) and Acmon Blues (*Icaricia acmon*) work the plants on the slope while an occasional Western Tiger Swallowtail (*Papilio rutulus*) soars in and out of the dappled sunlight.

At about two miles (or less) after leaving your car, the rocky slope becomes perpendicular to the trail—severely so. (If you've gotten to the raised, single-file bridge that crosses back over the creek, you've gone too far. Turn around.) You'll see the host plant of the butterfly before you'll see the butterfly itself. Canyon liveforever (*Dudleya cymosa*) hangs on in the cracks and crevices of the chert cliffs. (It's a species of succulent that is currently under threat from people pulling them out of the native soil for their gardens—a shameless enterprise that should bring a curse upon your entire family!) Shimmering like a piece of insane blue foil, the Sonoran Blue (*Philotes sonorensis*) butterfly, about the size of a dime, appears around your feet here in an alarming, matter-of-fact way. The males, with a single red dot on their forewings, patrol up and down the gulleys looking for females. These females have four red dots (one on each wing) and are considered by many to be the Most Beautiful Small Butterfly in the Country—a subjective superlative no doubt, but there is something truly jaw-dropping when one sees this butterfly for the first time: lipstick red against a cerulean blue. You'll notice its utterly unique flight as well, fluttering with many wingbeats as if it's trying out its wings for the first time and can't relax. Note the time of this walk on the calendar year. One of the earliest butterflies to emerge (not counting the overwintering species), it might even be on the wing in early January.

I used to feel it wasn't a new butterfly season till I made my annual pilgrimage here with my friend the late Bill Shepard. This man had so much glee for butterflies that he downright giggled when he saw them. It pleased him no end to point out this rare butterfly when people were just walking past it. "Wanna see something cool?" he would ask them with a big smile on his face. I like to think Bill is permanently there on a chair waiting patiently for the next group, then springing up with the same enthusiasm each time—a docent for the ages who instills sincere wonder.

On the hike back to the car there is a large, grand water fountain like only the Gilded Age could have created. Stop and treat yourself to just a sip of the "magical waters." I remember as a boy coming here on a class field trip and being told to "pinch my nose to get the water down." Unless your dream is to vomit with screaming children in a big yellow school bus, I'd advise you to forego this.

Warning: This trail is frequently closed due to washouts from winter and early spring rains. Call ahead to check for any trail closures. If you get halfway up the trail and a barricade stops you from proceeding on, no worries . . . the butterfly can still be seen on the high ridge trails above Sunol Regional Wilderness in Alameda County and Balconies Trail and Balconies Cliffs in Pinnacles National Park in San Benito County.

KENT PUMP ROAD

(behind Alpine Dam)
Marin County
When: June

I was assigned "behind Alpine Dam" one year by the leader Wendy Dreskin of the Marin Butterfly Count. My companions for this day were Bob Hall and Matthew Delaroca. The Kent Pump Road behind the dam was a path I'd taken many times, but this day left me with "This is a must for my top six list."

First, directions: Get yourself onto Bolinas Road, which you can find cutting through the quaint town of Fairfax. Stay on this road for a good fifteen to twenty miles. (It may be shorter than that, a windy road that is slow going. Be careful of the weekend bikers, and God help you if there is a race going on.) You'll stay on this until you begin to notice a large body of water appearing on your left side. That is Alpine Lake. You'll want to prepare to stop before the road makes a sharp left turn going over the dam. Try to park before this, near the gate or any of the pullouts along the lake there. The trailhead down the road is right behind the gate, and the butterflies usually start immediately there. Yes, this road is the trail, and you'll see mile signs there once you begin.

Pipevine Swallowtails (*Battus philenor*) dart among the cliffs above your parked car.

Acmon Blues (*Icaricia acmon*) were seen on count day dancing on the lower plants in the beginning of the road. The trail makes its way through completely shaded areas to open glades of sun. Bring lunch, as you will be spending a few hours walking about three miles. All level ground, no ups or downs—a dream hike.

Species likely to be seen: both Pale Swallowtails and Anise Swallowtails (four species of swallowtails total along with the Pipevine and Western Tiger), Margined Whites in the shade, Variable Checkerspots, Mylitta Crescent, and Umber Skippers. On my day observing with Bob and Matthew, at the two-mile mark a large, pale brown skipper began appearing. I thought at first it was the female Northern Cloudywing (*Thorybes pylades*), but the wings were held flat on the leaf. Cloudywings hold their wings slightly up and about a quarter open. When Bob stepped in for a better shot, the butterfly lifted its wings to reveal a large, prominent silver spot. No question now. And soon, we counted thirteen Silver-spotted Skippers (*Epargyreus clarus*) hanging off the yerba santa bushes; this is the largest skipper in the Bay Area and the only Bay Area population I know of other than that of Chews Ridge in Monterey County. I confirmed with Wendy Dreskin, the count leader, later that this population is seen annually on this count. A fantastic discovery so close to the San Francisco Bay.

You'll pass a fork in the road that splits toward Old Vee Fire Road. Do not take it—stay on the Pumphouse Road for about another mile. Unbeknownst to many, you are in the presence of a special plant, California false indigo (*Amorpha californica*) that can be seen along the road as a large bush. It is not only the host plant for that Silver-spotted Skipper, but also for another extraordinary butterfly here. Park yourself near some blooming bull thistles for lunch and let this other butterfly come to you. You might be blessed with the arrival of our official state insect, the California Dogface (*Zerene eurydice*) aka the Flying Pansy. The female is solidly lemon yellow (not like the orange yellow of the Orange Sulphur) and the male is outrageously fancy with black forewings that reveal the side profile of a dog's face bathed in a mauve, plum color. His hind wings are yellow. We had a male blast through lunch the day I joined for the count, which set off cheers among us.

GARRAPATA STATE PARK

Monterey County
When: April–June

As one passes some of the only native Monterey cypress trees at Point Lobos along Highway 1, the terrain changes from beaches to rocky coves and redwood watersheds. Beyond the enclave of Carmel Highlands is a state park that if you blink you might miss. What you *will* notice is the extreme left and short exit off the highway and toward a line of parked cars below the trees. Parking can be rough here on the weekends, as is true for many state parks in California. Welcome, you've arrived at Garrapata State Park. Bring lunch and plenty of water. The word *garrapata* is Spanish for "tick," and though I've never had any issues there, be forewarned.

I was assigned Garrapata one year for the Monterey Butterfly Count run by Chris Tenney, the count leader, which is when I discovered this walk. It starts easy enough through a flat, coastal scrub trail that is being invaded by a wall of non-native prickly pear cactus marching to the coast from inland. Rural Skippers

(*Ochlodes agricola*) and Variable Checkerspots (*Euphydryas chalcedona*) careen at your feet or land on some of the most robust native flowering plants I have ever seen. I've had both Two-banded Checkered Skippers (*Pyrgus ruralis*) and Mylitta Crescent (*Phyciodes mylitta*) along the creek there, both butterflies found near running water. Once one crosses over a few small creek bridges in the willows, the view opens up to what makes the place so incredible: the wildflowers. After the deluge of 2023, a superbloom made this place look like a flower stand on psychedelics. Yellows, whites, purples, and reds all competed for every inch of space. I felt like I was floating over a coral reef, or visiting some place over the rainbow that Dorothy skipped through.

One of the coolest aspects of Garrapata is the butterfly enthusiast's ability to see the endangered Smith's Blue (*Euphilotes enoptes smithi*) here along the Big Sur coast.

It's easy to key it away from the other small blue here, the Acmon Blue (*Icaricia acmon*) because the Acmon has an orange band (called the aurora) dorsally on both the female and male's hind wings. Smith's does not. Smith's has chunkier black spots on the underside.

The trail proceeds up the valley for a bit, and when it makes a sharp, right-hand turn, a new set of species appears: Silvery Blues (*Glaucopsyche lygdamus*), Margined Whites (*Pieris napi*), and Sara Orangetips (*Anthocharis sara*) bring light into this shady corner. Saras can be distinguished on the wing from Margineds with their orange-in-flight forewings. More Variable Checkerspots appear, but in among them I've spied Gabb's Checkerspots (*Chlosyne gabbii*) in a lazy flight along the trail. Check those spots on your checkerspots! Pale Tigers (*Papilio eurymedon*) and Western Tigers (*Papilio rutulus*) battle for platform rights on the Venus thistles.

The ecosystem then radically shifts to the understory of a redwood forest. The trail is well maintained, but some river crossings are a bit dicey, and from this point on the going starts going up. One of the few butterflies to penetrate this dark, shady world, Satyr Comma (*Polygonia satyrus*), will flash before you as a rust-colored leaf. A healthy population of the stinging nettle host plant it needs abounds in the glade openings. It mimics a dead leaf when at rest.

Many more sunny scrub openings along with shade continue for a while until you reach a junction—here you may see a sign that says the trail is closed from this point on. If it is open, know that it becomes extreme hiking, where you have to pull yourself up hand over hand in some areas. I personally have not made it to the summit, but the few times I've pressed on I have been rewarded with Lotus

Hairstreaks (*Callophrys dumetorum*) and a county record of the Marine Blue (*Leptotes marina*). If it is closed, no worries—I can guarantee you've already had a masterful day with your butterflies, and you get to do it all over again on your walk back to the car. Enjoy your stroll through Eden and be careful backing out of that parking-along-the-highway because it spits you out on it immediately and cars are flying past you.

There isn't a time I'm in Garrapata that I don't wonder why it isn't a national park. The scenery boggles the mind, and the butterflies never fail to come through.

MITCHELL CANYON TRAIL TO EAGLE PEAK

Contra Costa County
When: early May–late June

To hike this trail in Mount Diablo State Park, get yourself to Concord Avenue in the city of Concord and take it going toward the city of Clayton. Make a right on Mitchell Canyon Road, which will drop you into the northern section of the park. There's a parking fee of eight dollars. Warning: this is the most strenuous hike I recommend in the book. Bring lunch because it's going to be four hours (maybe five) easily round trip. And bring more water that you think you need. (Did I say more water?) And a wide-brimmed hat. It can be quite hot here during this time of year. Parking can be an issue on weekends.

The willows in the lower parking lot are worth checking out. It's the first place I ever saw the Dryope Hairstreak (*Satyrium dryope*). The park has a great visitor's center close to this entrance that can help you with butterfly identification lists. The demonstration garden behind the center will inevitably have your first California Ringlets (*Coenonympha california*) of the day. The trailhead for Mitchell Canyon is just beyond the garden.

The walk initially is flat and easy enough, taking one through a classic California oak forest. Gray Buckeyes (*Junonia grisea*) will bolt up to check you out as you enter males' territories along the path. The first few California buckeye trees (gas stations for butterflies when in bloom) might hold your first Variable Checkerspots (*Euphydryas chalcedona*), Lorquin's Admirals (*Limenitis lorquini*), and a tiny blue butterfly dancing along the canopy—the Echo Blue (*Celastrina echo*). You'll cross Mitchell Creek in about a mile. This is when the swallowtails should start appearing, and one has the distinct possibility of seeing five members of this family in a day, the largest of which, the Two-tailed Tiger Swallowtail (*Papilio multicaudata*), scales out as our largest butterfly in the Bay Area. All five—the

Anise, the Pipevine, the Pale, the Western Tiger, and the Two-tailed can be seen along the creek. Only one other in our Greater Bay Area, the Indra Swallowtail (*Papilio indra*), flies more to the north. It's rather easy to key the Two-tailed Tiger away from the other two yellow floating butterflies here: Two-tails are a lighter, lemon yellow on the wings than the more vibrant yellow of the Western Tiger, and Anise Swallowtails have black shoulders and are relatively smaller than the other two. The only other place one could see five swallowtails in a day is Mount Wanda at the John Muir National Historic Site in Martinez.

Lower Mitchell Canyon has so many butterfly species possibilities I'm bound to leave out a few. Northern Checkerspots (*Chlosyne palla*) will patrol the road before, and so will Mylitta Crescents (*Phyciodes mylitta*). A Mourning Cloak (*Nymphalis antiopa*) might be scared up out of the shade. Western Brown Elfins (*Callophrys augustinus*) will dart out from their perches to explore you. Don't forget to check flowers along the way. I've had Thicket Hairstreaks (*Callophrys spinetorum*) and Northern White Skippers (*Heliopetes ericetorum*) here many times.

A lovely picnic table greets you at the two-mile mark below an oak tree on the left. This is a good place to fuel up. Now, it's perfectly respectable to call it a day at this point and head back to the car. The strenuous part starts after this with a series of steep switchbacks. If you do decide to go on, you'll be rewarded with a whole new set of butterflies.

Around the third switchback and especially earlier in the spring, LuEsther's Checkerspot (*Euphydryas editha luestherae*) shows up. Smaller, redder, and with round forewings, this checkerspot looks much different than the ubiquitous Variable. This is a creature of the serpentine rock ecosystem, and the endemic type of wildflowers that grow in that challenging environment begin to appear. And yet another checkerspot, Leanira (*Chlosyne leanira*), begins to appear in the next few switchbacks. (Please drink more water at this point—the incline will kick your butt.) Its host plant is warrior's plume (*Pedicularis densiflora*), and the checkerspot's eye-catching spider web pattern from below makes it utterly unique. Golden Hairstreaks (*Habrodais grunus*) perch on the golden oak trees just below Deer Flat—a good place to rest or lunch. Comstock's Callippe Silverspot (*Argynnis callippe comstocki*) can be seen sometimes gliding over the grasses at Deer Flats.

Proceeding on to Eagle's Peak, the creek sometimes streams over the path and offers the opportunity to see butterflies at the mud. Decades back I saw the unnamed Moss Elfin (*Callophrys mossii*) from the mountain lapping up the minerals there. If there is a California Tortoiseshell (*Nymphalis californica*) movement, you'll be sure to see them at the mud as well. A rather steep meadow appears on the right just below the summit of the peak. Here I always seem to find Gorgon Coppers (*Lycaena gorgon*) on the yellow flowers; their host plant coast buckwheat (*Eriogonum latifolium*) abounds in every direction. The trail straddles quite treacherously both sides of the ridgeline before hitting the actual summit. Many of the butterflies I mentioned in lower Mitchell Canyon will also be up here, looking for mates. Thicket Hairstreaks (*Callophrys spinetorum*) sit in the lower branches of the gray pine trees, their silhouettes, the size of a nickel, easily seen against the blue sky. If you missed the Leanira Checkerspot earlier, you have a second chance up here, as it is a strong hill-topper. Also, perhaps you will see a handful of Duskywing species. I was once very privileged to have found a Great Blue Hairstreak (*Atlides halesus*) sitting on the highest point on the highest bush possible, watching me catch my breath after my ascent. Once you get here, drink some water, and congratulations for climbing one of the highest peaks in the park. Rest. The good thing is, it truly is all downhill from here. On the descent I've seen Hedgerow Hairstreaks (*Satyrium saepium*) in the chamise bushes and Common Wood Nymphs (*Cercyonis pegala boopis*) where the trail opens up and then deposits you once again at the trailhead.

CHEWS RIDGE

Monterey County
When: June–July

Within the fog-shrouded mountains of the Santa Lucia Range in Monterey County lies a butterfly place of lore. Jim Kruse took me for the first time to Chews Ridge early in my journeyman years. We were on our way to the Hastings Natural History Reservation in Carmel Valley. "You won't believe this place," I remember him saying. "Butterflies just keep coming at you, and it can be a bit overwhelming." I have been there many times since, and it lives up to the banter among lepidopterists. In fact, Chews Ridge might be my number one recommendation out of all six destinations for butterflies.

To get there, take Carmel Valley Road from Highway 1 to the east. Go a good twenty miles until you see a right-hand turn onto Tassajara Road—take that turn. Now, the summit of Chews Ridge is approximately nine miles from this junction. Set your odometer and don't become frustrated, as a great deal of it is slow going, but even along the road there are a handful of spots worth checking out. The road crosses Conejo Creek at Cachagua Road. Pull out there and look for Sylvan Hairstreaks (*Satyrium sylvinus*) in the willows along the creek, and Lorquin's Admiral (*Limenitis lorquini*) will be among those willows too. A California Sister (*Adelpha californica*) might flap-flap-glide its way over to you from the oak trees.

The asphalt becomes a gravel washboard right around the Tassajara Buddhist retreat, and you have to really slow down. At some point the entire road is covered in a canopy of trees—pull out along the side (trust me you'll probably have the place to yourself, and the road is wide enough to double-park). Below the canopy, a large gathering of the California false indigo bushes lines the roadside. It's quite possible to see a butterfly that hosts on it along the sides of this alley, mainly California Dogfaces (*Zerene eurydice*). Remember the girl is a pale, lemon yellow in flight, and the boy has black forewings. Just when it feels like this road is never going to end, you arrive at the summit. Constantine and Nellie Chew homesteaded 315 acres here in the late nineteenth century. It is now a part of the Los Padres National Forest. You are allowed to carry a net here and collect if one wishes to do that.

As you pull up to the summit, park on the right-hand side and start on the trail going west. You'll immediately notice the fritillaries blasting about. Park yourself near one of the blooming Venus thistles for the best photographic opportunities—but not too close. There are three species here of Silverspots that are rather easy to distinguish from one another: the Coronis or Crown Fritillary (*Argynnis coronis*)—big and orange with oblong silver spots ventrally; the Callippe Fritillary (*Argynnis callippe*)—tawny and sandy brown with silver spots ventrally; and finally the star of Chews Ridge, the butterfly that draws folks here from all over the country if they are keeping a life list: the Unsilvered Fritillary (*Argynnis adiaste clemencei*). (I'm using a subspecies trinomial here to distinguish it from the population in Santa Cruz County, which is referred to as *A. adiaste adiaste*.) And why is it so easy to differentiate it from the other two on the wing here? Because it has no silver spots on the underside. This California endemic has one of the most narrow ranges of any of our state's butterflies.

The trail meanders and is sometimes hard to follow, but just use the worn path going from hilltop to hilltop. Pale Tiger Swallowtails (*Papilio eurymedon*) chase one another and are a supreme example of hill-topping behavior. It's a good spot for finding Gabb's Checkerspot (*Chlosyne gabbii*), *especially* since there are no Northern Checkerspots (*Chlosyne palla*) in Monterey County to confuse it with. When in full bloom, ceanothus and coffeeberry blossoms are magnets to an array of hairstreaks: Hedgerow (*Satyrium saepium*), Mountain Mahogany (*Satyrium tetra*), Golden (*Habrodais grunus*), California (*Satyrium californica*), and Gray Hairstreaks (*Strymon melinus*) are all possible. One of the more prominent lycaenids here is the rich, sable Western Brown Elfin (*Callophrys augustinus*).

Next, head back to your car and cross the street, then go up the road behind the gate. After an easy hike, go up to the fire tower that will be revealed on your right—the flowers below the tower will have lots of butterfly activity. It's a good place to work on your Duskywings (*Erynnis* spp.)—you may find Mournful (*E. tristis*), Propertius (*E. propertius*), and Pacuvius (*E. pacuvius*) and even recently a Sleepy Duskywing (*E. brizo lacustra*). They'll be virtually impossible to key apart on the wing as they all dance madly, looking for the right mates. Below the back side of the tower in the dirt has been for many years a reliable spot for a Columbian Skipper (*Hesperia columbia*). Follow the slope down behind the tower, make your way through an opening in the hedges, and the place opens up again from denser vegetation. This is a great spot for Silver-spotted Skippers (*Epargyreus clarus*). As you are walking back to the car, take in the spectacular view from Chews Ridge and start dreaming about your next visit.

If you go there a little earlier, say April or May, you might come across the Arrowhead Blue (*Glaucopsyche piasus*), a species reported from here but in need of more sightings. Be sure to get that all-important photograph of the underwings—we need to see arrowheads, folks, for positive identification, since Boisduval's Blue (*Icaricia icariodes*) flies here as well, and one could easily confuse the two. Post your shot to iNaturalist, and you will actually be making a little history. The challenge is on.

PINNACLES NATIONAL PARK

San Benito County
When: May–July

Do you know what *xeric* means? Siri will tell you it's an ecological adjective describing a place that contains little moisture; very dry: "xeric conditions." The elements converge to give one a perfect example of this in Pinnacles National Park in San Benito County, a place that is so much a part of my butterfly adventure, I added this entire county just so I could write about the park. Created by Teddy Roosevelt in 1908 as a national monument, it was bumped up in status in 2013 to a national park. Visitors from all over the country are now discovering this unique oasis that many of us in the Bay Area have known about all along.

Paul Johnson has been a park ranger there for decades. I first met Paul back in the 1990s at the Big Creek Butterfly Count in Monterey, when he showed Jerry Powell and me his lepidoptera inventory he had just begun. It was beautifully pinned in boxes, and it was clear this guy was serious. A quiet and whip-smart man, Paul began the Pinnacles Butterfly Count at my suggestion, and it has grown into a swell event that kick-starts the butterfly count season on the first weekend in June of each year. I suggest you camp out the night before, as Paul gives a wonderful slideshow in an amphitheater campground that is the perfect primer for the novice.

The campground is one of my favorite places to look for butterflies. Blooming buckeye tree and California buckwheat blossoms there hold many of the Lycaenidae one could see throughout the day: San Bernardino Blues, Great Coppers, Gorgon Coppers. I've even had a Tailed Copper land on the asphalt before me there. (Within the scope of this book there is no greater place to observe your coppers than at Pinnacles.) There is a trail that goes east a little beyond the campgrounds called South Wilderness. Take it as far as you can toward the boundary of the park, and you should be lucky enough to see Northern White Skippers.

Make your way up to the parking lot next to the visitor's center at Bear Gulch. Among the Acmons, Clemence's, and San Bernardino Blues one can experience a maelstrom of hairstreaks—Hedgerow, Mountain Mahogany, and Gold-hunter's—there. It can be somewhat overwhelming. I suggest you take a minute and look up here at the magnificent geology shroud in colors that only Cézanne, the French Impressionist, could have painted. There is almost a blue tinge to them.

Over seventy species of butterflies have been recorded by Paul in the park over the years.

Going in the earlier spring may reveal Sonoran Blues on the High Balconies Trail. The primary skipper, the Rural Skipper, darts about with its heavily teethed forewings. Be prepared that one can hit it on a scorcher of a day. Bring and carry on the trail more water than you think you need.

At the end of the day in Pinnacles, the senses can be overloaded. Your list of species you saw will be bigger than you had expected. Treat yourself to some ice cream or a cold drink in the Pinnacles Park Store back at the campground. You might even run into Paul there. He loves to hear about what you've seen and will gladly help you with any sightings you might have struggled with.

It's the perfect Butterflyland.

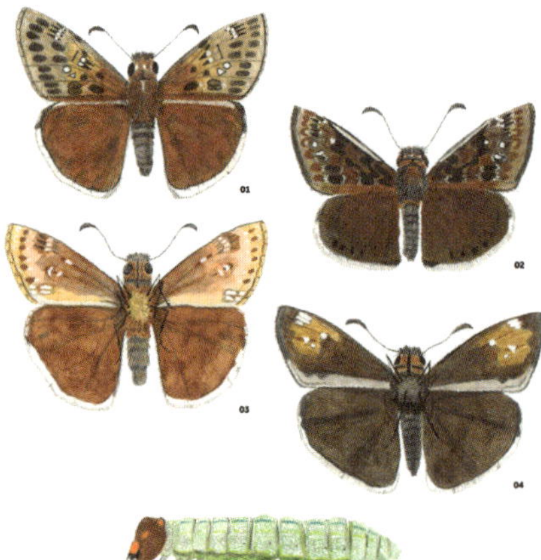

01: dorsal ♀
02: dorsal ♂
03: ventral ♀
04: ventral ♂
05: mature larva

MOURNFUL DUSKYWING

Boisduval 1852

Erynnis tristis

One of the easier ones to learn when you are first starting out because of the white fringe it has along its hind-wing border. Our other common Duskywing, Propertius, does not have white fringe. The only other Duskywing you could confuse it with that has the white fringe is the Funereal, which shows up to our area from time to time in abundance. Here's an easy rule of thumb: if you are in an area vastly different from an oak woodland and you have a Duskywing with white fringe, you are probably looking at a Funereal.

Habitat: Oak woodlands
Host Plants: Coast live oak (*Quercus agrifolia*) and others in the genus *Quercus*
Life Phases: Pupa olive gray. In adults: the general field (or ground color) to the female is a sable brown and the male is a dark, mud brown. Both sexes have a distinct pattern to their forewings. Much like the Propertius Duskywing but multivoltine: (many flights) March–October.
Counties They Fly In: All except San Francisco (though occasional vagrants have been verified there along the county line)
Great Places to See Them: Literally any place that has an oak forest. Our most ubiquitous Duskywing, which, when on the wing, can be quite widespread. Sierra Azul Open Space Preserve (Santa Clara County); Marsh Creek Regional Trail, Mount Diablo State Park (Contra Costa County).

The Structure of These Entries

I've arranged each entry according to the following structure: an introduction to the butterfly, its habitat, its host plant or plants, life phases, the county or counties the butterfly flies in, and great places to see them. I didn't want to bog down the learning aspect with a lot of verbiage but rather wanted to focus on the paintings and the physical details of each species. Each entry begins with the common and scientific name, a silhouette depicting the actual size of an adult, and the person who brought it to science. After sharing cool facts about the creature, I break down information about each species into the following sections:

Habitat: The ecosystem of which the butterfly is a part, usually due to a unique host plant which is part of that community. In an effort to streamline things, I don't go into great detail about each of these communities (chaparral, oak and willow riparian, et cetera), but each of them is present in the glossary of terms. Some of them, like "vacant lot," "forest clearings," and "open meadows," are general enough to be self-explanatory.

Host Plant: You might be surprised to find that wonderful backdoor butterflies are a wonderful back door into learning your botany. Simply put, the host plant is the plant or family of plants that the female lays her eggs on. Mastering the host plants will definitely give you an advantage in finding butterflies. Some butterflies use the host to either sit on or patrol for mates (Mission Blues on lupines, Coastal Green Hairstreaks on coast buckwheat) for most of their flight. Others dance about the canopy of their host tree (yes, trees can be hosts), like Echo Azures on California buckeye and Golden Hairstreaks around golden oak and tan oak. I'd say seven out of ten times when you find the host plant the butterfly is not far off.

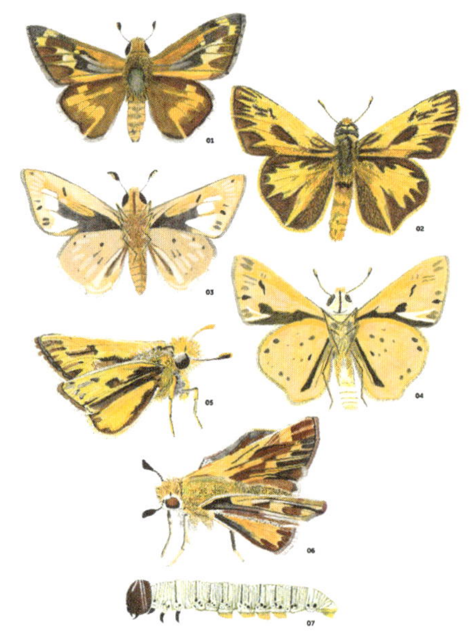

01: dorsal ♀
02: dorsal ♂
03: ventral ♀
04: ventral ♂
05: bombardier ♀
06: bombardier ♂
07: mature larva

FIERY SKIPPER

Drury 1773

Hylephila phyleus

This flash of yellow orange zipping about neighborhoods makes this creature quite well known. Sometime between being a toddler and kindergartener, I remember lying on the cool grass and watching them. Identifying this butterfly can be a head-scratcher for the novice because the female and male have two very different and distinct underside patterns: the female has a complicated field of beige spots while the male has been described as having a "pepper grinder" smattering of dark spots against a tawny field. Both have heavily teethed forewings.

Habitat: Urban areas, city parks, and coastal marshes
Host Plants: Bermuda grass (*Cynodon dactylon*), alkali grass (*Distichlis spicata*), and the American Lawn (a composite of grass species)
Life Phases: Pupa yellow brown. Larvae have adapted to the constant mowing of the American Lawn by placing their shelters horizontally at the base of the grass (Howe.1975.490). Adults have short antennae. Multivoltine (many flights): March–November.
Counties They Fly In: All
Great Places to See Them: Almost any suburban subdivision or county park where large swaths of lawn occur. At the Great Lawn in Golden Gate Park (San Francisco County) or out at India Basin and Heron's Head Park one can observe them using their original host, the alkali grass, along the seashore.

Life Phases: I explain how the butterfly appears in each of the four life cycles. You'll quickly notice that I make almost no visual reference to the pupae, chrysalises or cocoons, or the eggs. That is my fault, folks. Something had to go; otherwise I'd have been painting for another few years, and it's something I hope to rectify in a later edition. Lucky for us most of them are brown or green, and some indeed are quite spectacular. I've painted a few here. I've tried to stick to concise bullet points for each of the larvae and adults (also called the imago). A butterfly either has a single flight per season: univoltine; two flights: bivoltine; or many flights all year long: multivoltine. Life spans are in reference to how long the adult butterfly lives.

Counties They Fly In: At the time of publication of this book, this represents the most up-to-date knowledge of the actual presence of a species within a county, compiled from iNaturalist postings, personal and others' observations, and the work of John Steiner. I'm not big on range maps, maps that guesstimate where a butterfly should be. Authors use known range maps from one another, thereby compounding wrong data, and it goes on and on. One book has the California Dogface (*Zerene eurydice*) listed as "common" in the county of San Francisco. I wish.

Great Places to See Them: This is the cornerstone of the book and a primary reason why I wanted to write it. It's a wonderful level of knowledge when you are able to identify a butterfly on the wing; it's a whole other level to seek out a particular species at a certain locale. I've collected these special places over decades. No doubt I'll receive pushback from those who think these special places should remain a secret or God forbid the dreaded collector go on a frenzy there. You'll just have to trust me, but in the thirty years of doing field work I have never run across this person. I'm not saying they don't exist, but most places I'm sending you to are places where you can't wield a net. The fear is not in proportion to reality. Get your photograph and leave no mark.

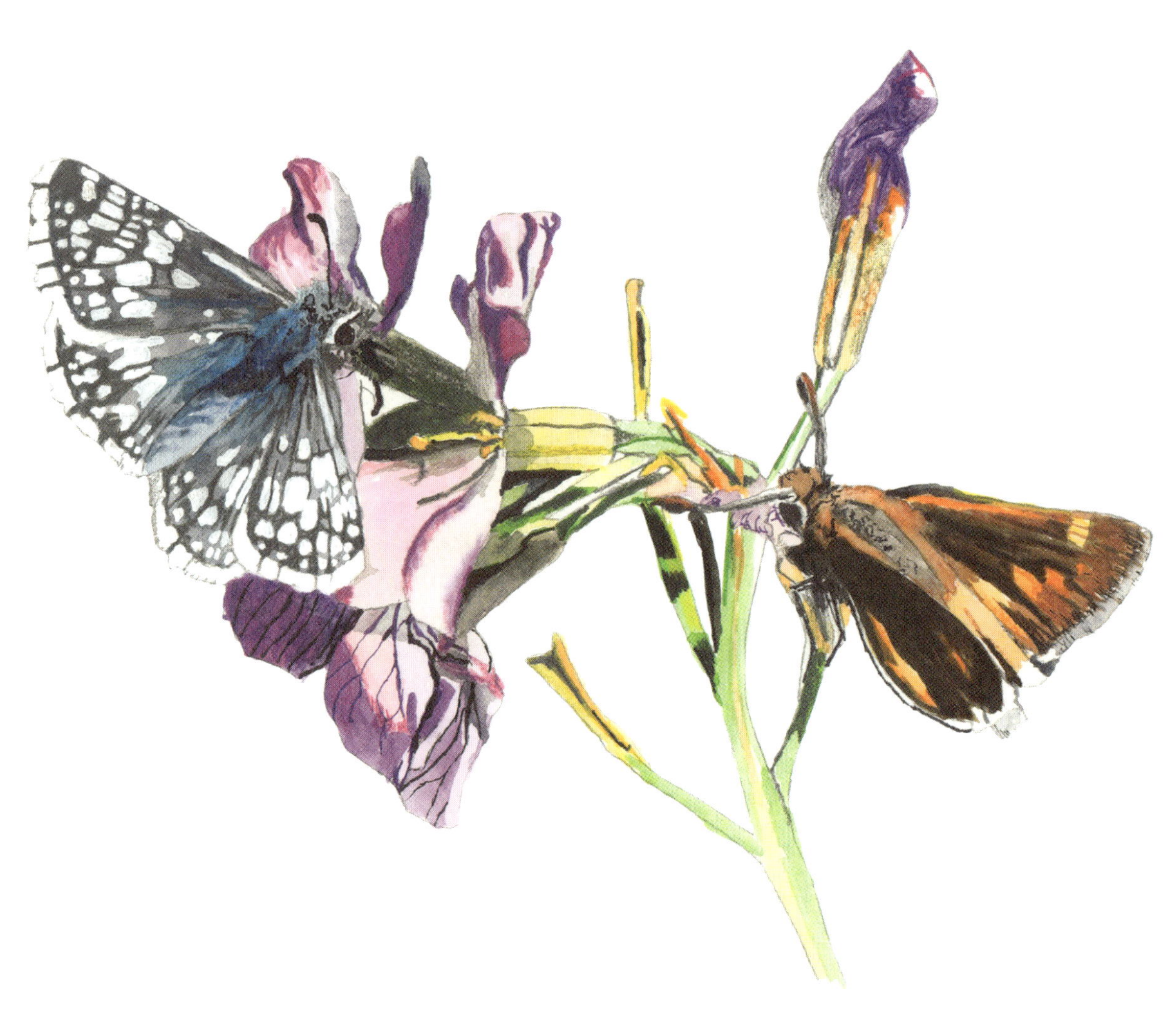

The Skippers (*Hesperiidae*)

Once thought to be halfway between a butterfly and a moth, skippers are diurnal (day-flying) and named for their quick, darting movements. Most have clubbed antennae tips and lack the wing-coupling structure of moths. The greatest diversity of species of skippers is found in the Neotropics, more than 3,500 in all worldwide.

Skippers are in the Order Lepidoptera (moths and butterflies) within the Family Hesperiidae, which now belongs to the Superfamily Papilionoidea—the superfamily of butterflies. Previously Hesperiidae was placed in its own superfamily, Hesperioidea, but that has been revised: skippers are butterflies. They have wide heads, and their wings are triangular.

We have about thirty species in the Greater Bay Area, each belonging to one of the three subfamilies: the spreadwings (Pyrginae), which hold their wings flat to the ground for the most part; the branded or grass skippers (Hesperiinae), which sit in what is referred to as the bombardier position—forewings up hind wings out like little jets on an aircraft carrier; and a lone representative within the skipperlings group (Heteropterinae).

One of my favorite new facts that I learned while researching this book: all skipper larvae have different "faces" on their caterpillar heads. No two are alike. Usually exaggerated with large black spots for eyes (their actual eyes being much smaller), they come in an array of unique patterns.

Umber and Common Checkered Skippers

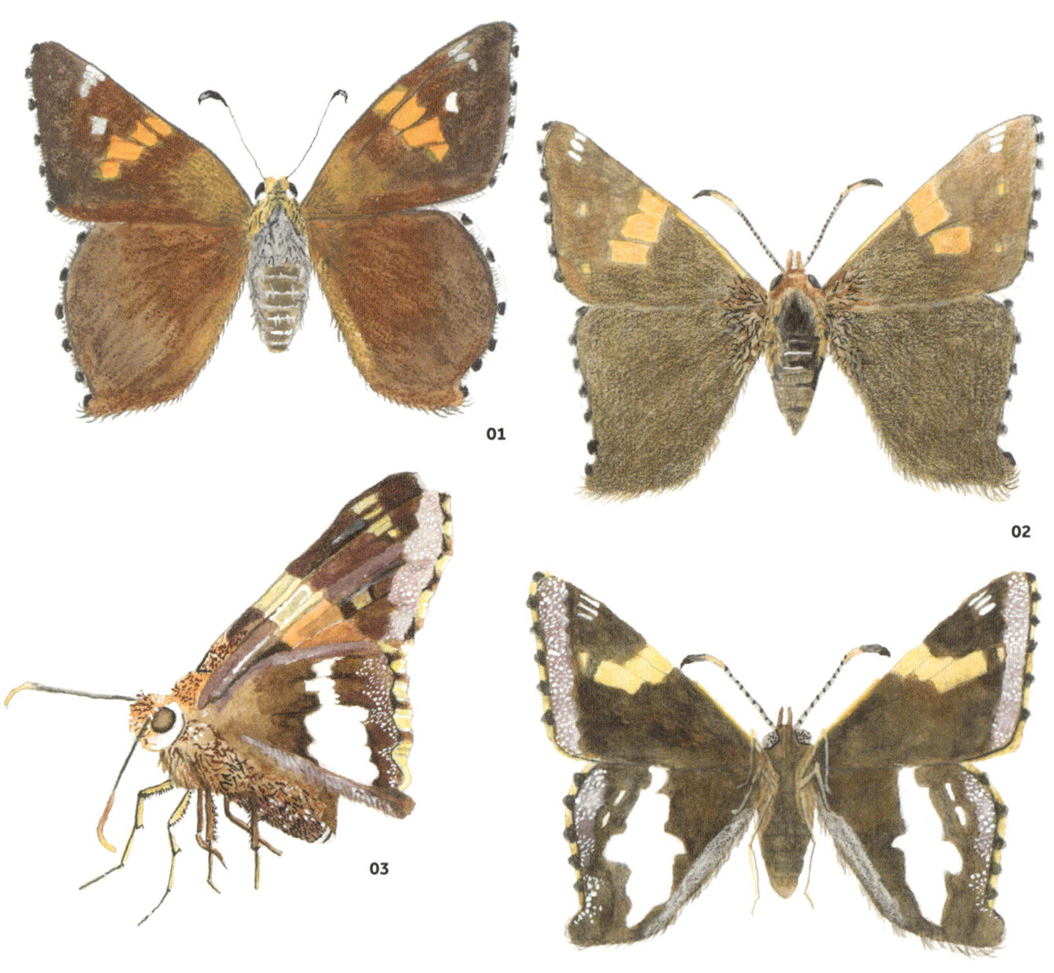
01
02
03
04
05

SILVER-SPOTTED SKIPPER

MacNeil 1975

Epargyreus clarus

The most recognizable and largest skipper in our area. Strangely, it was not named till the seventies yet flies all over our country. For sheer heft, there is nothing else like it. Females are slightly bigger. I love how it hangs upside down from flowers to nectar.

Habitat: Riparian, chaparral
Host Plants: False indigo (*Amorpha californica*), locust (*Robinia* spp.), and pea (*Lathyrus* spp.)
Life Phases: Eggs have a red dot on top with greenish ribbing alongside. First instar makes shelter from leaves. Pupa dark brown. Mature larvae overwinter in shelters. Univoltine: May–early July.
Counties They Fly In: Marin, Monterey, Napa, Solano, and Sonoma
Great Places to See Them: Chews Ridge, Carmel Valley (Monterey County)—see **Best Butterfly Walks** for directions; Bothe-Napa Valley State Park, Calistoga (Napa County); Los Padres National Forest (Monterey County); Glen Ellen (Sonoma County); Rincon Valley Community Park, Santa Rosa (Sonoma County). My favorite new place to see them occurred in the spring of 2023—Kent Pump Road (Marin County). Walk a good two miles back there and you should start seeing them.

01: dorsal ♀
02: dorsal ♂
03: in situ
04: ventral
05: mature larva

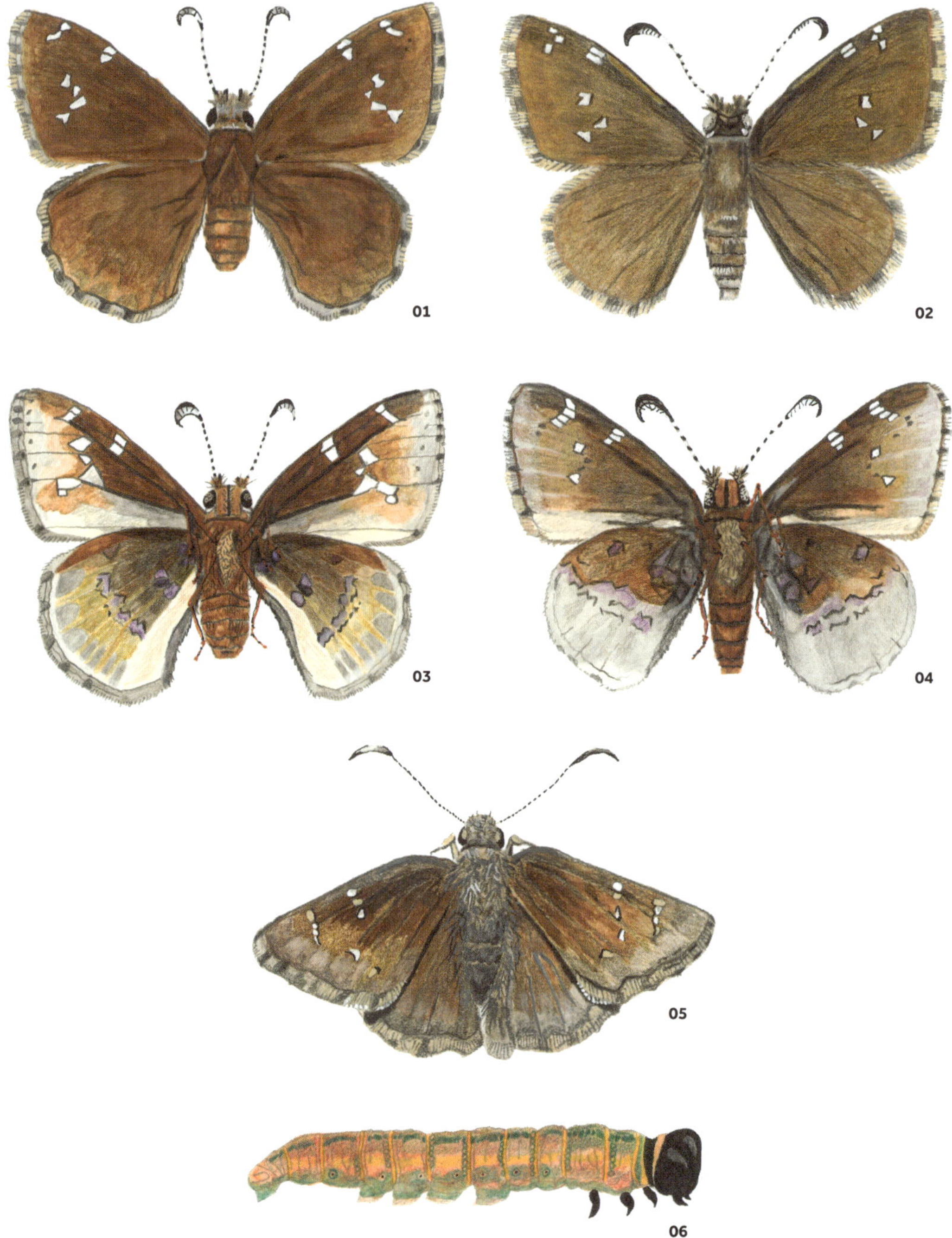
01
02
03
04
05
06

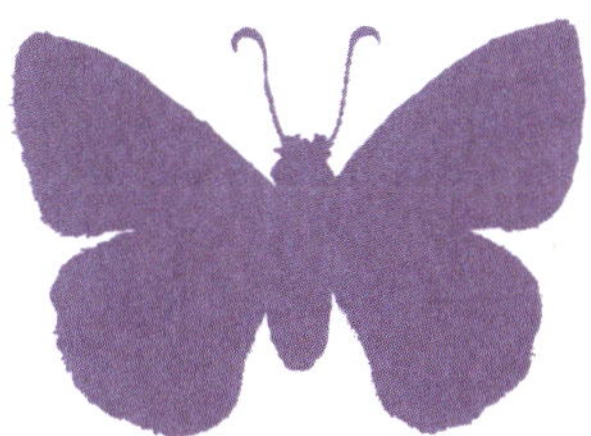

NORTHERN CLOUDYWING

Scudder 1870

Thorybes pylades

You won't find a more widespread member of the genus *Thorybes* in North America than this species. The ventral side was a revelation to paint, holding spots of lavender here and there. Enjoy the simplicity of the Cloudywing here, folks, because you'll want to pull your hair out over the next group—the Duskywings.

Habitat: Riparian, oak woodland, chaparral, forest clearings, canyons, and open meadows
Host Plants: Various clover (*Trifolium* spp.), alfalfa (*Medicago sativa*), false indigo (*Amorpha californica*), bird's-foot trefoil (*Lotus* spp.), and locoweed (*Astragalus* spp.)
Life Phases: Overwinter as mature larvae, building a nest from leaf litter. Pupa greenish brown. As an adult, the sable brown dorsal side has white dots in rows of twos and threes toward the apex of the forewing. Univoltine: March–July.
Counties They Fly In: All except San Francisco
Great Places to See Them: Hastings Natural History Reservation, Carmel Valley (Monterey County); Chews Ridge, Carmel Valley, (Monterey County); Silver Peak Wilderness, Big Sur (Monterey County); along Calistoga Road, Santa Rosa (Sonoma County); Oat Hill Mine Trail (Napa County); Skyline Wilderness Park (Napa County); Baldy Ryan Creek, Morgan Hill (Santa Clara County). One was seen on San Bruno Mountain along the Ridge Trail in the 2023 season (San Mateo County).

01: dorsal ♀
02: : dorsal ♂
03: ventral ♀
04: ventral ♂
05: wings flat
06: mature larva

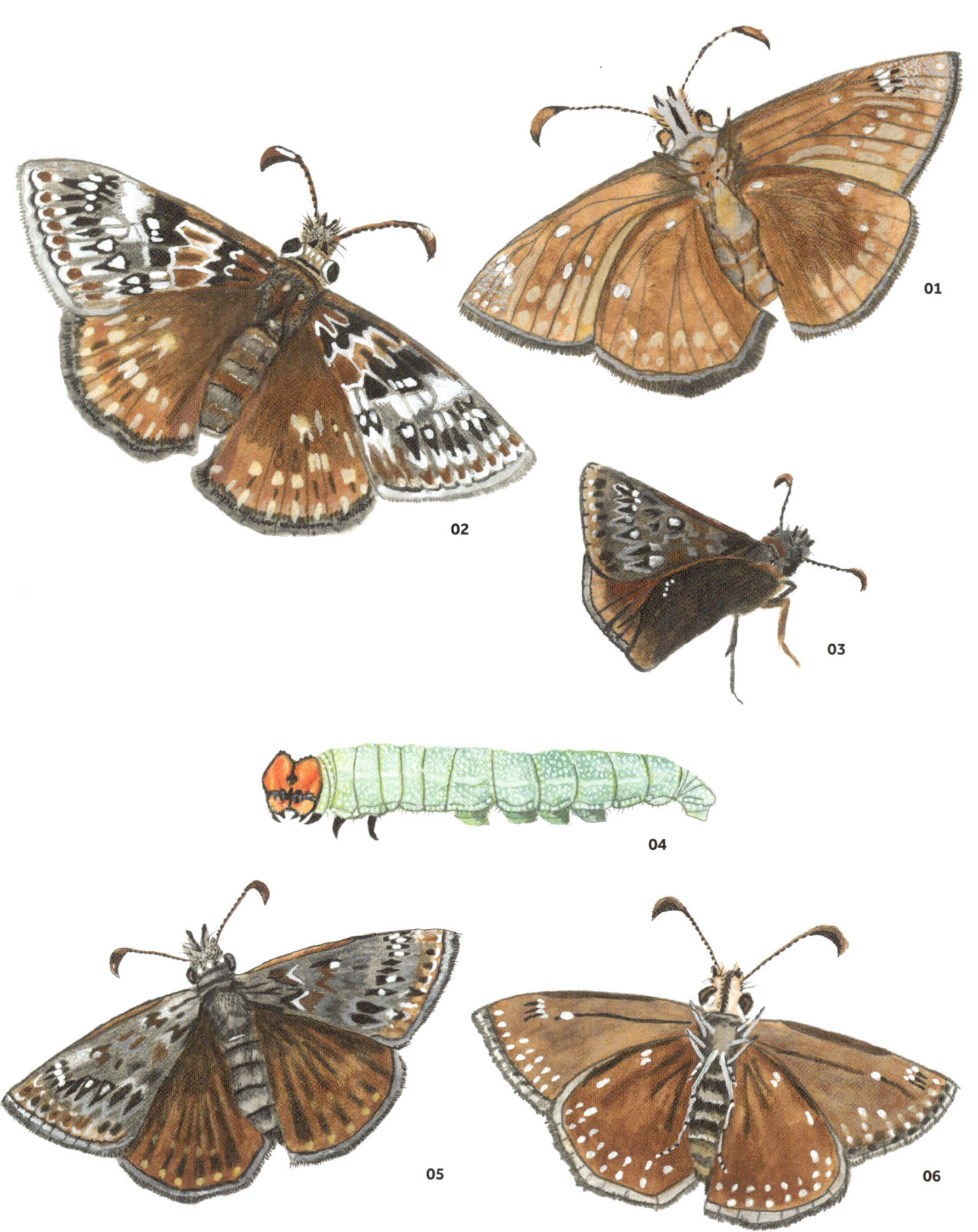
01
02
03
04
05
06

PROPERTIUS DUSKYWING

Scudder & Burgess 1870

Erynnis propertius

This is a spring butterfly that sits with its wings open like other Duskywings. There are glassy spots on the forewings among the brown and black brindled pattern. No white fringe on Propertius, which, when flying with the Mournful Duskywing, is a quick way to key them apart. Adults sit with wings half up.

Habitat: Oak woodlands
Host Plant: Coast live oak (*Quercus agrifolia*)
Life Phases: Eggs are laid on new growth. Five instars (or molts) with the fifth one overwintering. Pupa wing case is whitish with overall green to it. Univoltine: March–June.
Counties They Fly In: All except San Francisco
Great Places to See Them: Sugarloaf Ridge State Park (Sonoma County); Chews Ridge, Carmel Valley (Monterey County); Pinnacles National Park (San Benito County); Jasper Ridge Biological Preserve (San Mateo County)

01: ventral ♀
02: dorsal ♀, wings flat
03: wings half up ♂
04: dorsal ♂, wings flat
05: mature larva
06: ventral ♂

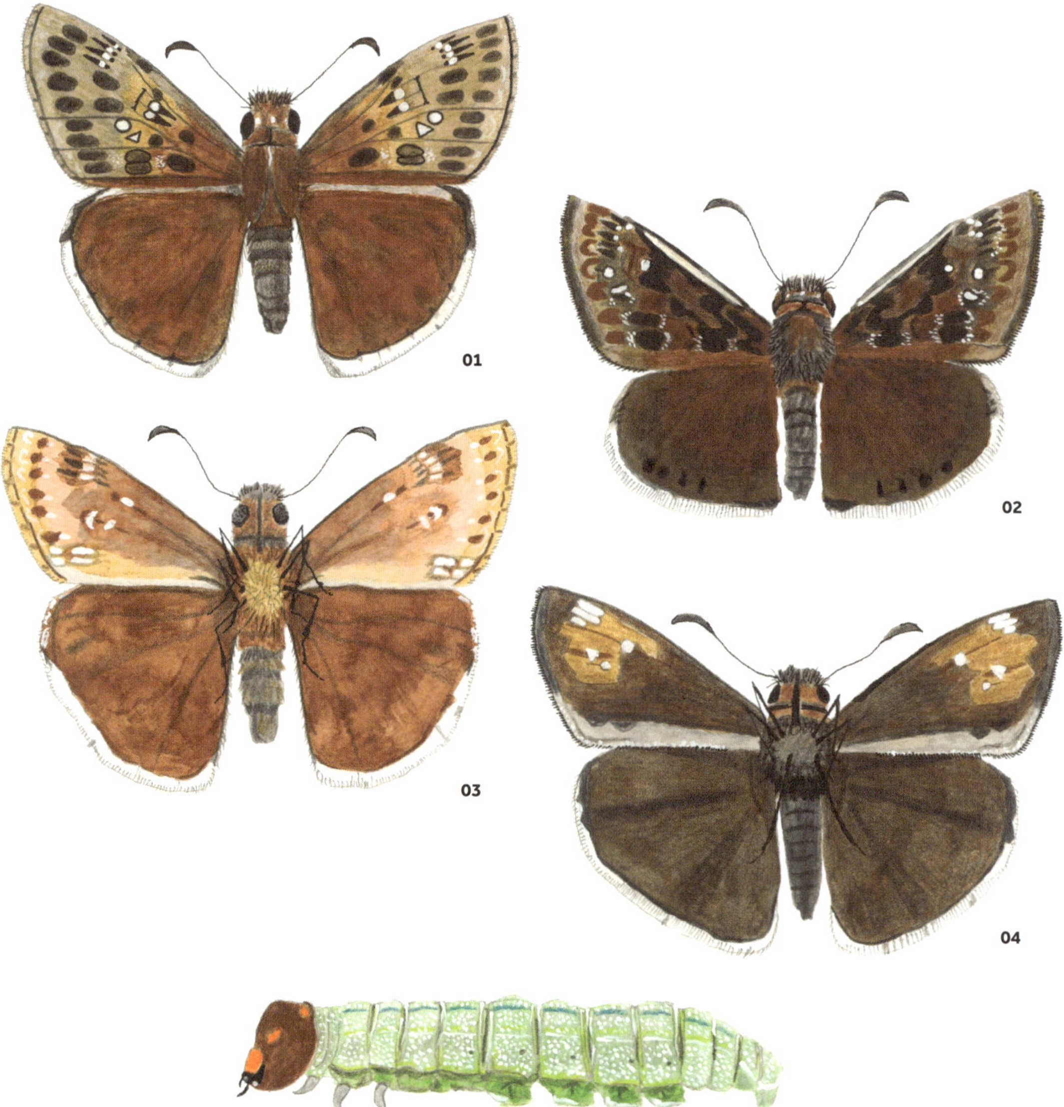
01
02
03
04
05

MOURNFUL DUSKYWING

Boisduval 1852

Erynnis tristis

One of the easier ones to learn when you are first starting out because of the white fringe it has along its hind-wing border. Our other common Duskywing, Propertius, does not have white fringe. The only other Duskywing you could confuse it with that has the white fringe is the Funereal, which shows up to our area from time to time in abundance. Here's an easy rule of thumb: if you are in an area vastly different from an oak woodland and you have a Duskywing with white fringe, you are probably looking at a Funereal.

Habitat: Oak woodlands

Host Plants: Coast live oak (*Quercus agrifolia*) and others in the genus *Quercus*

Life Phases: Pupa olive gray. In adults: the general field (or ground color) to the female is a sable brown and the male is a dark, mud brown. Both sexes have a distinct pattern to their forewings. Much like the Propertius Duskywing but multivoltine: (many flights) March–October.

Counties They Fly In: All except San Francisco (though occasional vagrants have been verified there along the county line)

Great Places to See Them: Literally any place that has an oak forest. Our most ubiquitous Duskywing, which, when on the wing, can be quite widespread. Sierra Azul Open Space Preserve (Santa Clara County); Marsh Creek Regional Trail, Mount Diablo State Park (Contra Costa County).

01: dorsal ♀
02: dorsal ♂
03: ventral ♀
04: ventral ♂
05: mature larva

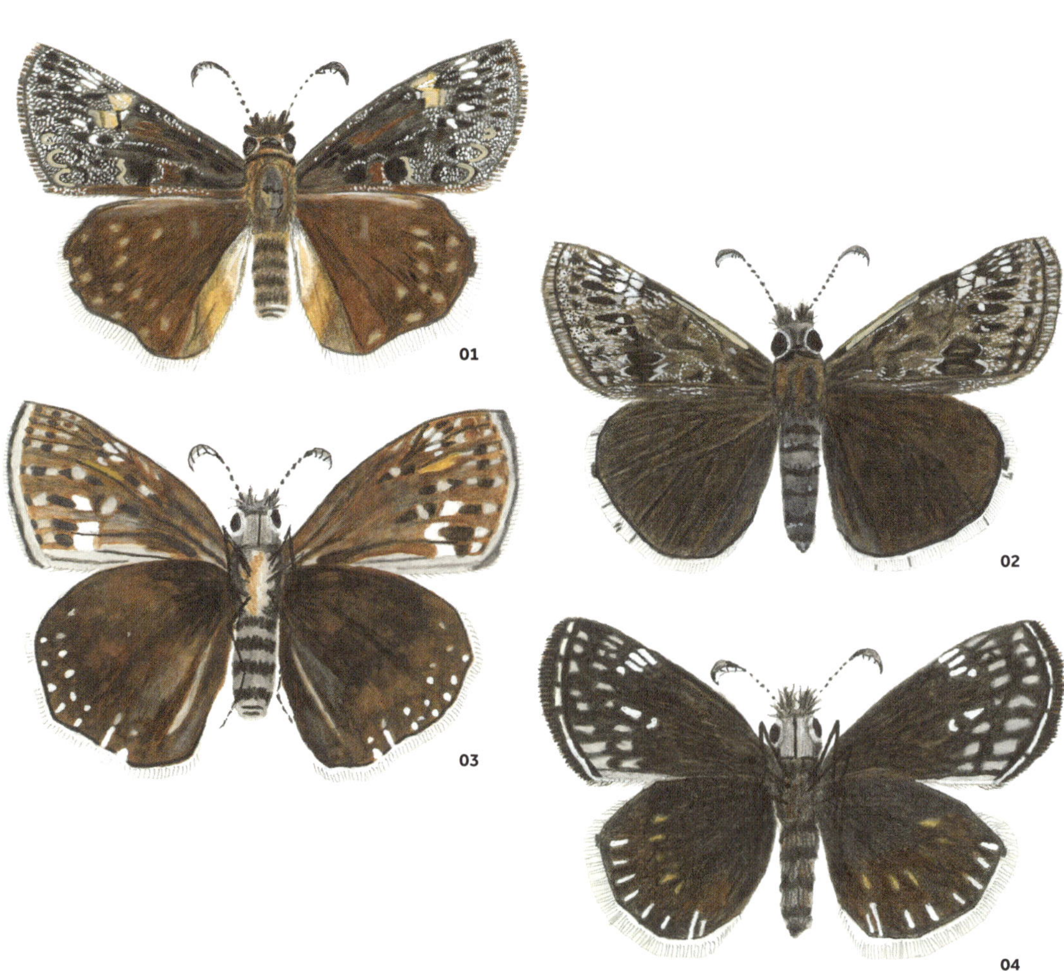

05

FUNEREAL DUSKYWING

Scudder & Burgess 1870

Erynnis funeralis

A butterfly that pushes into our area from the south every few years. There is a trick one learns to key this species away from the Mournful Duskywing (both have white fringe on their hind wings). The Funereal has elongated forewings and on them each has a large gray patch around the discal cell. The patch is more cream-colored in the females. In most years one won't have to struggle between the two identifications because the Funereal isn't around—when they are here, keying the two apart in the field can be a head-scratcher. Unlike Mournful, this butterfly is not known to hill-top.

Habitat: Many—from pristine native ecosystems to weedy vacant lots
Host Plants: Bird's-foot trefoil (*Lotus* spp.), deerweed (*Acmispon glaber*)
Life Phases: Eggs are laid on new plant growth. Fifth instar overwinters. Pupa is green. Adult female Funereals have a brown hind wing with muted spots. The adult male's hind wings are muddy brown. Both have a frosted forewing dorsally. Multivoltine: April–September.
Counties They Fly In: It can literally show up anywhere. I've seen it twice in San Francisco—once at Heron's Head Park and once at the Lobos Dunes of the Presidio.
Great Places to See Them: The best that I can do here is give you places it has shown up in the past, with no guarantees it'll be there again. Mount Diablo State Park (Contra Costa County); Briones Regional Park (Contra Costa County); Pescadero Creek Park (San Mateo County); the city of Brisbane (San Mateo County).

01: dorsal ♀
02: dorsal ♂
03: ventral ♀
04: ventral ♂
05: mature larva

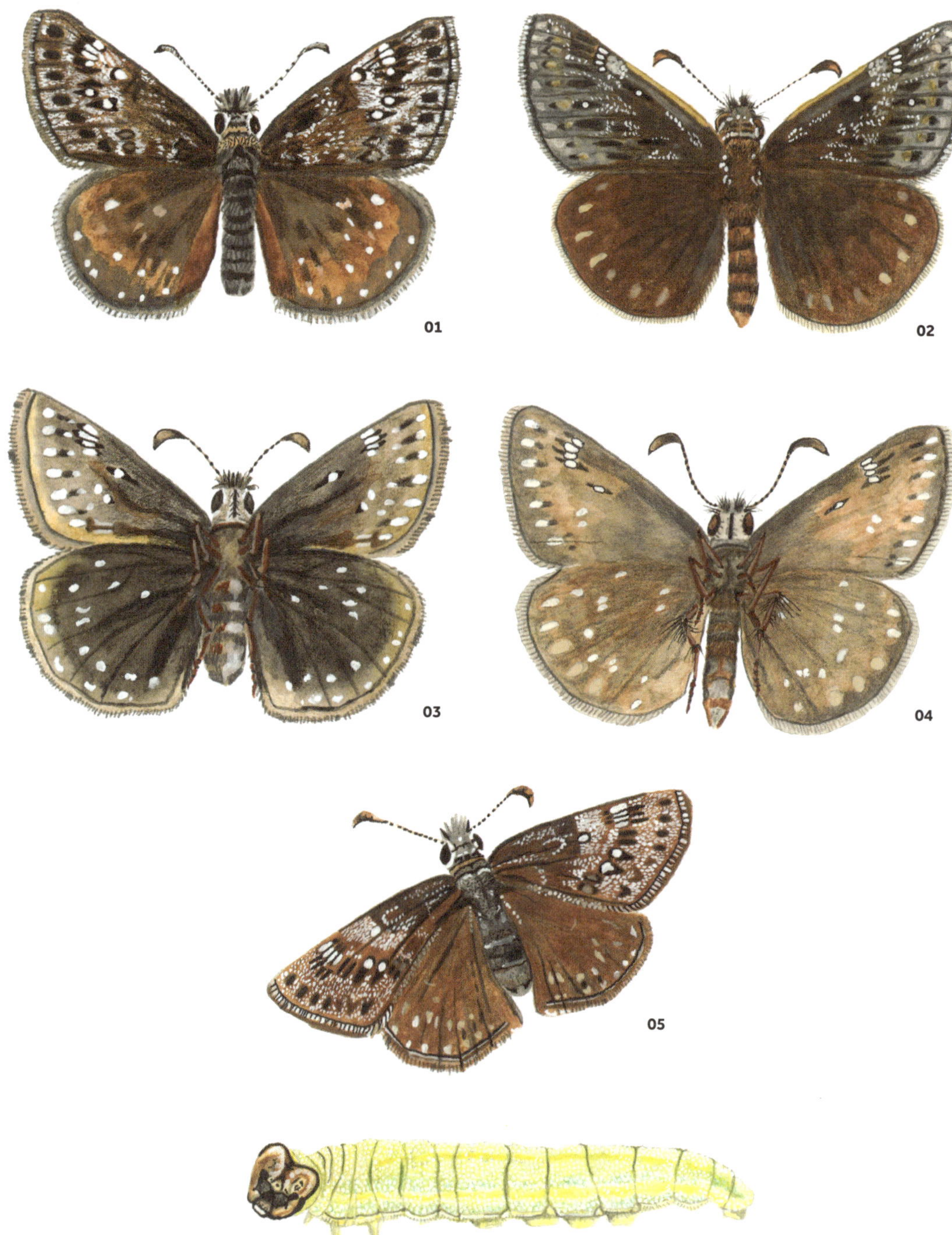
01
02
03
04
05
06

PERSIUS DUSKYWING

Scudder 1863

Erynnis persius

"A very difficult species to recognize" (Garth & Tilden.1986.171). Adults are strong mud-puddlers and hill-toppers. I've heard it is also called the Hairy Duskywing. To compare with the Pacuvius Duskywing, see **Appendix B, Tableau 1**.

Habitat: Riparian, chaparral, and evergreen forests
Host Plants: Willow (*Salix* spp.), cottonwood (*Populus* spp.)
Life Phases: Overwinters as last instar larva. Pupa similar to Funereal. Adult males have tufts of black hairs on their tibias (Pacuvius does not). Persius has a large, faded brown patch near the forewing discal cell (Pacuvius does not). Lots of white hairs on the forewing male giving him a rather hairy appearance. Persius has an extra hyaline spot on the forewing that Pacuvius does not. Univoltine: mainly in June.
Counties They Fly In: Contra Costa, Napa, San Mateo, Santa Clara, Santa Cruz, Solano, and Sonoma
Great Places to See Them: Cascade Canyon Preserve (Marin County); the Grove of the Stout-Hearted Wonders (Santa Cruz County); intersection of Gazos Creek Road and Cloverdale Road (San Mateo County)

01: dorsal ♀
02: dorsal ♂
03: ventral ♀
04: ventral ♂
05: wings flat ♀
06: mature larva

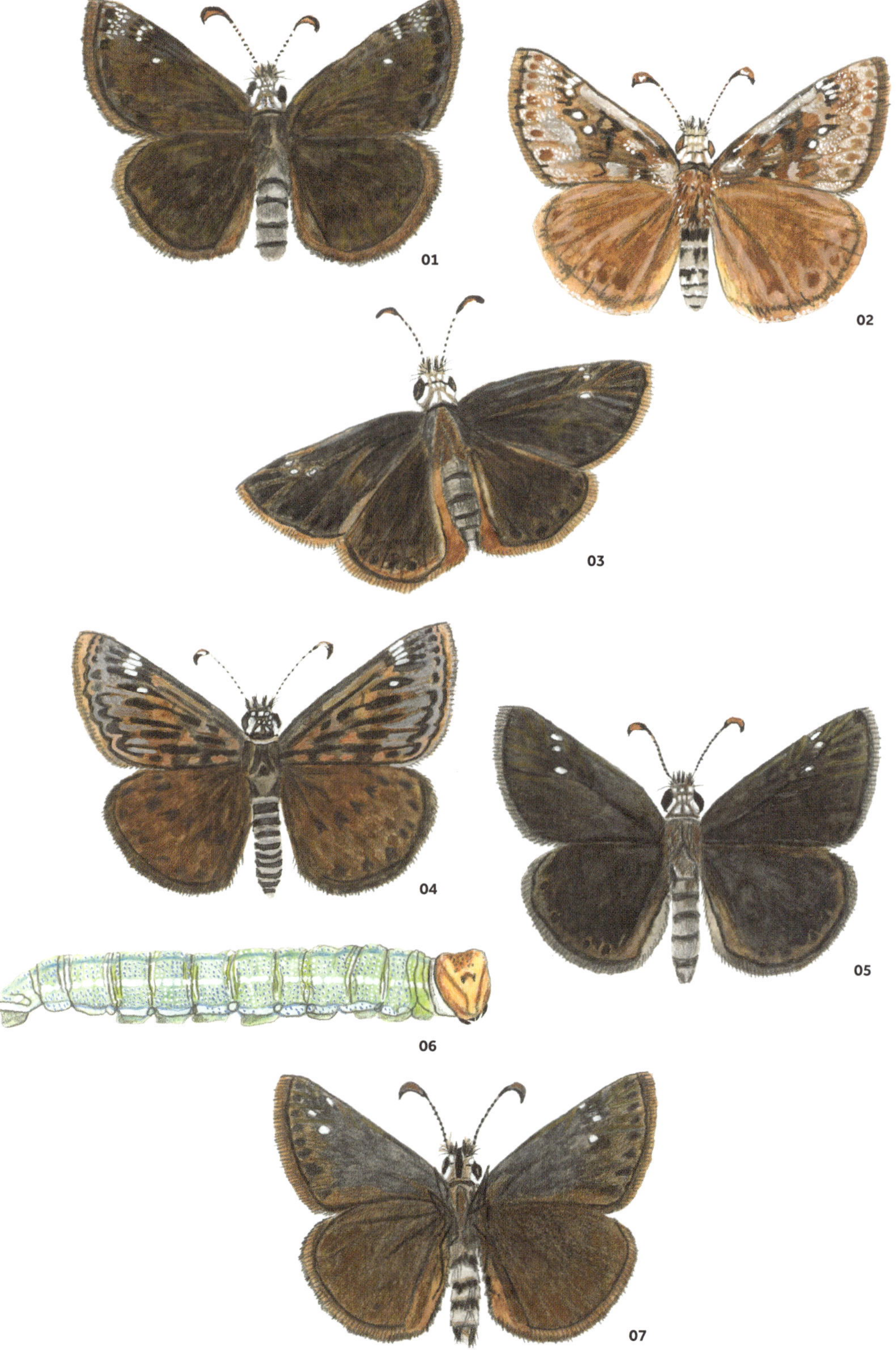
01
02
03
04
05
06
07

PACUVIUS DUSKYWING

Grinnel 1905

Erynnis pacuvius

01: dorsal ♀, dark Central California subspecies
02: dorsal ♀, Mendocino County
03: dorsal ♂, wings flat
04: dorsal ♂, dark Central California subspecies, variation one
05: dorsal ♂, dark Central California subspecies, variation two
06: mature larva
07: ventral ♂, dark Central California subspecies, variation two

Like some in the genus *Erynnis*, these butterflies sit with their wings held in a "V" pattern. Adults are strong mud-puddlers and hill-toppers. There are two forms (or variations) in our area: the more inland brown male and female, and a subspecies (*Erynnis pacuvius pernigra*) found along our coast with an extremely dark form (almost black). It's the darkest Duskywing we have. To compare with the Persius Duskywing, see **Appendix B, Tableau 1**.

Habitat: Chaparral, coastal scrub
Host Plant: California lilac (*Ceanothus* spp.)
Life Phases: Eggs are pale green and do not change color (Scott.1986.489). Pupa is dark greenish brown. Adult males lack the tufted tibia seen in Persius. In most places this is the smallest duskywing. Univoltine: May–June.
Counties They Fly In: Marin and Monterey.
Great Places to See Them: May be difficult to find. Chews Ridge (Monterey County); Goat Campground (Monterey County); China Camp Campground (Marin County)

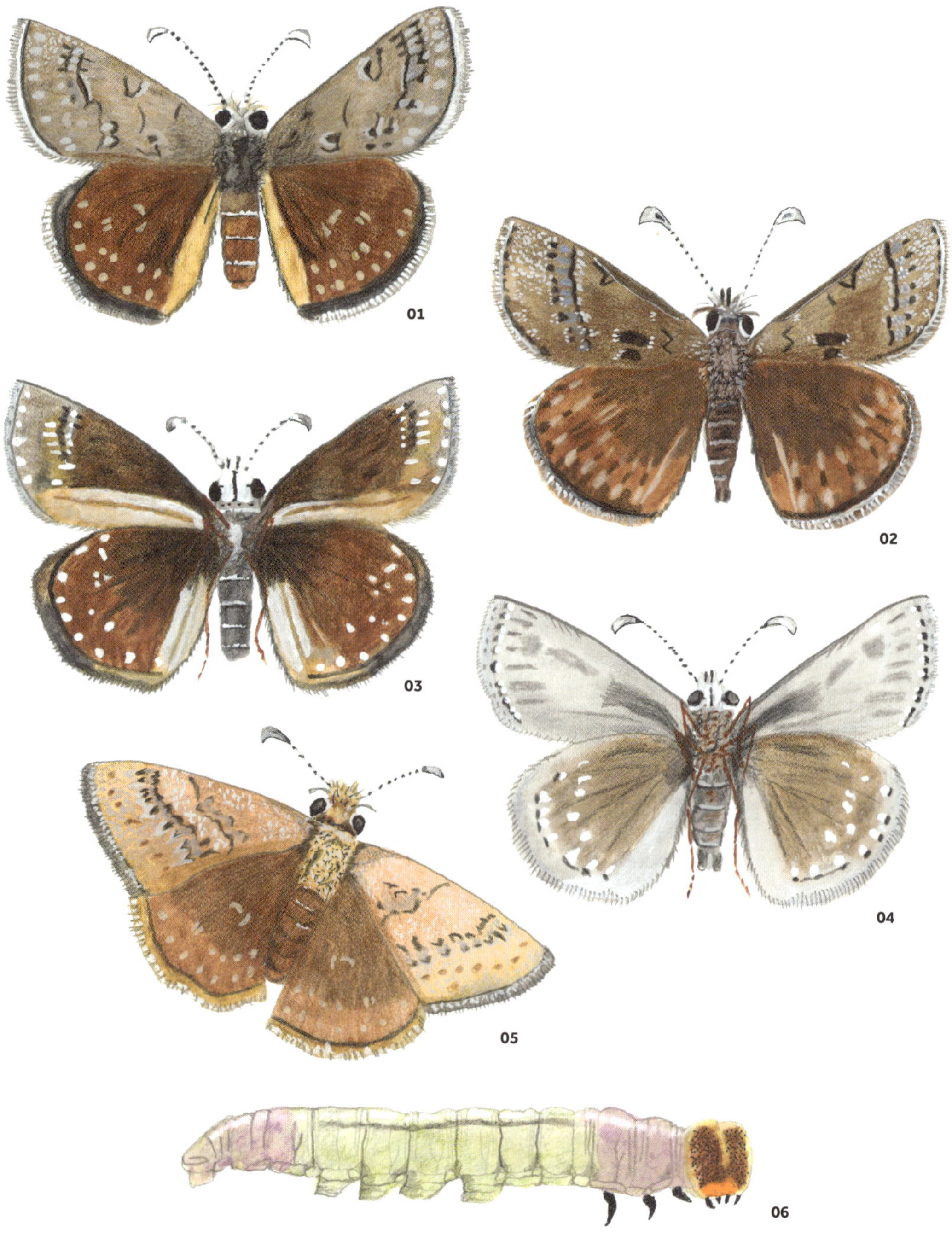
01
02
03
04
05
06

SLEEPY DUSKYWING

Wright 1905

Erynnis brizo lacustra

This is a butterfly of the serpentine community in our Greater Bay Area. An early emergent, "Sleepys" are strong hill-toppers. The females are only slightly lighter dorsally than the males. The spots on the female can have a hint of yellow to them.

Habitat: Chaparral, serpentine
Host Plants: Scrub oak (*Quercus durata*)—a chaparral plant confined to serpentine soils in Central California (Howe), Nuttall's scrub oak (*Quercus dumosa*)
Life Phases: Much like Funereal. Pupa green. Adults are uniformly gray above except for the dark, discal marks like a chain on the forewing. Univoltine: March–May.
Counties They Fly In: San Benito, Monterey, and Sonoma
Great Places to See Them: Pine Flat Road (Sonoma County); Hope Road, Healdsburg (Sonoma County); San Benito Mountain Research Natural Area (San Benito County)—this area needs a permit for entrance and requires a vehicle with high clearance.

01: dorsal ♀
02: dorsal ♂
03: ventral ♀
04: ventral ♂
05: wings flat ♀
06: mature larva

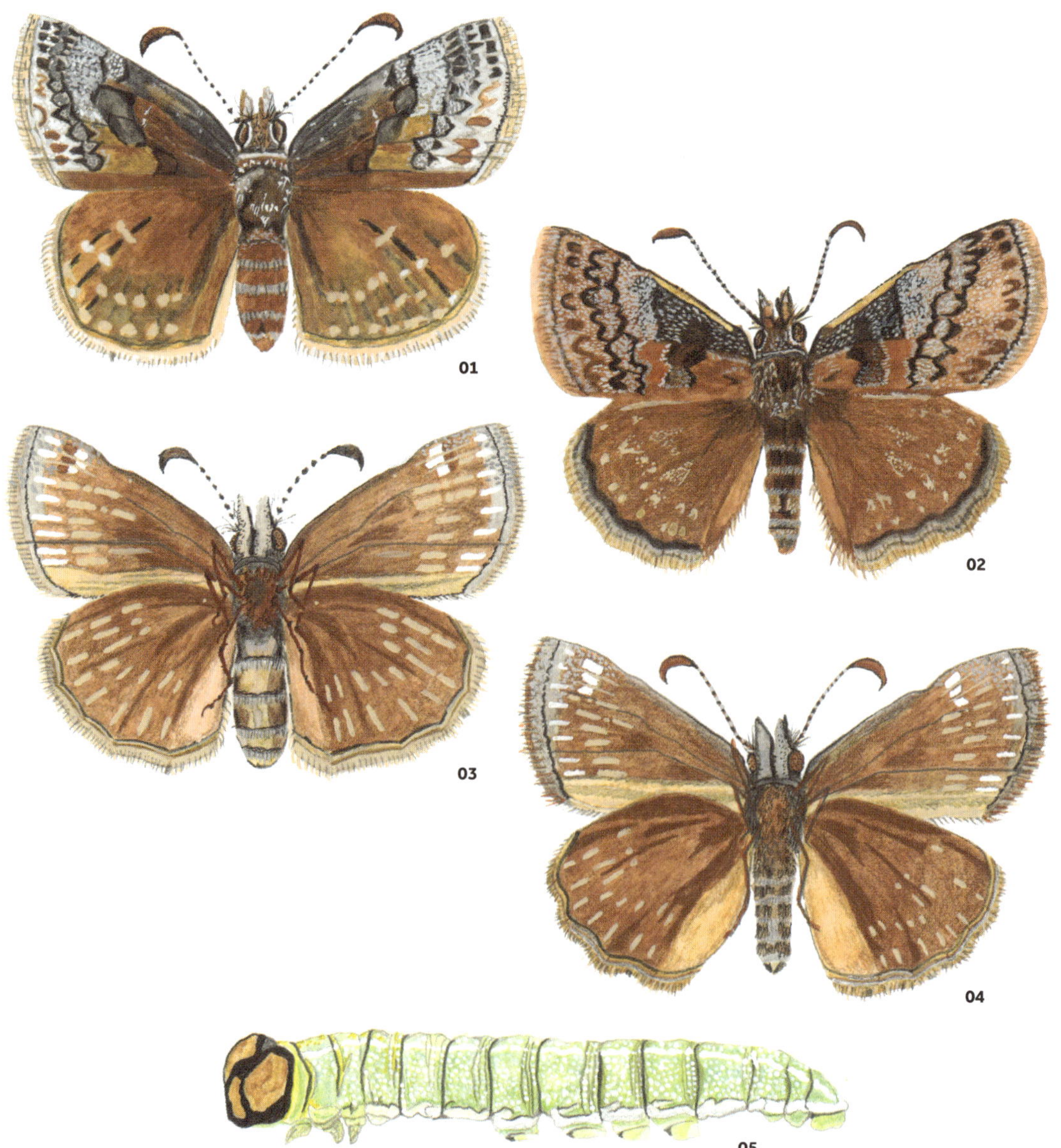
01
02
03
04
05

DREAMY DUSKYWING

Scudder and Burgess 1870

Erynnis icelus

Aside from one historic sighting, there are no modern records of this butterfly in any of the counties covered by this book. I am hesitant to even include it. I shied away from including vagrants (butterflies that have been seen only a handful of times in any of the counties) because the Duskywings we do have are tough enough to learn. But I already painted it . . . so might as well. It's a beautiful little butterfly. Dr. Art Shapiro is a firm believer that this butterfly has gone underreported in our area (Shapiro & Manolis.2007.223). Community scientists and iNaturalist users go forth! Find it and take some pictures of this bug.

Habitat: Riparian, glade openings in forests, creeks

Host Plants: Primarily willows (*Salix* spp.), cottonwood (*Populus* spp.)

Life Phases: Eggs are laid on the very tip of new growth of the host tree. Overwinter as larvae. Pupa is reddish or yellowish brown. In adults, no glassy white hyaline spots on forewings. A chain of white spots outlined in black running vertically through each of the forewings is a defining characteristic of the Dreamy. Univoltine: March–July.

Counties They Fly In: A single specimen was recorded by J. W. Tilden in the last century in Sonoma County.

Great Places to See Them: Unknown

01: dorsal ♀
02: dorsal ♂
03: ventral ♀
04: ventral ♂
05: mature larva

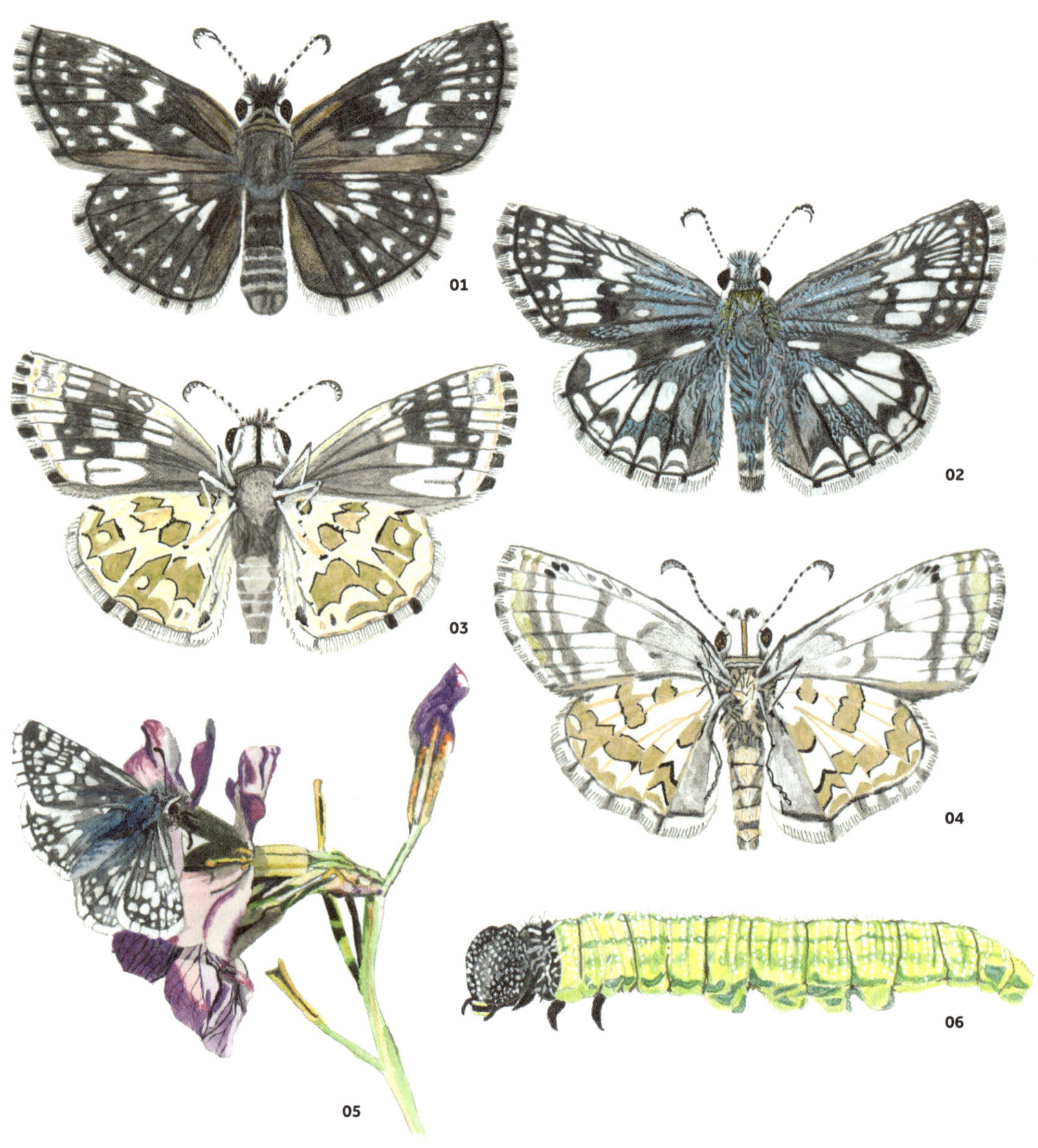
01
02
03
04
05
06

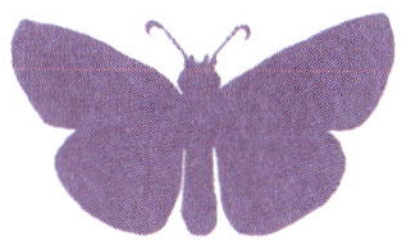

COMMON CHECKERED SKIPPER

Grote 1872

Burnsius communis

You'd be hard pressed not to see them everywhere—it's usually the first species I see in January for the New Year. In San Francisco this butterfly used to be univoltine, making one flight in the spring and laying eggs on our only native mallow—checkerbloom—as its host. Then a new, non-native mallow arrived: cheeseweed—a catch-all name for a handful of exotic mallows (*Malva* spp.), and because she now has new options to lay her eggs on, both he and she fly twelve months a year there, an evolution of multivoltine behavior. Other parts of the Greater Bay Area, the butterfly flies less than year round. This *Burnsius* as its genus is rather new. Older books will refer to it as *Pyrgus communis*. Many novices mistake this butterfly for a blue from the Lycaenidae family because of all the blue hairs all over it. I got called once to a neighborhood dispute on Potrero Hill in San Francisco over a vacant lot being developed. The anti-development folks were insistent they'd seen the endangered Mission Blue there (a butterfly that only flies on Twin Peaks in the city—it does not fly along its streets). "We've seen a blue butterfly!" The lot was resplendent with Common Checkered Skippers.

Habitat: From pristine native ecosystems to lowlands, vacant lots, and wastelands

Host Plants: Mallows (*Malva* spp.), dwarf checkerbloom (*Sidalcea malviflora*)

Life Phases: Eggs are a turban shape. The larvae roll a shelter out of the leaves they are feeding on. Pupa light green. The adult female is primarily dark gray with white checkerboarding. The male is primarily white with an abundance of sky-blue hairs on its wings, thorax, and abdomen. Multivoltine: (many flights) year-round.

Counties They Fly In: All

Great Places to See Them: Literally anywhere a butterfly flies: disturbed areas where *Malva* grows, vacant lots, city parks, residential yards, trailside, et cetera

01: dorsal ♀
02: dorsal ♂
03: ventral ♀
04: ventral ♂
05: wings flat ♂
06: mature larva

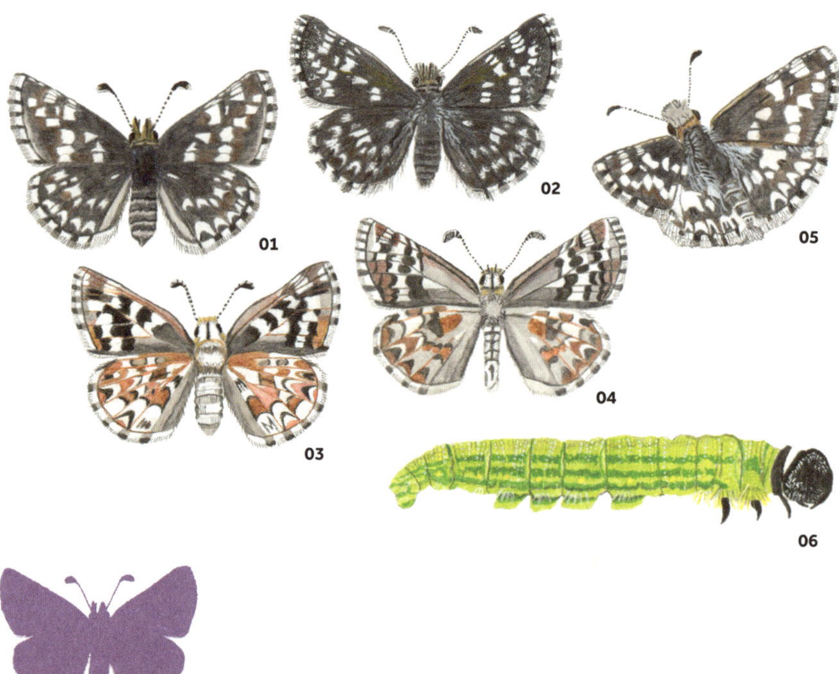

TWO-BANDED CHECKERED SKIPPER

Boisduval 1852

Pyrgus ruralis

Like the Mylitta Crescent (*Phyciodes mylitta*), this small butterfly with white checkered fringe patrols creeks and streams in pursuit of its mate. It is noticeably smaller than the Common Checkered Skipper (*Burnsius communis*) and is never found in any abundance in any area.

Habitat: Water concourses (streams, creeks, rivers) and coastal scrub
Host Plants: Cinquefoil (*Drymocallis glandulosa*), California horkelia (*Horkelia californica*)
Life Phases: Eggs laid on host plant individually. Pupa green and browner toward tip. The checkerboard pattern is primarily gray with white squares, the gray being more prominent on the female. Univoltine: May–June.
Counties They Fly In: All except Contra Costa, San Francisco, and Solano
Great Places to See Them: Garrapata State Park (Monterey County); San Vicente Trail, Moss Beach (Santa Cruz County); Burleigh H. Murray Ranch State Park (San Mateo County); Mount Madonna County Park (Santa Cruz County); Sanborn County Park (Santa Clara County); Gualala Point Regional Park Campground (Sonoma County)

01: dorsal ♀
02: dorsal ♂
03: ventral ♀
04: ventral ♂
05: wings half-open ♂
06: mature larva

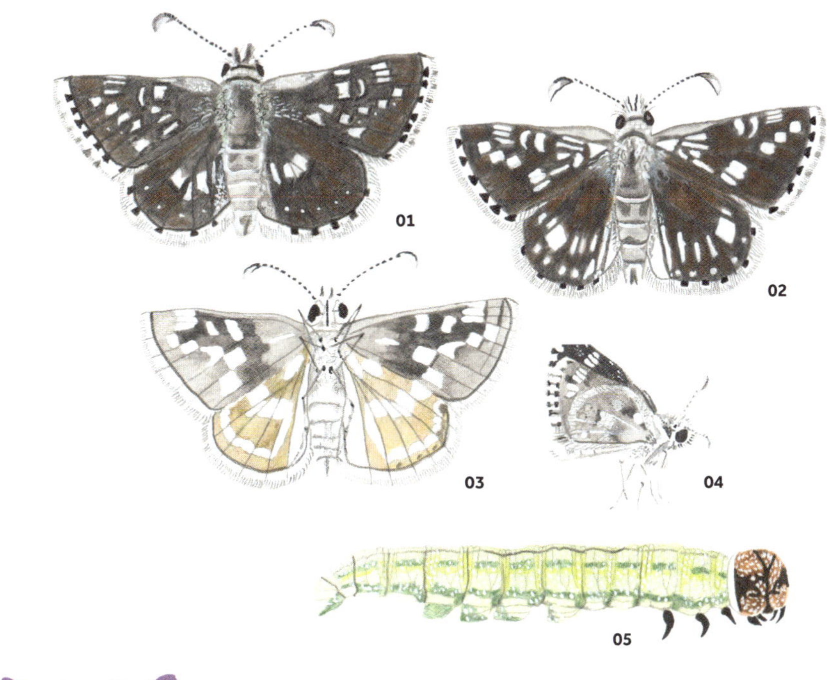

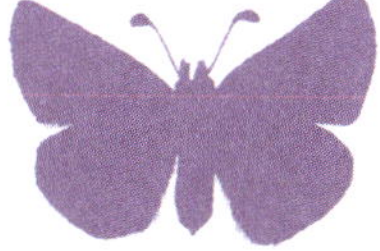

01: dorsal, wings flat, ♀ and ♂, summer
02: ventral, ♀ and ♂, summer
03: dorsal, ♀ and ♂, spring
04: ventral, wings up
05: mature larva

SMALL CHECKERED SKIPPER

Boisduval 1852

Pyrgus scriptura

The Small Checkered Skipper, like the Two-banded Skipper, holds its wings one-quarter open while at rest. Never far from its host, this butterfly for me has always presented itself while sitting on the host. Adults feed on mud, flowers, and dung.

Habitat: Salt marshes, disturbed areas and fields
Host Plant: Alkali mallow (*Malvella leprosa*)
Life Phases: Eggs laid singly on the host plant. Adults are incredibly small (think dime size). Males lack a costal fold. Females are the color of dirt. Multivoltine: March–November with fall population most visible.
Counties They Fly In: Alameda, Contra Costa, Napa, San Mateo, Santa Clara, Solano, and Sonoma
Great Places to See Them: Big Break Regional Trail (Contra Costa County); Corteva Wetlands Preserve, Worth Shaw Community Park, Antioch (Contra Costa County); Black Diamond Mines (Contra Costa County)

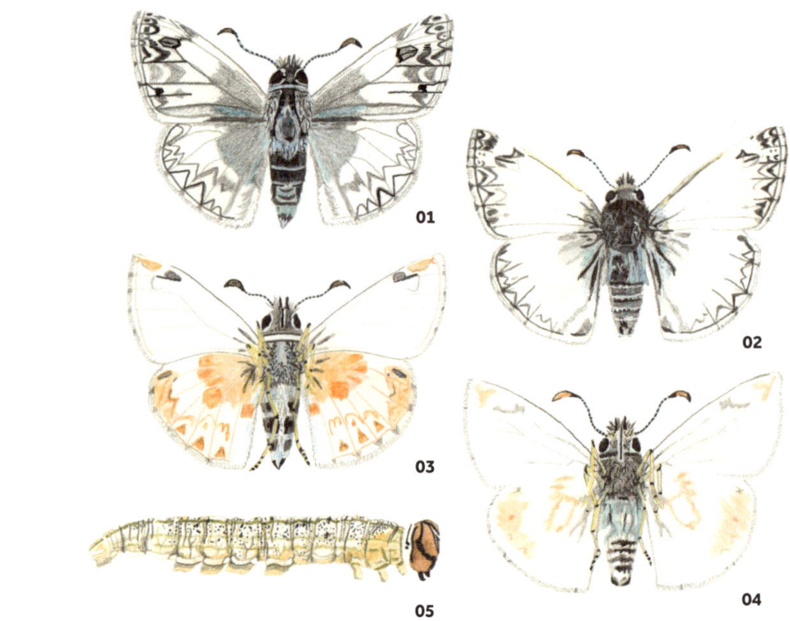

NORTHERN WHITE SKIPPER

Boisduval 1852

Heliopetes ericetorum

The butterfly is usually sitting on or patrolling the host. Paul Johnson, a park ranger at Pinnacles, told me that this butterfly only moved in after a fire scorched an eastern section of the park—*Malacothamnus* came up, and then the butterfly arrived (Johnson.pers.comm). Coming from points beyond, how do they find these plants?

Habitat: Hot dry chaparral where the host plant has established itself
Host Plant: Bush mallows in the genus *Malacothamnus*—a plant that follows fire
Life Phases: Eggs spherical with cross-ribbing. Pupa yellow or brown throughout with a pinkish center. Adult females resemble the Common Checkered Skipper (*Burnsius communis*)—a much smaller butterfly. Bivoltine (two flights): spring and fall.
Counties They Fly In: Found in most counties covered by this book
Great Places to See Them: Pinnacles National Park (San Benito); Mount Diablo State Park (Contra Costa County)

01: dorsal ♀
02: dorsal ♂
03: ventral ♀
04: ventral ♂
05: mature larva

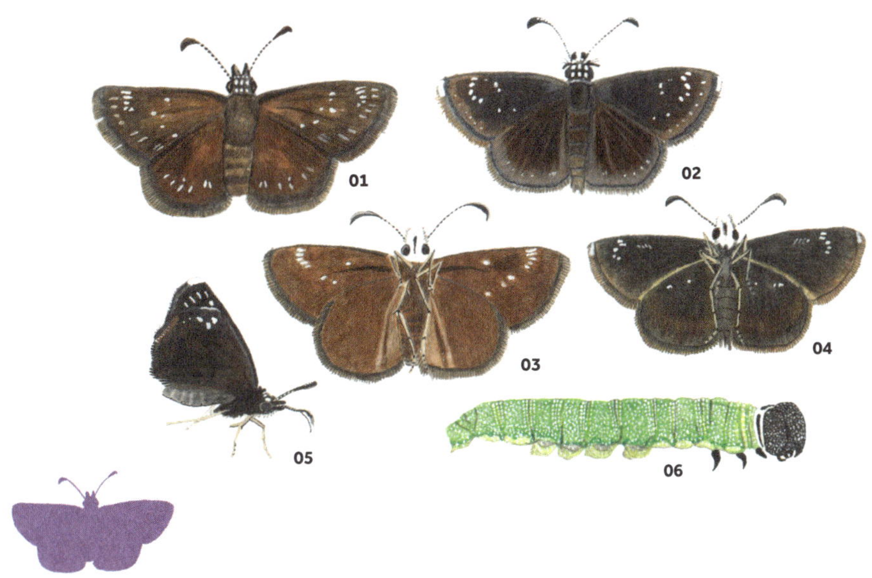

COMMON SOOTYWING

Fabricius 1793

Pholisora catullus

This richly brown or black, small butterfly flies low to the ground and is getting harder and harder to find (which is weird considering its hosts are still in many vacant lots). All the postings on iNaturalist give me hope it's not completely blinking out in our area. Breaking my own book's parameters, I'd say check all the open spaces in western Sacramento throughout the late spring.

Habitat: Sand dunes, disturbed areas, rural
Host Plants: Lamb's quarters (*Chenopodium* spp.), ragweed (*Ambrosia* spp.), pigweed (*Amaranthus* spp.)
Life Phases: The small, pale yellowish-brown eggs are conical and strongly ribbed (Dornfeld.1980.114). The larvae are night-grazers and camouflage among the leaves during the day. Pupa greenish to yellow brown. Multivoltine (many flights): May–August.
Counties They Fly In: Contra Costa and Solano
Great Places to See Them: Start with the sides of levees and vacant lots in Solano County's Delta Country. There are lots of historic records on John Steiner's species distribution map for Contra Costa County (Steiner.1980.204). I used to see it regularly when I did the Antioch portion of the Mount Diablo Butterfly Count, and all of a sudden, years back, it just stopped being seen. iNaturalist has postings for Pittsburg (Contra Costa County); Contra Loma Regional Park, Antioch (Contra Costa County); Big Break Regional Park, Oakley (Contra Costa County)

01: dorsal ♀
02: dorsal ♂
03: ventral ♀
04: ventral ♂
05: wings up ♂
06: mature larva

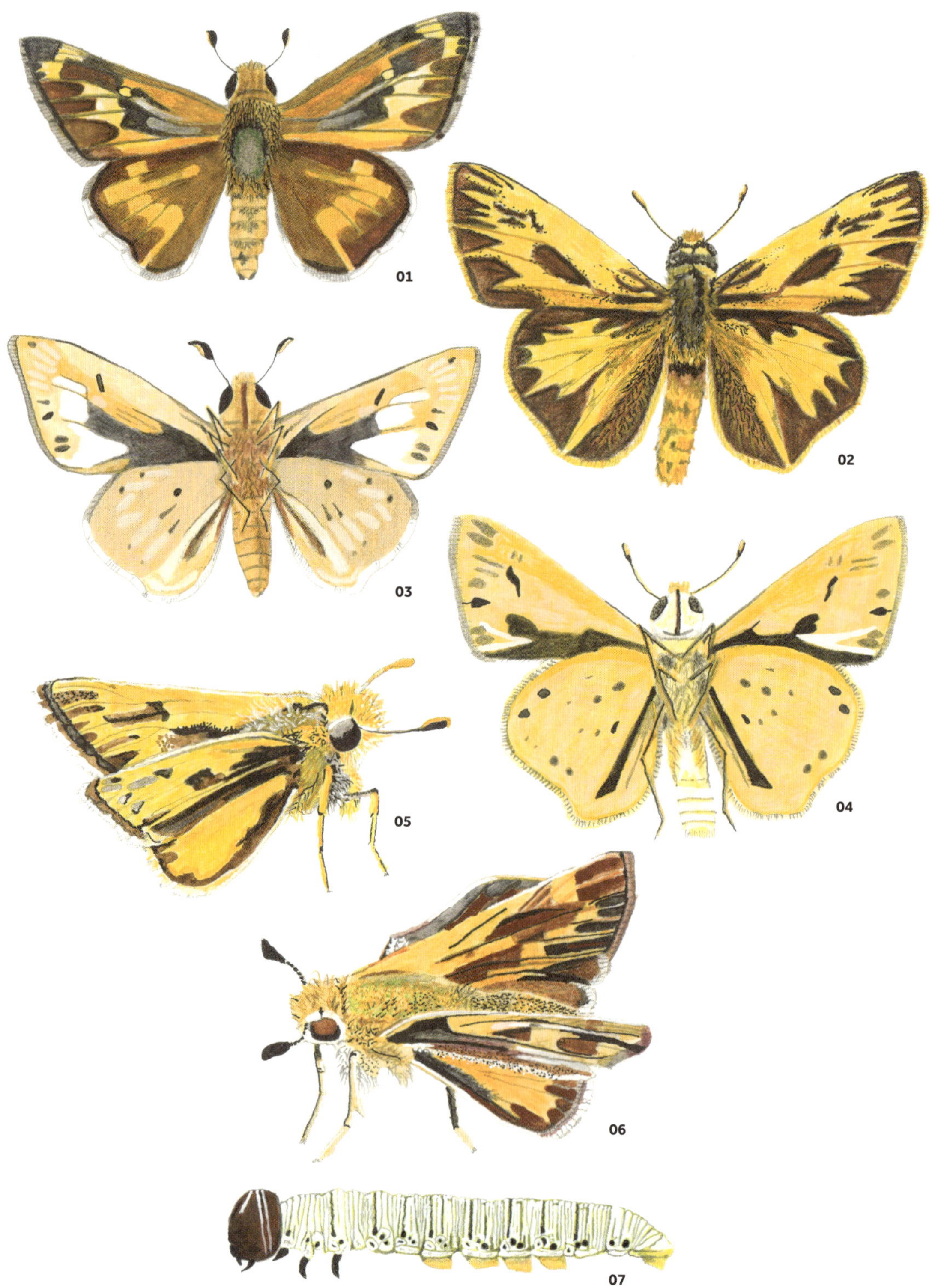
01
02
03
04
05
06
07

FIERY SKIPPER

Drury 1773

Hylephila phyleus

This flash of yellow orange zipping about neighborhoods makes this creature quite well known. Sometime between being a toddler and kindergartener, I remember lying on the cool grass and watching them. Identifying this butterfly can be a head-scratcher for the novice because the female and male have two very different and distinct underside patterns: the female has a complicated field of beige spots while the male has been described as having a "pepper grinder" smattering of dark spots against a tawny field. Both have heavily teethed forewings.

Habitat: Urban areas, city parks, and coastal marshes
Host Plants: Bermuda grass (*Cynodon dactylon*), alkali grass (*Distichlis spicata*), and the American Lawn (a composite of grass species)
Life Phases: Pupa yellow brown. Larvae have adapted to the constant mowing of the American Lawn by placing their shelters horizontally at the base of the grass (Howe.1975.490). Adults have short antennae. Multivoltine (many flights): March–November.
Counties They Fly In: All
Great Places to See Them: Almost any suburban subdivision or county park where large swaths of lawn occur. At the Great Lawn in Golden Gate Park (San Francisco County) or out at India Basin and Heron's Head Park one can observe them using their original host, the alkali grass, along the seashore.

01: dorsal ♀
02: dorsal ♂
03: ventral ♀
04: ventral ♂
05: bombardier ♀
06: bombardier ♂
07: mature larva

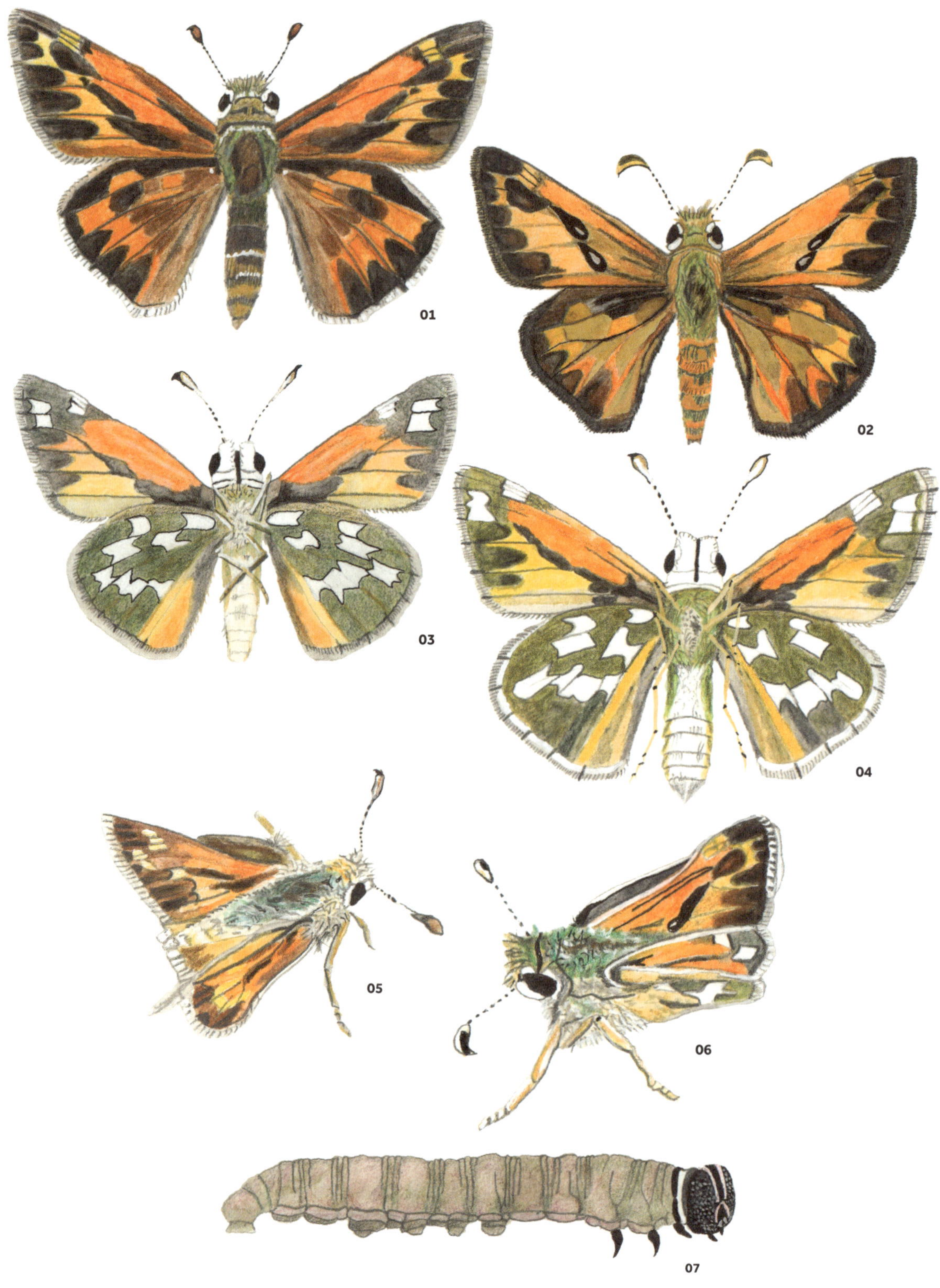
01
02
03
04
05
06
07

JUBA SKIPPER

Scudder 1872

Hesperia juba

The only other butterfly this one could be confused with for sheer size is Lindsey's Skipper (*Hesperia lindseyi*), but the greenish hue of Juba's underside should easily key the two apart. This butterfly flies in abundance in the higher and colder elevations of the Sierra Nevada. This range shift may be another impact of global warming—perhaps the coast is becoming too warm for it now. Males like to sit in depressions in the ground waiting for females.

Habitat: A vast array of ecosystems
Host Plants: Bunchgrass, Kentucky bluegrass (*Poa pratensis*)
Life Phases: Each of the five instars collectively last for two weeks (Steiner.1980.36). The field or ground color on the ventral side of both adults is olive green. Pupa is light brown. Bivoltine: April–June then again August–early October.
Counties They Fly In: Historic records from all except Marin, Monterey, San Benito, San Francisco, and Solano
Great Places to See Them: Sonoma Coast State Park (Sonoma County) is the only place this bug has been seen in recent times.

01: dorsal ♀
02: dorsal ♂
03: ventral ♀
04: ventral ♂
05: bombardier ♀
06: bombardier ♂
07: mature larva

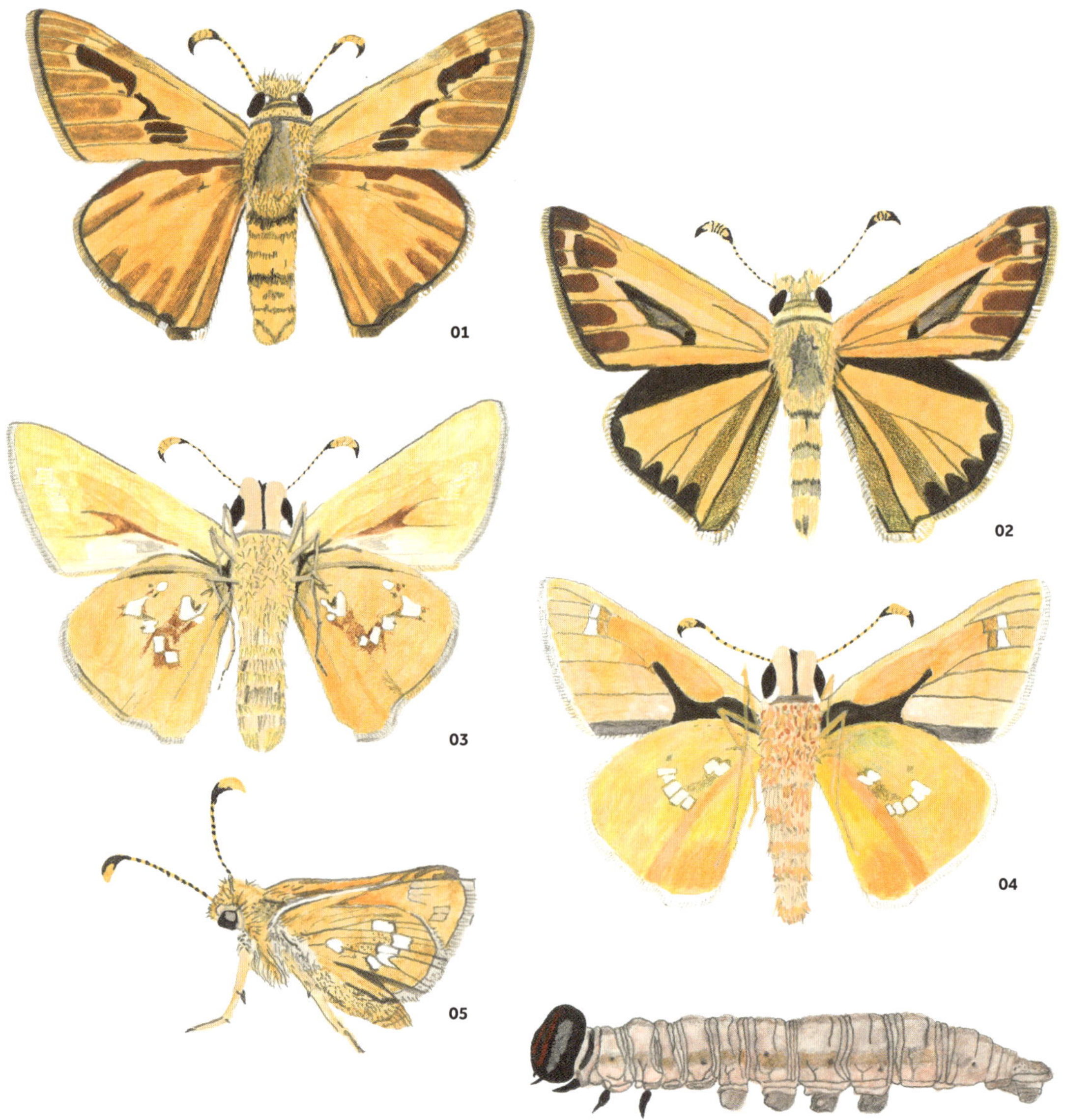
01
02
03
04
05
06

COLUMBIAN SKIPPER

Scudder 1872

Hesperia columbia

The Columbian Skipper is a strong hill-topper, as testified to by the recommended places to see this butterfly. It will sit on the ground at these high points or a low bush. On Mount Tamalpais, get yourself to the top of the fire tower on the East Peak and hike around to the eastern side so you face the Bay. Make a slow boulder crawl down to the vegetation. Butterflies should be zipping and landing on rocks and bushes.

Habitat: Grasslands, chaparral, rocky outcrops at summits
Host Plants: June grass (*Koeleria macrantha*), bunchgrasses
Life Phases: The pale yellowish larvae have dark brown heads whose face is marked with creamy streaks (Dornfeld.1980.113). Pupa is pale light brown. Adult males have a broad stigma with gray and yellow at its center. Bivoltine: March–June then again September–November.
Counties They Fly In: Alameda, Contra Costa, Marin, Monterey, Napa, San Benito, Santa Clara, Solano, and Sonoma
Great Places to See Them: The summit of Mount Tamalpais (Marin County); the summit of Mount Hamilton (Santa Clara County); Indian Valley Open Space, Novato (Marin County); Robert Ferguson Observatory (Sonoma County); Landels-Hill Big Creek Reserve (access by permission), Big Sur (Monterey County)

01: dorsal ♀
02: dorsal ♂
03: ventral ♀
04: ventral ♂
05: wings up ♀ and ♂
06: mature larva

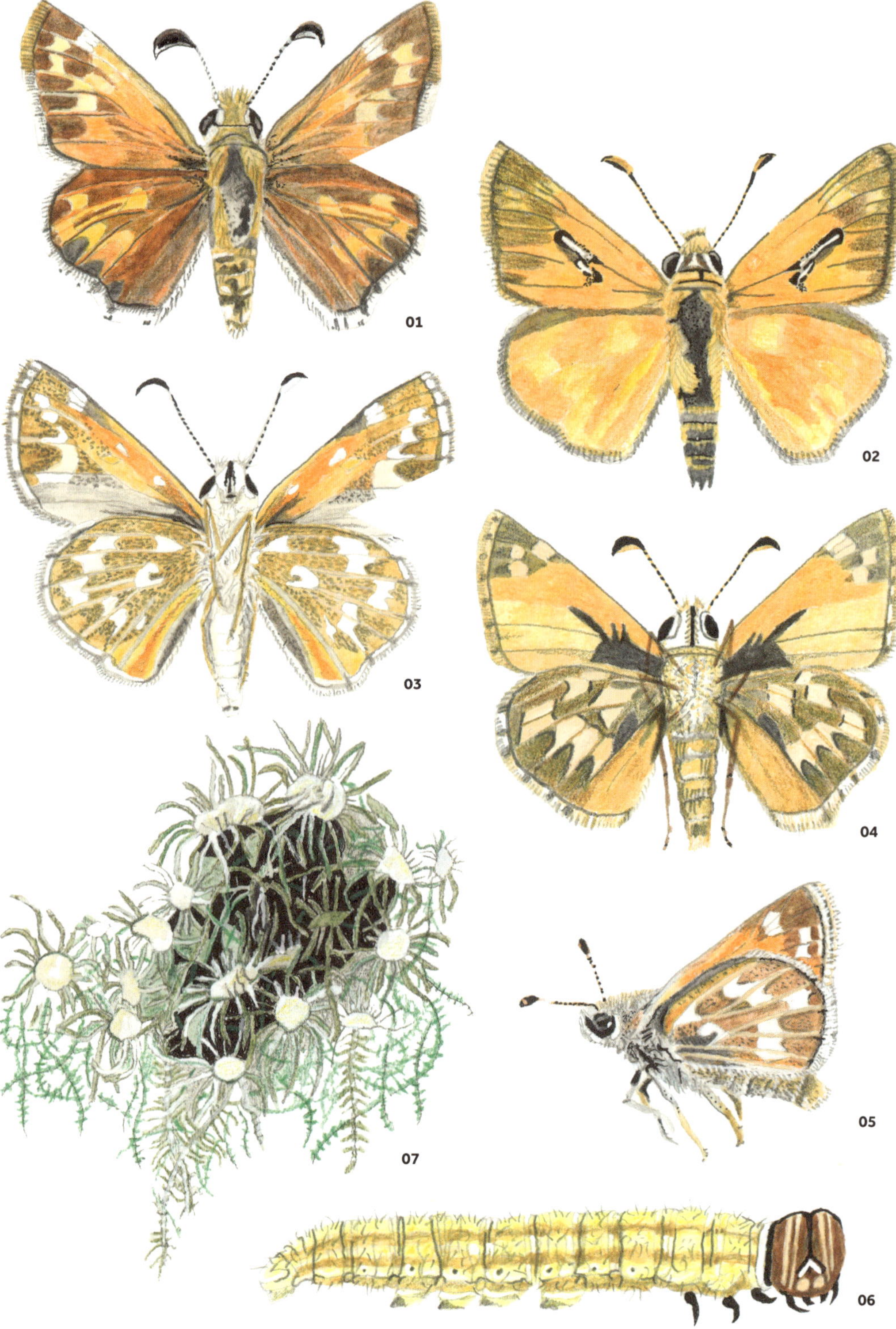

01
02
03
04
07
05
06

LINDSEY'S SKIPPER

Hesperia lindseyi

Holland 1930

Endemic to the state of California, Lindsey's Skipper was separated from the Common Branded Skipper complex and given full species status awhile back. The unique behavior of laying its eggs on or near the lichen now named *Usnea intermedia*, western bushy or Arizona bushy lichen (Johnson, J.2023.per.comm) was discovered by C. Don MacNeill as a grad student at UC Berkeley in 1964 (Pyle.2002.86). The larvae do not eat the lichen but have to go looking for their grass hosts after hatching. I went to MacNeill's study site Azalea Hill in 2022. The scrub oaks there are festooned with other *Usnea* spp., but I could not find that particular lichen—a showy one with spidering apothecia, though the genus *Usnea* is a tough group to key apart (Sharnoff.2014.210). I wondered, could MacNeill have gotten the species of lichen incorrect? Lots to still investigate within this phenomenon.

Habitat: Grasslands, chaparral

Host Plants: Idaho fescue (*Festuca idahoensis*), oat grass (*Danthonia californica*)

Life Phases: Eggs whitish, laid singly, mainly on tree lichens at one site and haphazardly at others (Scott.1986.443). Pupa green or brown. Overwinters as an egg. Adult males will have white felt within their stigma. Univoltine: June–July.

Counties They Fly In: All except San Mateo, San Francisco, Santa Cruz, and Solano

Great Places to See Them: Azalea Hill (along Bolinas Road and across from the Pine Crestline Trailhead, Fairfax (Marin County); Clear Creek Management Area (San Benito County); Indian Tree Open Space Preserve (Marin County); Mendenhall Springs (Alameda County); Mine Road (Contra Costa County)

01: dorsal ♀
02: dorsal ♂
03: ventral ♀
04: ventral ♂
05: wings up ♀
06: mature larva
07: Arizona bushy lichen

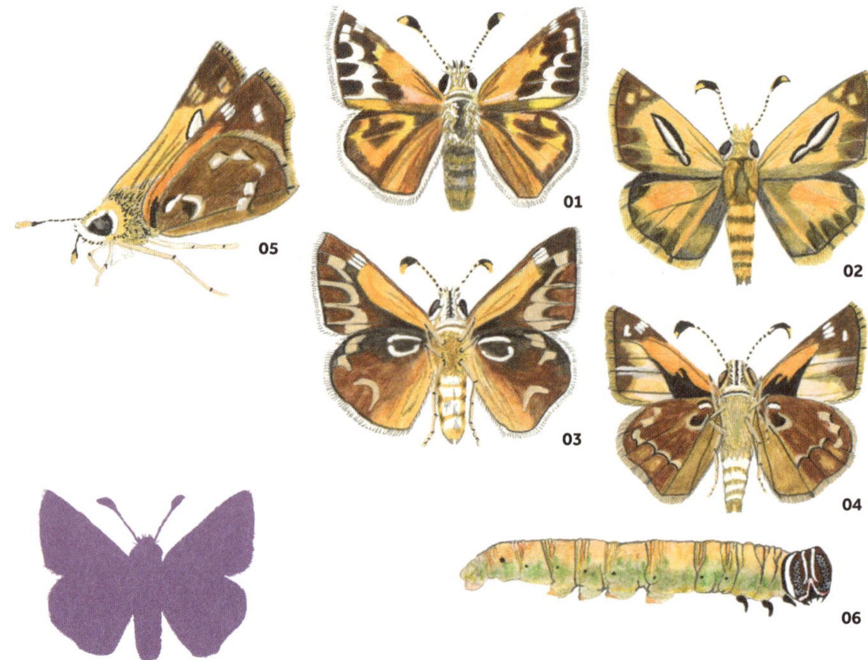

DODGE'S SKIPPER

Hesperia colorado dodgei

Bell 1927

It's been hard for me in years past to key these out in the field. Hopefully now that I've painted it, some semblance of knowledge will stick. Overall, I can say they're more of a dark brown than Tilden's (*Hesperia colorado tildeni*) from below when they are flying. They can be quite abundant. To compare with Tilden's Skipper, see **Appendix B, Tableau 2**.

Habitat: Coastal chaparral, grasslands, glades in forests
Host Plants: Red fescue (*Festuca rubra*) and other grasses (Poaceae family)
Life Phases: Larvae go through six instars (Shapiro & Manolis.2007.239). Overwinters as first instar with the egg. Pupa is light brown. In adults: females have a complicated pattern from above. Males have a long, thin stigma with white felt within it. Head-scratching patterns continue below. Univoltine: June–October.
Counties They Fly In: In the fog-belt zones of Contra Costa, Marin, San Mateo, and Santa Cruz (but not in San Francisco or Sonoma)
Great Places to See Them: Tomales Bay Ecological Reserve near Point Reyes Station in August (Marin County); Long Ridge Open Space Preserve off of Skyline Boulevard/Highway 35 (San Mateo County); UC Santa Cruz campus (Santa Cruz County)

01: dorsal ♀
02: dorsal ♂
03: ventral ♀
04: ventral ♂
05: wings up
06: mature larva

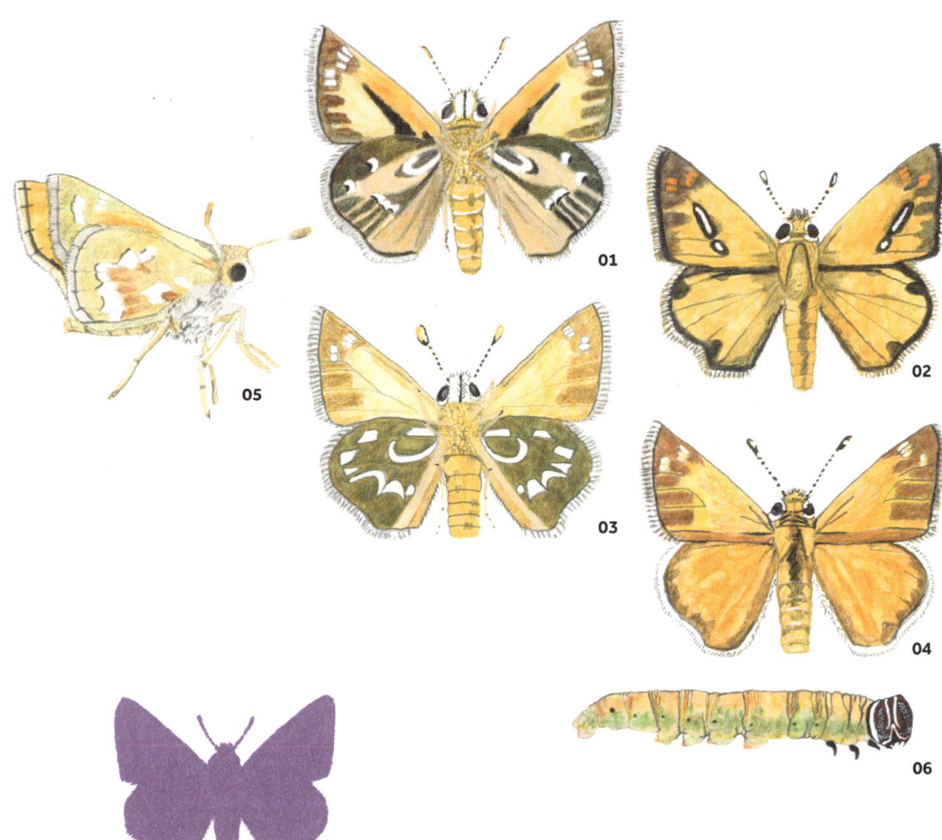

TILDEN'S SKIPPER

Freeman 1955

Hesperia colorado tildeni

Can easily be confused with Lindsey's Skipper where the two fly together. *Hesperia lindseyi* is quite large (think quarter size), whereas Tilden's is more like a nickel. Also Lindsey's flies earlier than Tilden's. To compare with Dodge's Skipper, see **Appendix B, Tableau 2**.

Habitat: Grassland
Host Plants: Grasses (Poaceae family)
Life Phases: Larva similar to Dodge's, though Tilden's has only five instars. Pupa is light brown. Adults have a greenish field or background ventrally on both wings. Univoltine: June–October.
Counties They Fly In: Alameda, San Benito, Santa Clara, and Sonoma
Great Places to See Them: Mount Hamilton in the Diablo Range (Santa Clara County); the town of Idria (San Benito County); Pepperwood Preserve (Sonoma County)

01: dorsal ♀
02: dorsal ♂
03: ventral ♀
04: ventral ♂
05: wings up
06: mature larva

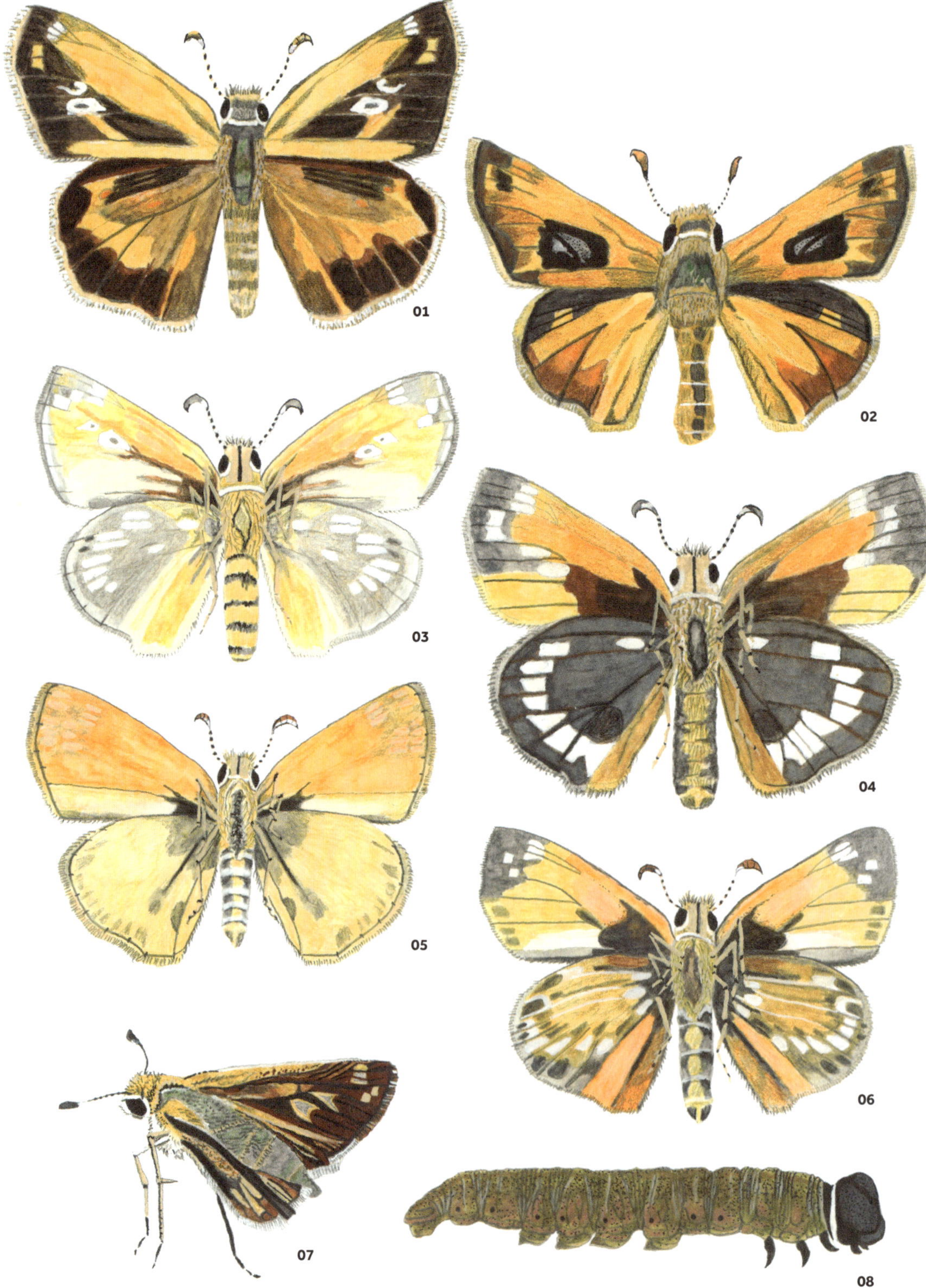
01
02
03
04
05
06
07
08

SACHEM

Boisduval 1852

Atalopedes campestris

Because it is a colonial butterfly (like Marine Blues, Eastern Tailed Blues, and Funereal Duskywings), Sachem has explosive years of being everywhere in our area and barren years during which one cannot find them at all. I had a great day on Tyler Island in Solano County at the junction of River Road and Theater Street seeing many with my fellow butterfly enthusiasts Erica Harris and Cat Chang. A huge hedge of lantana wraps around a gas station on the corner. More Sachems and Eufala Skippers (*Lerodea eufala*) were nectaring and flying than we could count.

Habitat: Riparian woodlands, marsh communities, open roads, and disturbed habitats
Host Plants: Oat grass (*Danthonia californica*), saltgrass (*Distichlis spicata*)
Life Phases: Eggs cream or white. Larvae live in nests of leaves. Pupa is blackish brown with white dots at thorax. Adult females are brown dorsally with many glassy spots at the outer margin of each forewing. The male from above has an incredibly large and prominent black stigma patch (the largest of any of our grass skippers). This patch is a determining factor in identifying this species. Both females and males come in a lighter and darker form ventrally—the darker form being more inland. Bivoltine: June–October.
Counties They Fly In: All except San Francisco (where I'm bound and determined to find it someday)
Great Places to See Them: They can be rather elusive. They have been spotted in Indian Tree Open Space Preserve, Novato (Marin County); Laguna Wetlands Preserve (Sonoma County); Tyler Island, Walnut Grove (Solano County).

01: dorsal ♀
02: dorsal ♂
03: ventral ♀, light form
04: ventral ♀, dark form
05: ventral ♂, light form
06: ventral ♂, dark form
07: bombardier ♀
08: mature larva

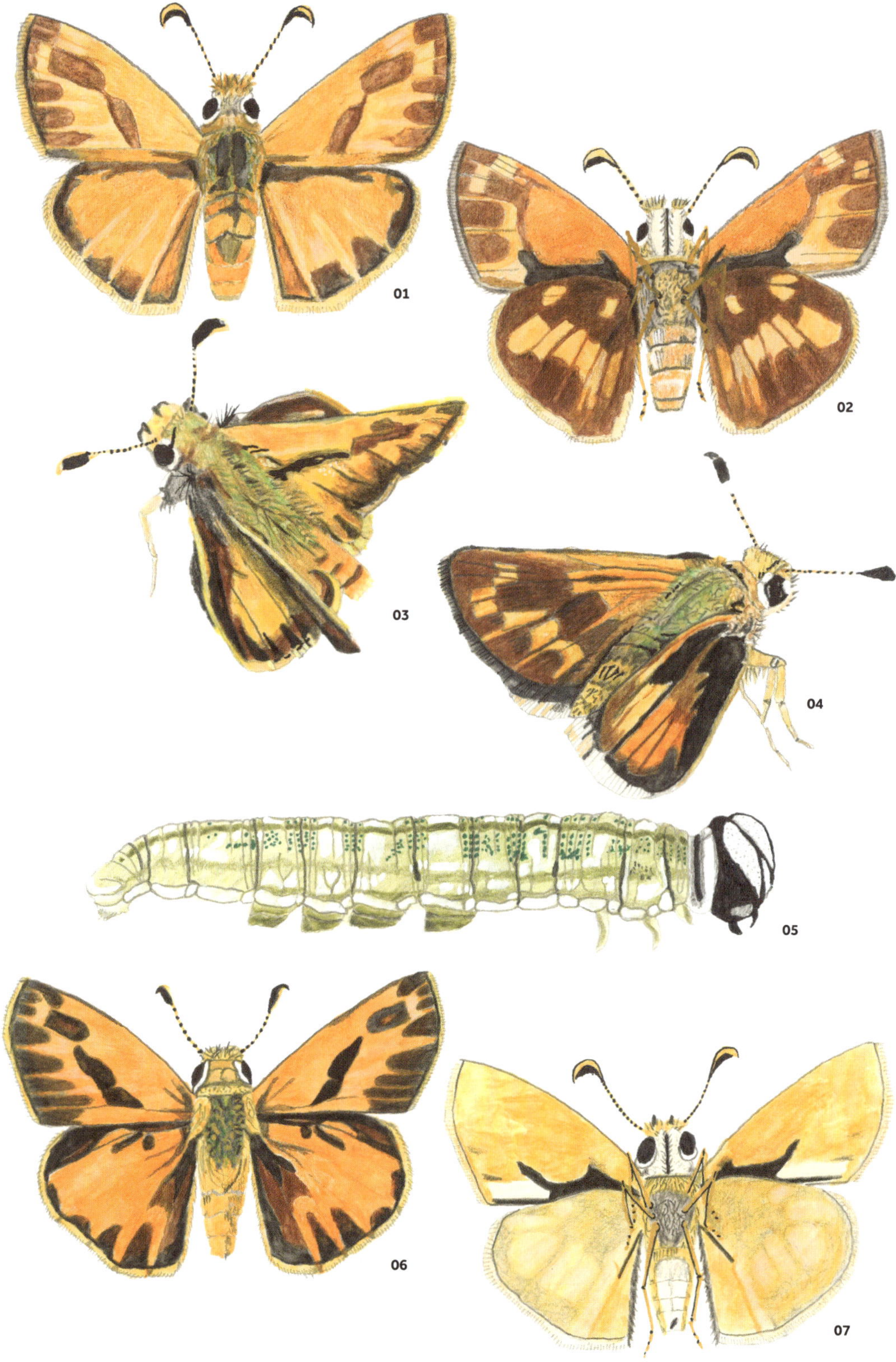

01
02
03
04
05
06
07

WOODLAND SKIPPER

Boisduval 1852

Ochlodes sylvanoides

A sage lepidopterist once told me the two *Ochlodes*—the Woodland and the Rural—have flights that never overlap. Yet, while inventorying Yerba Buena Island in the middle of the San Francisco Bay, I personally saw the two flying together. The Rural is distinctively smaller and bright tawny orange, and the Woodland is bigger with bold, blockish pale markings from below. In general. however, one would normally see the Rural worn with a fresh Woodland. To compare with the Rural Skipper, see **Appendix B, Tableau 3**.

Habitat: Chaparral, sagebrush, oak woodlands
Host Plants: *Phalaris* spp., *Elymus* spp., and other grasses (Poaceae)
Life Phases: Eggs creamish, pupa light brown with a whitish bloom. Pupa has an overall gray coating on a light brown ground color. Adults come in two different ventral forms; the coastal form is darker than the lighter inland form, both having hind-wing spot bands composed of chunky, squarish spots. Where the Rural does not occur, Woodlands tend to fly much earlier. Univoltine: June–October.
Counties They Fly In: All
Great Places to See Them: Marsh Creek Regional Trail, Mount Diablo State Park (Contra Costa County); Garin Regional Park, Carmel Valley (Monterey County); Sierra Azul Open Space Preserve, Mount Umunhum (Santa Clara County); Pinnacles National Park (San Benito County)

01: dorsal ♀
02: ventral ♀
03: bombardier ♂
04: bombardier ♀
05: mature larva
06: dorsal ♂
07: ventral ♀ and ♂, inland form

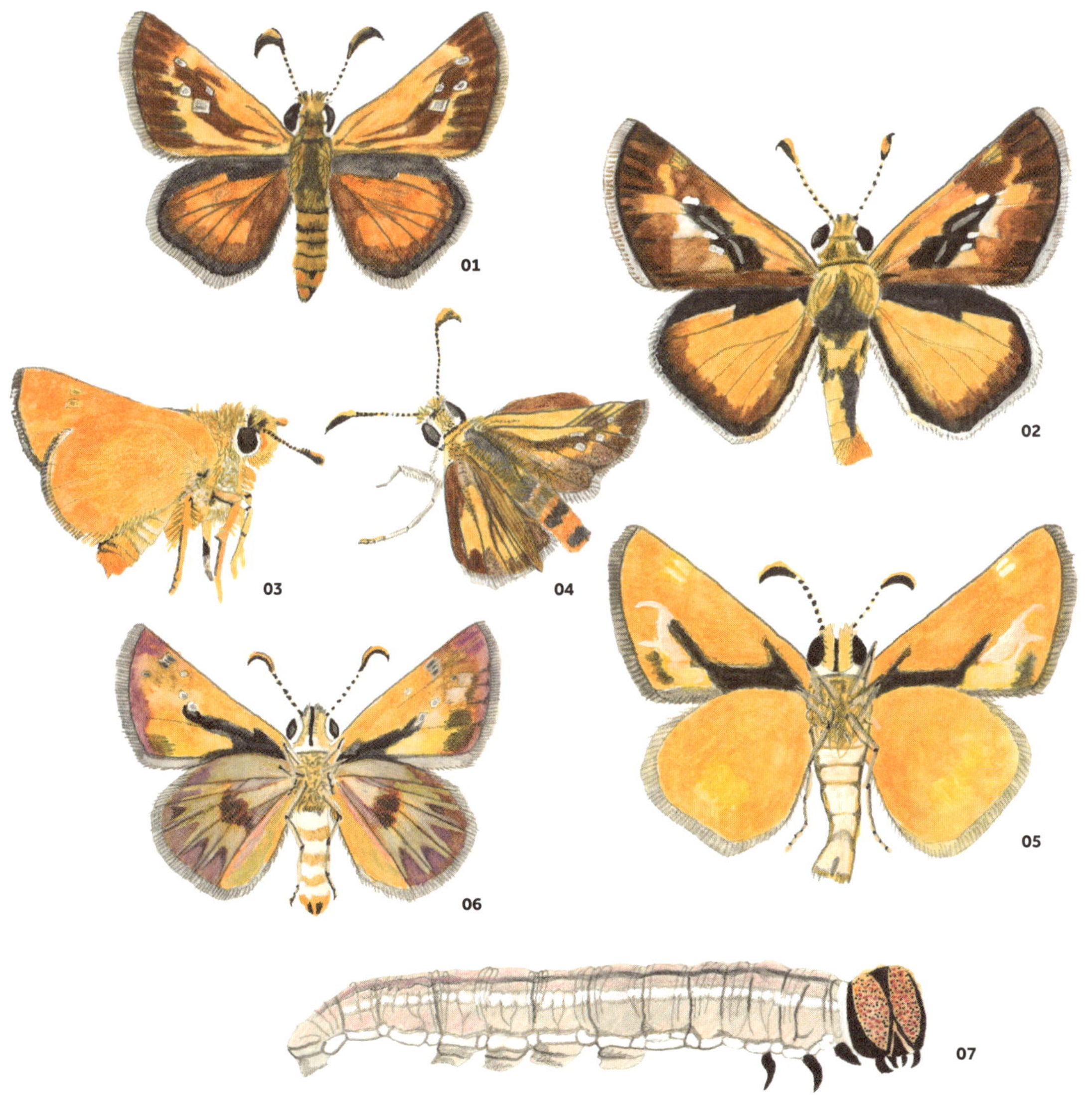
01
02
03
04
05
06
07

RURAL SKIPPER

Boisduval 1852

Ochlodes agricola

also known as **THE FARMER**

Adults show a deep appreciation for the flowering California buckeye tree at which they nectar and find mates (Opler.1999.449). Like those of the Woodland, Rural forewings are heavily teethed along the margins. It's easy to confuse these two skipper species. To compare with the Woodland Skipper, see **Appendix B, Tableau 3**.

Habitat: Grassland, streams, glade openings in forests
Host Plants: Grasses (Poaceae family), potentially oniongrass (*Melica bulbosa*)
Life Phases: Eggs green evolving to a pale white. In adults: females underwings orange or brown to slightly purplish, males golden orange with virtually no markings below. Univoltine: May–July. Emerging earlier than Woodland.
Counties They Fly In: All
Great Places to See Them: All the places recommended for its doppelganger, the Woodland Skipper. The Rural is the ubiquitous skipper in all the side canyons of San Bruno Mountain State and County Park (San Mateo County)

01: dorsal ♀
02: dorsal ♂
03: wings up ♂
04: bombardier ♀
05: ventral ♀
06: ventral ♂
07: mature larva

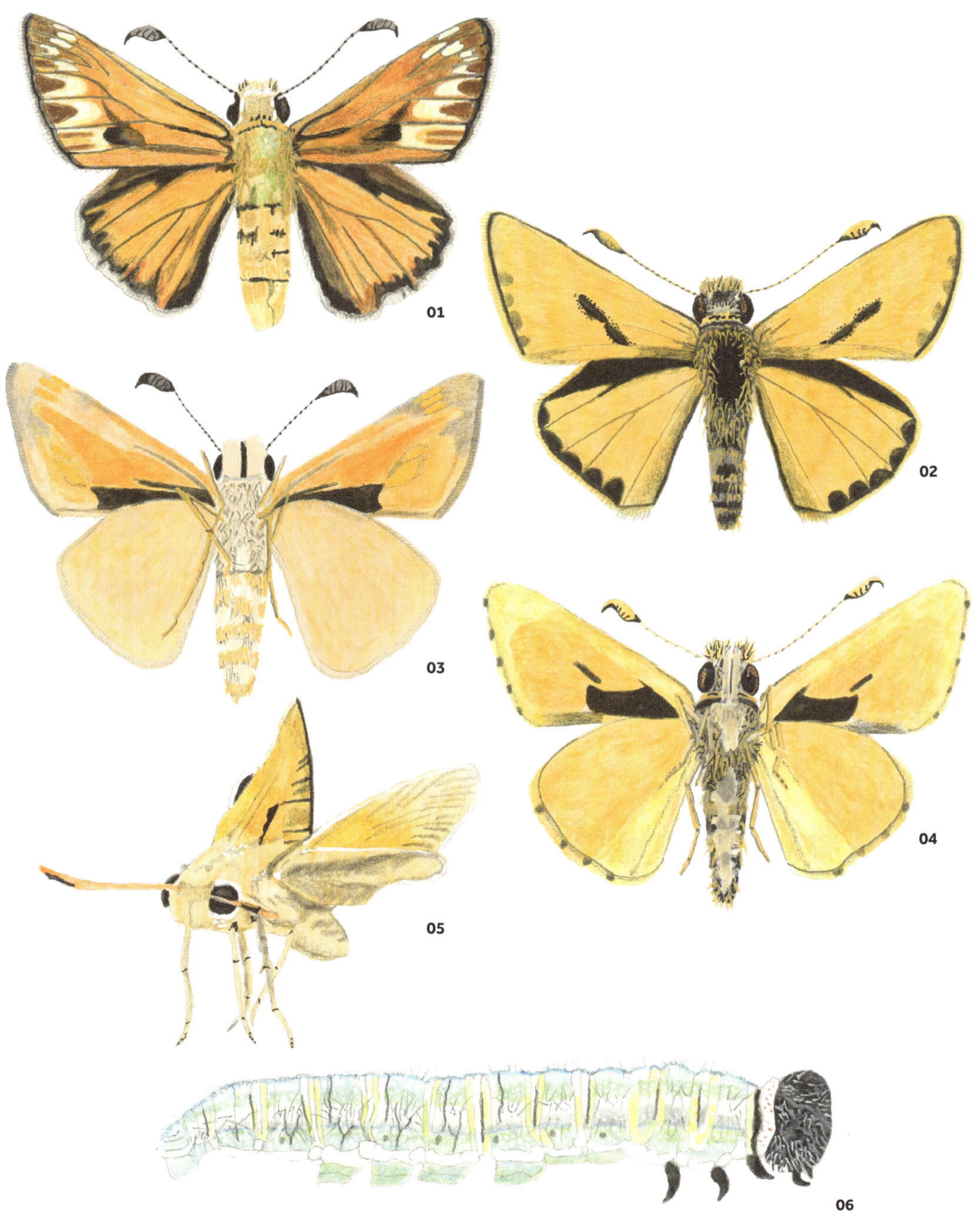
01
02
03
04
05
06

YUMA SKIPPER

W. H. Edwards 1873

Ochlodes yuma

Our largest grass skipper. They bullet about at insane speeds but become rather approachable for photography once they've landed. Female Yumas have a series of pale chevrons running along their forewing margins. Males are solidly orange yellow with a long, thin black stigma. They use blades of grass to create nest shelters.

Habitat: Salt marshes, sloughs, and canals
Host Plant: Common reed (*Phragmites australis*)
Life Phases: The pale greenish eggs are large and somewhat hemispherical (Howe.1975.449). Pupae are chocolate brown. Bivoltine: June then again August–September.
Counties They Fly In: Contra Costa and Solano
Great Places to See Them: Plantings along the jogging path on the levee below the Benicia–Martinez Bridge (Benicia/Solano County side). Butterfly nectars on any available flower and stays out of the wind in the gullies below the levee. At the point of the Antioch Marina (Contra Costa County); Grizzly Island (Solano County); Big Break Regional Shoreline, Oakley (Contra Costa County)

01: dorsal ♀
02: dorsal ♂
03: ventral ♀
04: ventral ♂
05: bombardier
06: mature larva

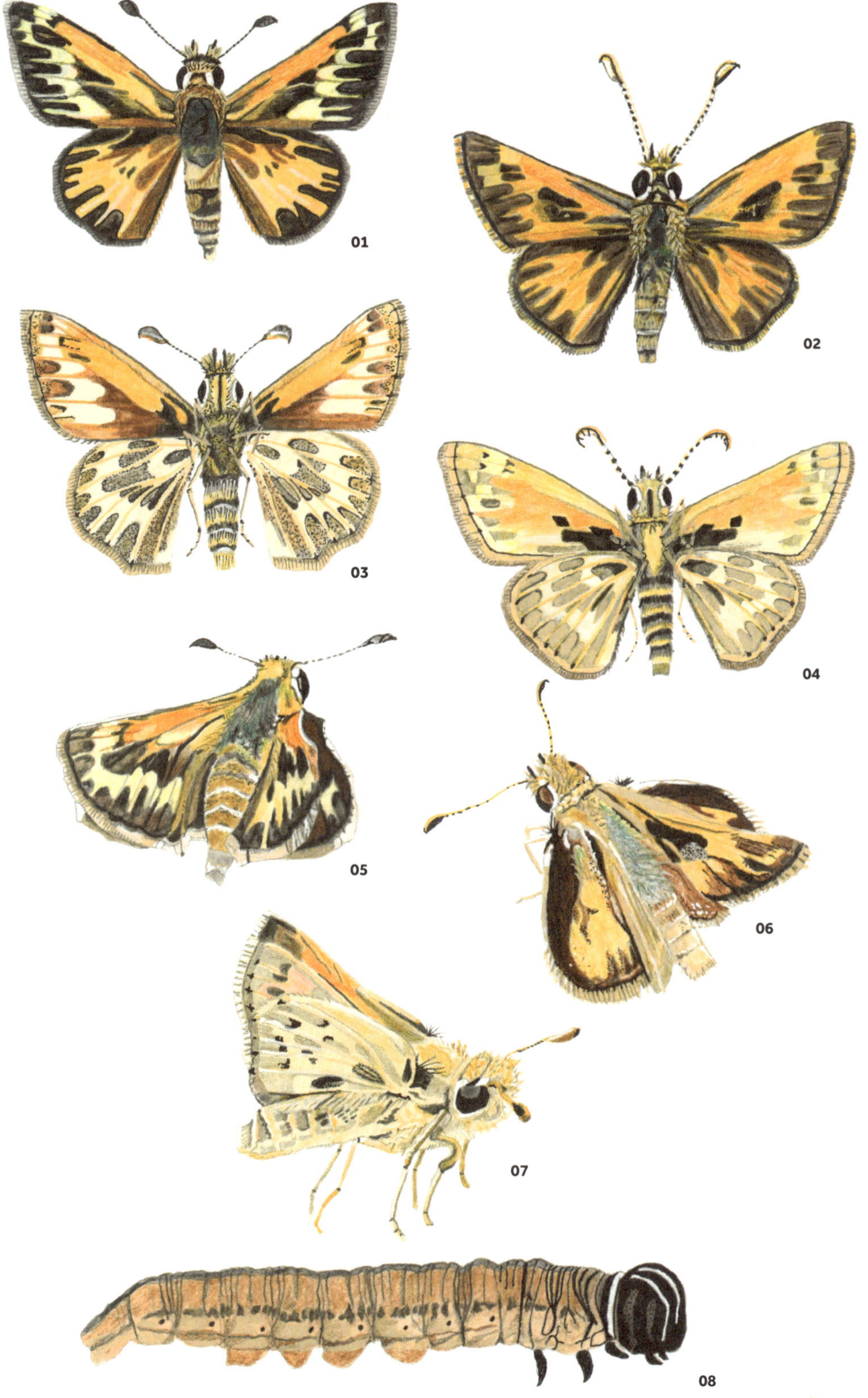
01
02
03
04
05
06
07
08

SANDHILL SKIPPER

Boisduval 1852

Polites sabuleti

When this mania took hold a few decades back, I remember a seasoned lepidopterist named John DeBenedictis teaching me quite a valuable party trick in identifying this one. On the ventral side of the hind wing, look for a sideways version of the international symbol of "flipping the bird" to someone. This helped tremendously early on keying it away from others. To put it more . . . scientifically . . . I quote Robert Michael Pyle here: "ventral hind wing has a yellowish or ivory-colored spot/band and a basal crescent with elongated middle spots that run out along the veins" (Pyle.2002.86). Found mostly in tidal marshes but regularly seen far inland. My friend Cat Chang has seen one in her garden in Emeryville, just a few miles from the San Francisco Bay shoreline; its host plant saltgrass is still in many people's backyards along the bay.

Habitat: Coastal scrub, salt marshes, urban parks, and residences
Host Plants: Saltgrass (*Distichlis spicata*), Bermuda grass (*Cynodon dactylon*)
Life Phases: Overwinters in the larval stage. Pupa green with a line across the front. Both sexes of the adults are heavily teethed on each upper and lower wing. Bivoltine: April–October.
Counties They Fly In: All
Great Places to See Them: Hayward Regional Shoreline (Alameda County); Candlestick Point State Recreation Area (San Francisco County); Coyote Hills Regional Park (Alameda County)

01: dorsal ♀
02: dorsal ♂
03: ventral ♀
04: ventral ♂
05: bombardier ♀
06: bombardier ♂
07: wings up, very light form
08: mature larva

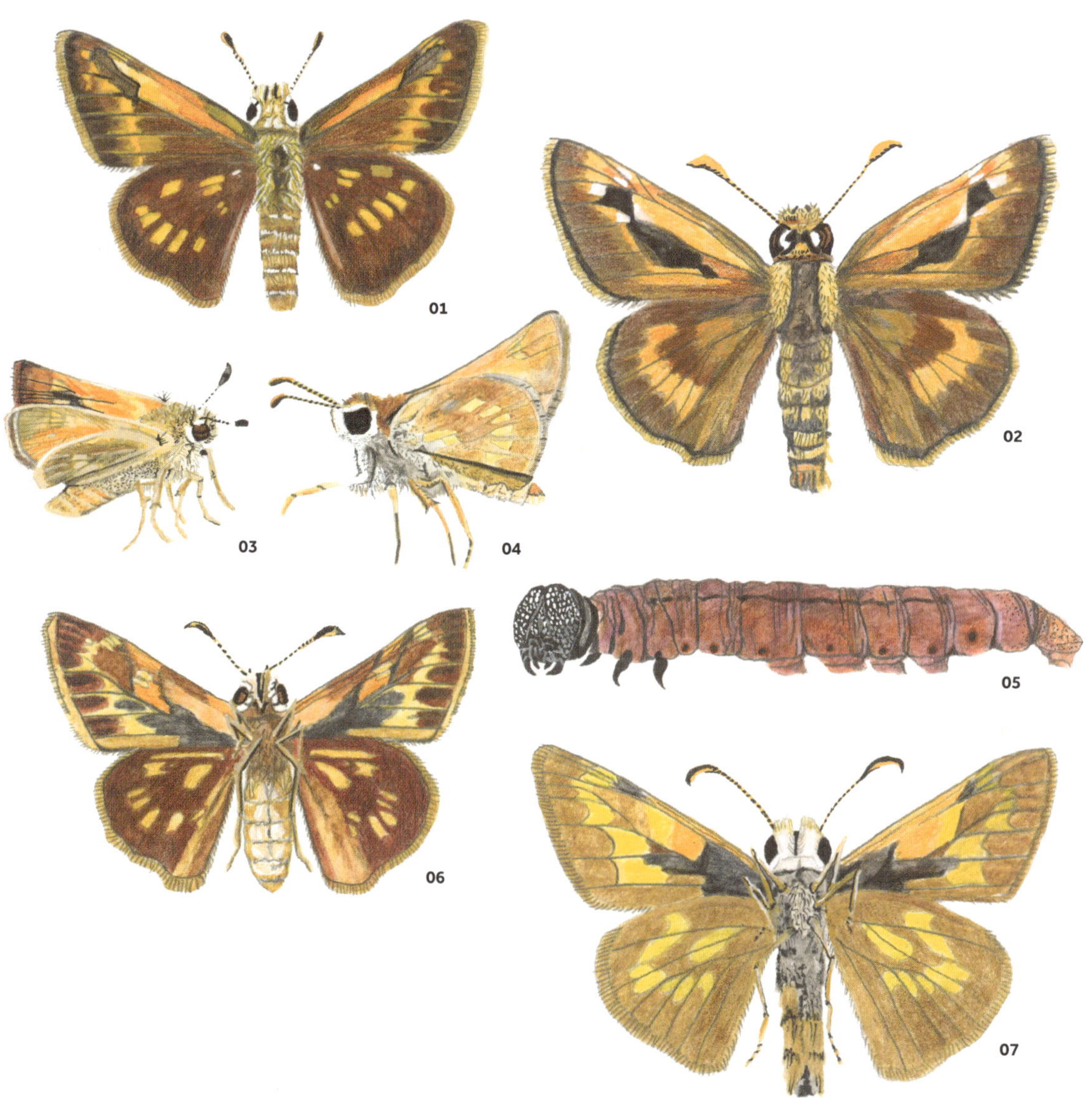
01
02
03
04
05
06
07

DOGSTAR SKIPPER

W. H. Edwards 1881

Polites sonora siris

This is a different subspecies from populations widespread in the High Sierra. The Dogstar is listed with a vulnerable status on iNaturalist. Flowers are used as rendezvous platforms during courtship.

01: dorsal ♀
02: dorsal ♂
03: bombardier ♀
04: bombardier ♂
05: mature larva
06: ventral ♀
07: ventral ♂

Habitat: Pine and Douglas fir forest understories, wet meadows
Host Plants: Grasses (Poaceae family), Idaho fescue (*Festuca idahoensis*)
Life Phases: Eggs are light green (Crabtree.1998.18). The cream-colored spot band on the adult ventral hind wing creates an inverted C shape along the basal disc. The field ventrally is tan brown as well. Univoltine: May–August.
Counties They Fly In: Northern Sonoma
Great Places to See Them: Salt Point State Park (Sonoma County)

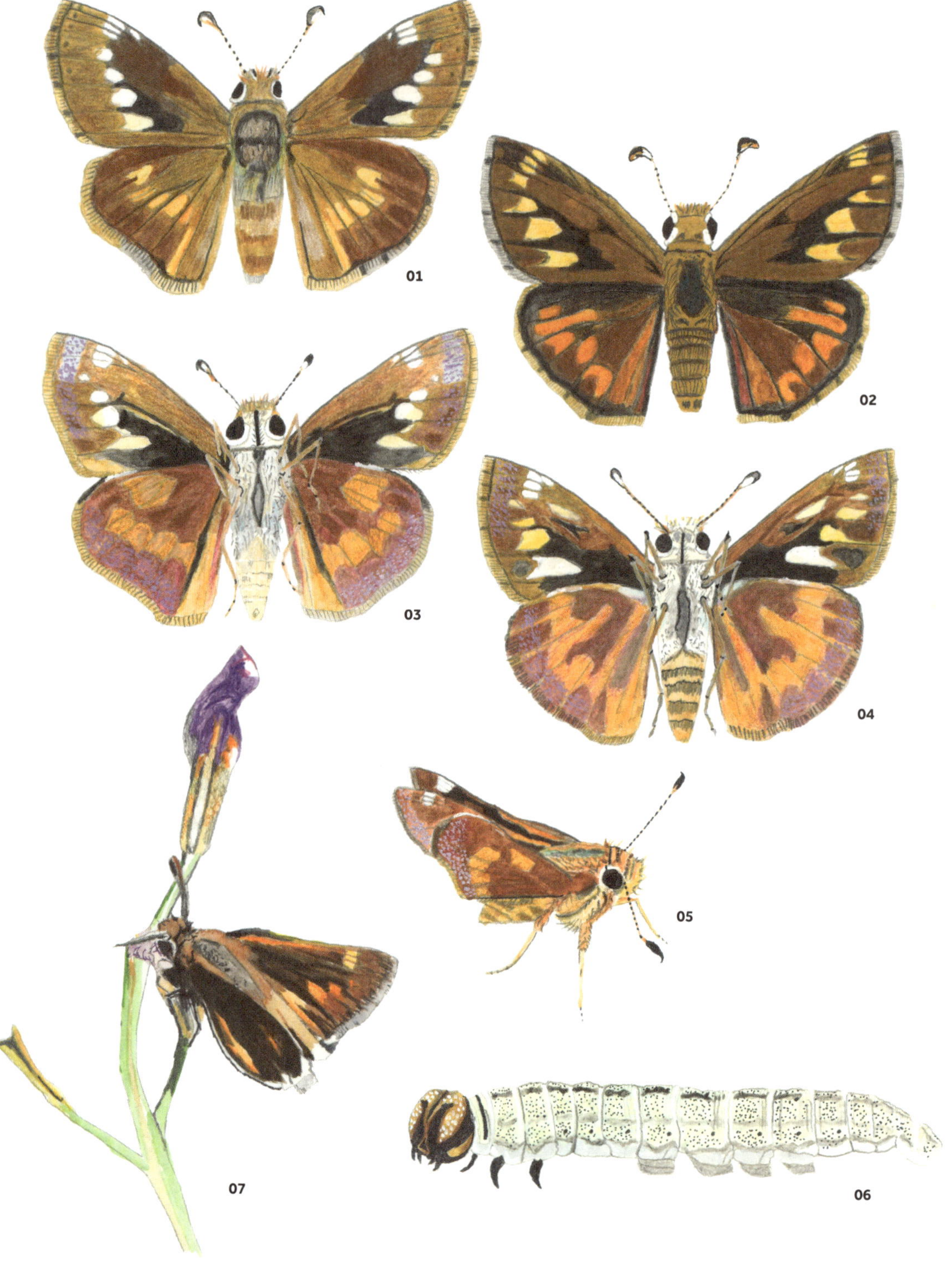
01
02
03
04
05
06
07

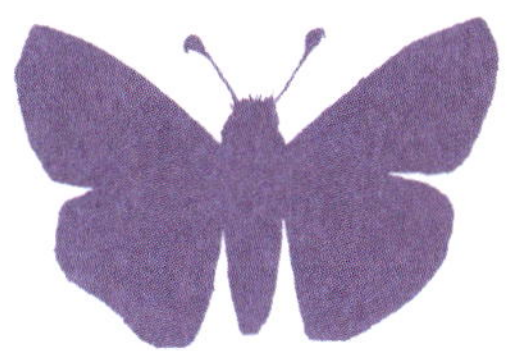

UMBER SKIPPER

W. H. Edwards 1869

Lon melane

One of the three most common skippers in San Francisco (the others being Fieries and Common Checkereds), this creature is in every part of the city. I am still amazed by that. I always find them in the dappled sunlight of an ecosystem's edge. Himalayan blackberry (*Rubus armeniacus*) blossoms are a good place to start looking for them in any park.

The Umber Skipper is a rather straightforward butterfly for the novice to learn. It's a rich sable brown overall with reddish brown or creamy yellow spots. Upper forewing spots help determine the sex of this bug—reddish in males, white in females. The underside has a plum hue along both the fore- and hind-wing borders.

Habitat: Many native ecosystems, disturbed areas, vacant lots
Host Plants: Bermuda grass (*Cynodon dactylon*), red fescue (*Festuca rubra*), California brome grass (*Bromus carinatus*)
Life Phases: Larvae live in nests of rolled or tied leaves (Scott.1986.451). Pupa light tan. Adult males lack stigma. Multivoltine: April–October.
Counties They Fly In: All
Great Places to See Them: City parks, botanical gardens, nurseries, friends' backyards

01: dorsal ♀
02: dorsal ♂
03: ventral ♀
04: ventral ♂
05: wing up ♀
06: mature larva
07: bombardier ♂

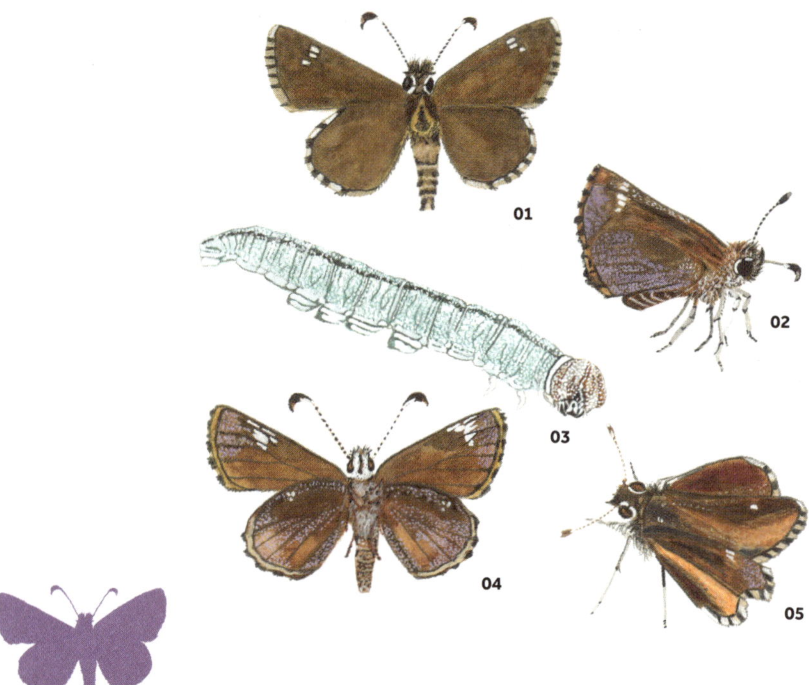

COMMON ROADSIDE SKIPPER

W. H. Edwards 1862

Amblyscirtes vialis

Though Opler refers to it as our "most widespread roadside skipper" (Opler.1999.464), it is clearly rare in our area. I believe it's part of a group of skippers (Arctic, Dun, Dogstar) that generally reflect that the Sonoma Coast is a disjunct relict of the Sierra Nevada. Even the wildflowers there (as pointed out to me by Cat Chang on an exploratory visit) are most like the ones seen in the High Sierra. It, too, like the skippers mentioned earlier, can be found near creeks and streams and clearings in the woods.

Habitat: Woodland, both riparian and evergreen
Host Plants: Blue grass (*Poa* spp.), oat grass (*Avena* spp.)
Life Phases: Larvae are covered in numerous setae, giving them the appearance of having been showered in powdered sugar. Pupa tip and tail reddish, midsection green. Overwinters in the larval stage. The adult butterfly's wings are bordered with checkered fringe, and both sexes have a metallic plum sheen to their undersides. Univoltine: April–May.
Counties They Fly In: Napa and Sonoma; one historic record in Alameda
Great Places to See Them: The only record within the scope of this book is an iNaturalist posting in North Asbury Creek (Sonoma County).

01: dorsal ♀ and ♂
02: wings up ♀ and ♂
03: mature larva
04: ventral ♀ and ♂
05: bombardier ♀ and ♂

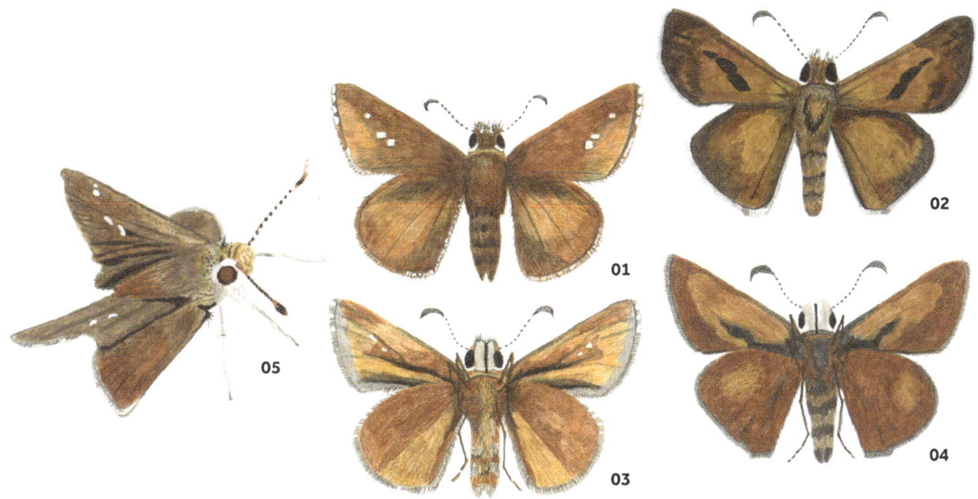

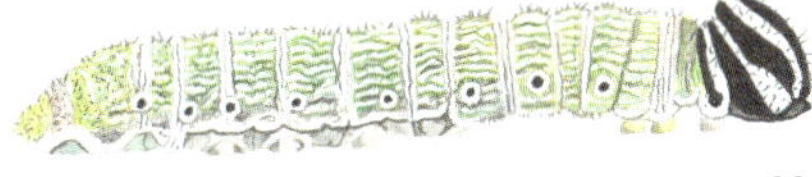

DUN SKIPPER

Linter 1878

Euphyes vestris

Whenever I've seen a Dun Skipper, this plain brown or black butterfly is either sitting on a blade of grass or down in a gulley in an opening in a forest. The first time I saw this species was along the coastal trail in Sea Ranch. Adults are known to feed on animal dung. Many do, including Lorquin's Admirals, California Sisters, and Satyr Commas, in order to extract nutrients for procreation.

Habitat: Ecosystems associated with Douglas fir and redwood forests
Host Plants: Sedges (*Carex* spp.)
Life Phases: Five instars with the third one overwintering. Pupa can come in brown, yellow, or green. Univoltine: May–June.
Counties They Fly In: Santa Cruz and Sonoma
Great Places to See Them: Sea Ranch (Sonoma County); Church Street in Gualala (Sonoma County); Sugarloaf Ridge State Park (Sonoma County); Salt Point State Park, behind campsite 49 (Sonoma County); off of Bear Creek Road, near Boulder Creek (Santa Cruz County)

01: dorsal ♀
02: dorsal ♂
03: ventral ♀
04: ventral ♂
05: bombardier
06: mature larva

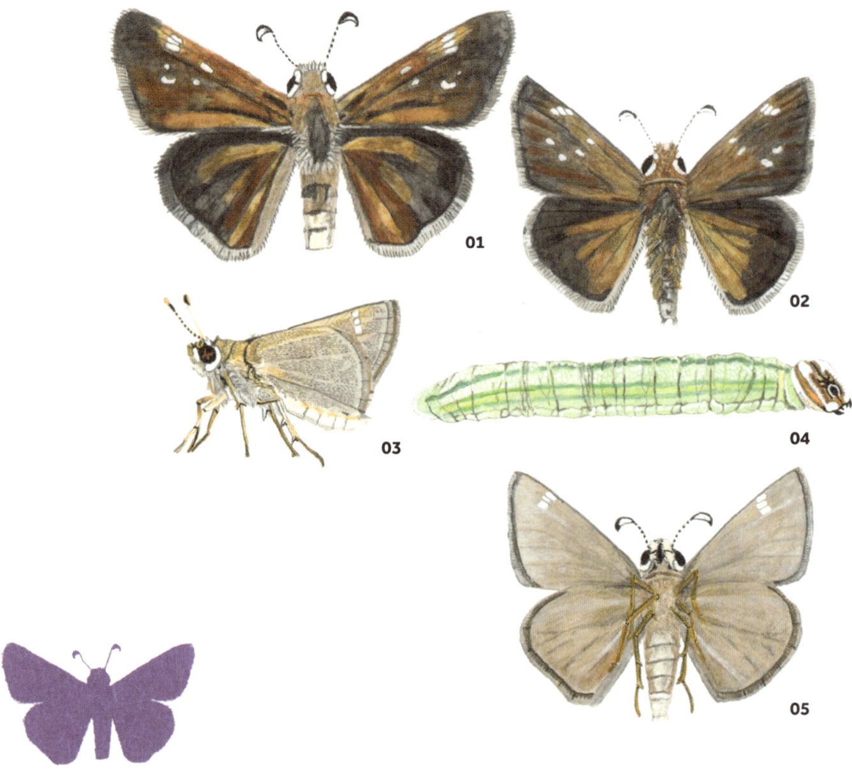

EUFALA SKIPPER

Edwards 1869

Lerodea eufala

This rather nondescript, small skipper can be easily overlooked, as the gray underside blends well with the ground. It has the overall color of dirty water. Larvae graze at night. It becomes more prevalent as one moves north through the delta country of the Sacramento River wetlands.

Habitat: Sand dunes, marshes, urban valleys, disturbed areas, moist spots along the Suisun Bay

Host Plants: Bermuda grass (*Cynodon dactylon*)

Life Phases: Eggs deposited on grass leaves. Pupa light green with stripes. Adults are uniformly gray below with white spots at the apex of their upper forewings. Multivoltine: June–November.

Counties They Fly In: Contra Costa, Napa, and Solano

Great Places to See Them: Seen along with Sachems (*Atalopedes campestris*) in Walnut Grove on Tyler Island (Solano County) (see Sachem); Madonna Drive, Suisun City (Solano County); Corteva Wetlands Preserve, Antioch (Contra Costa County)

01: dorsal ♀
02: dorsal ♂
03: wings up ♀ and ♂
04: mature larva
05: ventral ♀ and ♂

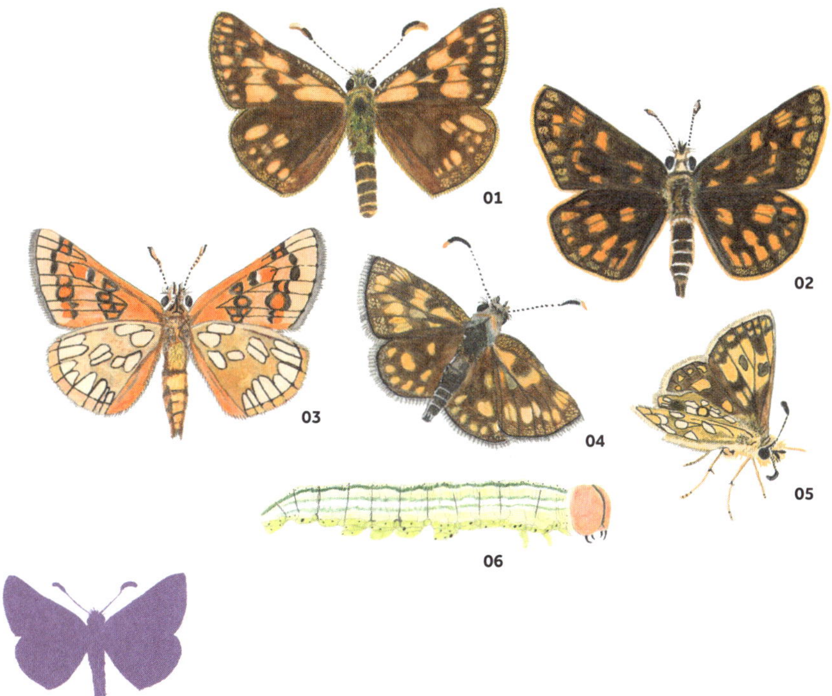

WESTERN ARCTIC SKIPPER

Matoon & Tilden 1998

Carterocephalus skada

More boreal than arctic, this butterfly flies among the understory of the forest. The Western Arctic Skipper is the only representative in the Bay Area of the Skipperling subfamily Heteropterinae. Adults are similar with their rich brown or black ground color and their gold checkerboarding. The female is larger and has more orange on her forewing. Wings are held flat when at rest. To see this butterfly, the following two spots in Salt Point are a sure bet. Get yourself to mile marker 39.17 along Highway 1 and pull off the road. Walk twenty feet up the spur trail to the main trail, make a left, then go fifty feet north to the spot. Cat Chang and I found this skipper immediately on the main trail. Also try Woodside Campground just up the road from this spot, one hundred yards behind campsite 49.

01: dorsal ♀
02: dorsal ♂
03: ventral ♀ and ♂
04: wings flat
05: wings up
06: mature larva

Habitat: Moist, boggy areas
Host Plant: Pacific reed grass (*Calamagrostis nutkaensis*)
Life Stages: Instars progressively live longer than the last (Garth & Tilden.1986.166). Pupa with a horn on its head. Univoltine: May–June.
Counties They Fly In: Coastal Sonoma
Great Places to See Them: Salt Point State Park (Sonoma County)

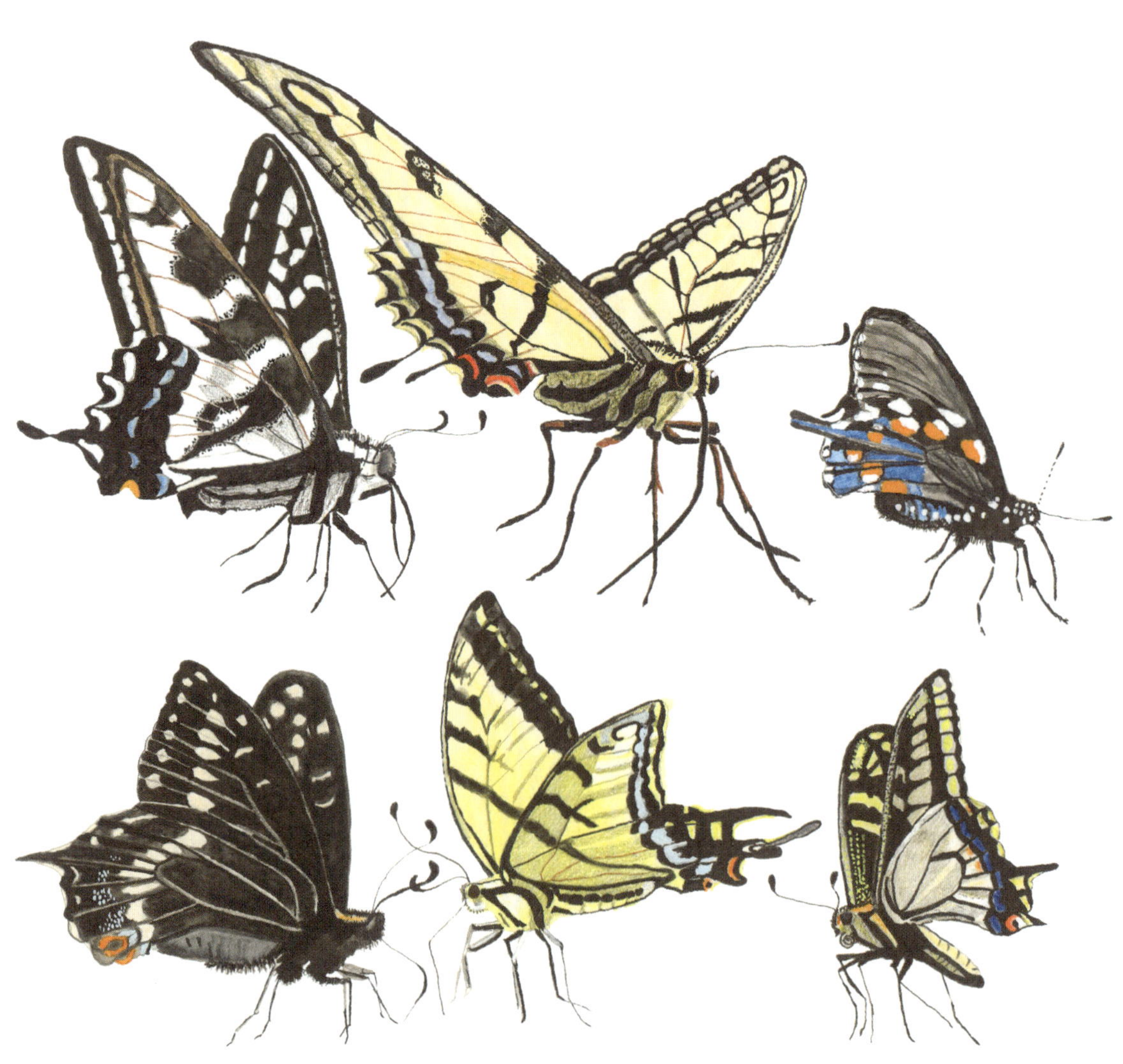

The Swallowtails (*Papilionidae*)

The swallowtails, so named for their tailed hind wings, are the largest and showiest of our local butterflies. The Two-tailed Tiger Swallowtail (*Papilio multicaudata*) is the largest butterfly in the United States. With about seven hundred species of this family known worldwide, these attractive beauties are divided into three tribes within the Subfamily Papilioninae: true swallowtails (Papilionini), of which we have five in this book; the poison eaters (Troidini), which are named for the *Aristolochia* toxins they eat at the larval stage, of which we only have one in this group—the Pipevine Swallowtail; and lastly the Kite Swallowtails (Leptocerini or Graphini)—this is a tribe that looks like our true swallowtails, but we actually have no representatives.

There is one other subfamily called the Parnassians, or Apollos, that are unlike all others. Covered in hairs and eye-catching because of their cherry spots, the Parnassians used to have a sole representative in the Bay Area up until the 1950s—the Clodius (*Parnassius clodius*). The populations in the Santa Cruz Mountains (called Strohbeen's Parnassian) and Sonoma went extinct, and the last one out of Marin County was collected by Robert Langston in the late 1950s. It would have been cool to have seen them on the wing around here.

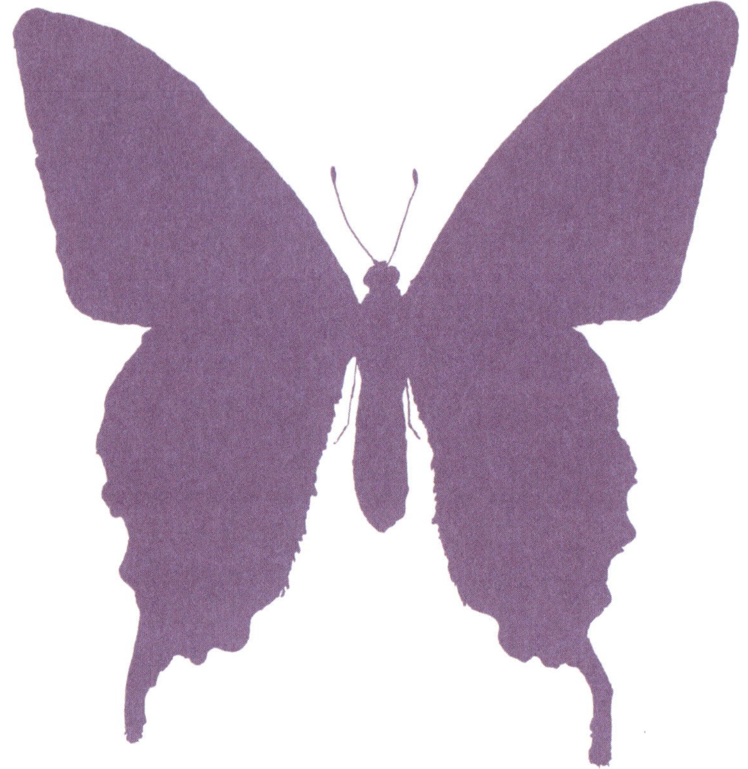

WESTERN TIGER SWALLOWTAIL

Linnaeus 1758

Papilio rutulus

At the base of each hind wing, a tail juts out with orange and blue metallic spots matching up to a "false head" (where the tail looks like antennae and the spots are eyes). This gives predators like birds something to aim for other than the real head.

Folks, this is the one that started it all for me. A decade or so later I collaborated with Amber Hasselbring of Nature in the City to bring the phenomenal story of this urban glider to the masses. The project is called Tigers on Market Street (see page 111 for the full story).

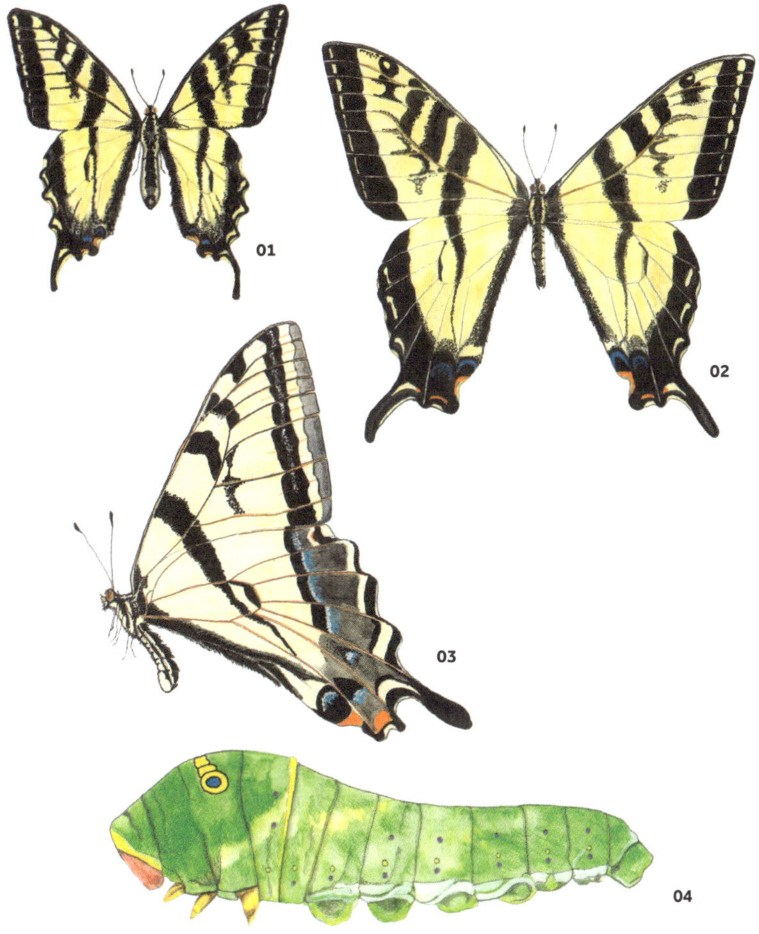

01: dorsal ♀
02: dorsal ♂
03: ventral
04: mature larva

Habitat: Riparian creeks and rivers, city and county parks
Host Plants: Primarily trees—willow (*Salix* spp.), sycamore (*Platanus* spp.), cottonwood (*Populus* spp.), alder (*Alnus* spp.), and elm (*Ulmus* spp.)
Life Phases: Caterpillar is like a big green cigar with prominent false yellow eyes. Pupa is light brown with gray markings. Overwinters as a pupa. Univoltine (May—September).
Counties They Fly In: All
Great Places to See Them: Many neighborhood streets in the Greater Bay Area are lined with the London plane tree (*Plantanus x acerifolia*)—one of the butterfly's hosts. Many people report seeing the butterfly float among its canopy and down city blocks. The butterfly is so widespread, I'd say first start with your local county parks. A more natural setting might be Mitchell Canyon Trail, Mount Diablo State Park (Contra Costa County), where you can work on distinguishing it from the other two swallowtails it flies with there, the Pale and the Two-tailed.

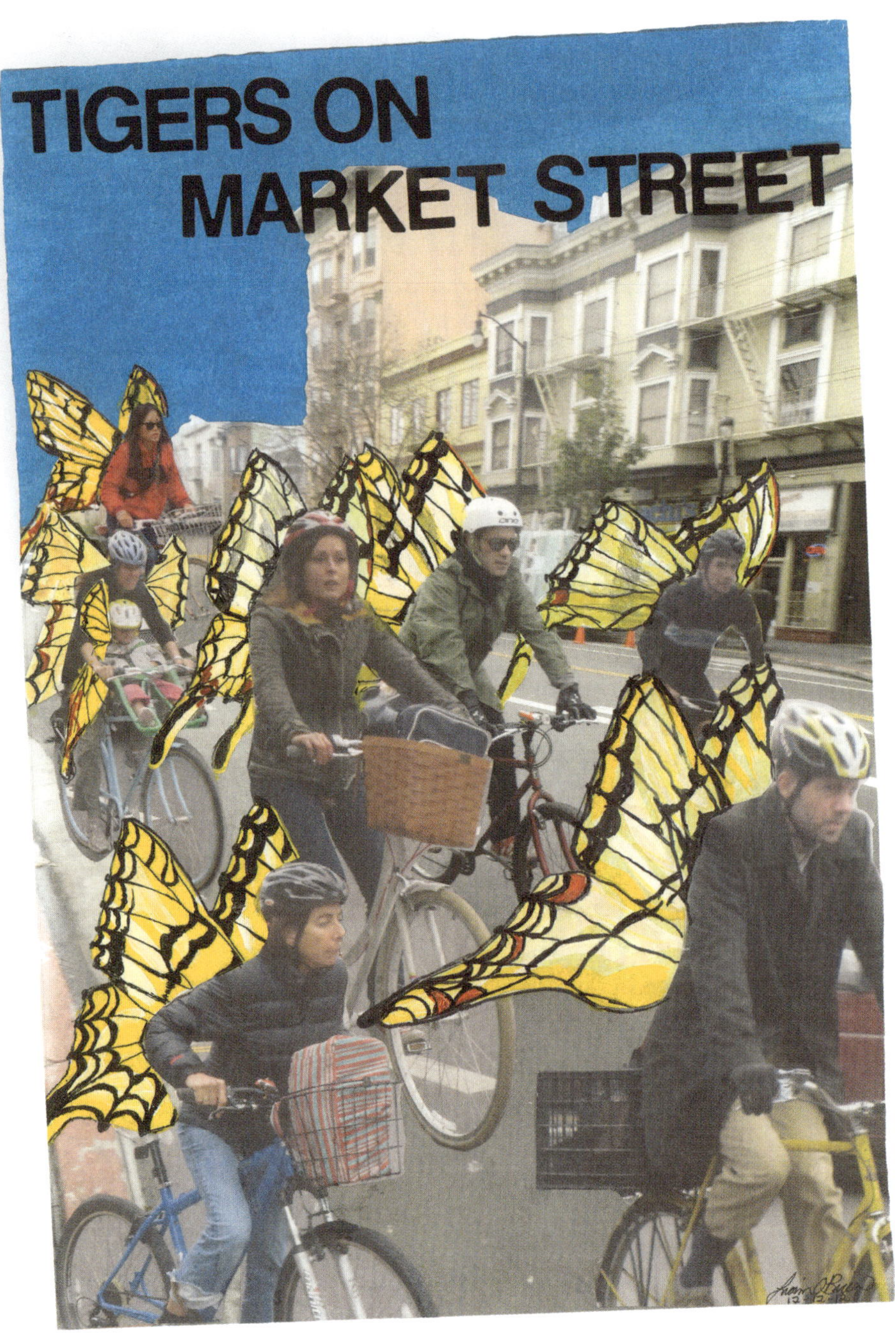
TIGERS ON
MARKET STREET

TIGERS ON MARKET STREET

The phenomenon was first noted by Harriet Reinhardt on the first Xerces Society's butterfly count in San Francisco (these are now run by the North American Butterfly Association): "In the highrise, downtown area," she wrote, "*P. rutulus*, the Western Tiger Swallowtail, is the main species seen, since the entire length of Market Street has been planted with sycamores" (Xerces Society. 1985.32).

I remember being young in the city back in my twenties, spending most days in the downtown area on some mindless temp job, either answering the phone in some office or folding, collating, and stapling six hundred reports for a law firm. I had no marketable skills. All my chips were pushed into the center of my dream to be an actor. I'd get a half hour for lunch wherever I worked, and that was spent on a wall outside eating from a brown bag. Had I known about the amazing story of this butterfly wafting over my head, I wonder if I would have cared then. Nature, it's easy to assume, is not on the mind of most techies walking briskly among the cement walls and sidewalks.

When I finally noticed this butterfly some thirty years later, I marveled that it could survive in this densely built city. The best way to understand why the Western Tiger Swallowtail (*Papilio rutulus*) haunts the hostile ecosystem of downtown San Francisco is to get yourself to the base of Market Street at Spear Street and look back up our most prominent grand boulevard. This is what you'd see: a long linear concourse between tall buildings that is lined with London plane trees, a hybrid between the American sycamore and the Oriental sycamore, and the most widespread urban tree planted in cities throughout the world. Their natural habit is to patrol canyons hollowed out by creeks and rivers with many varying hardwood trees on

Butterfly Brigade collage, 2010

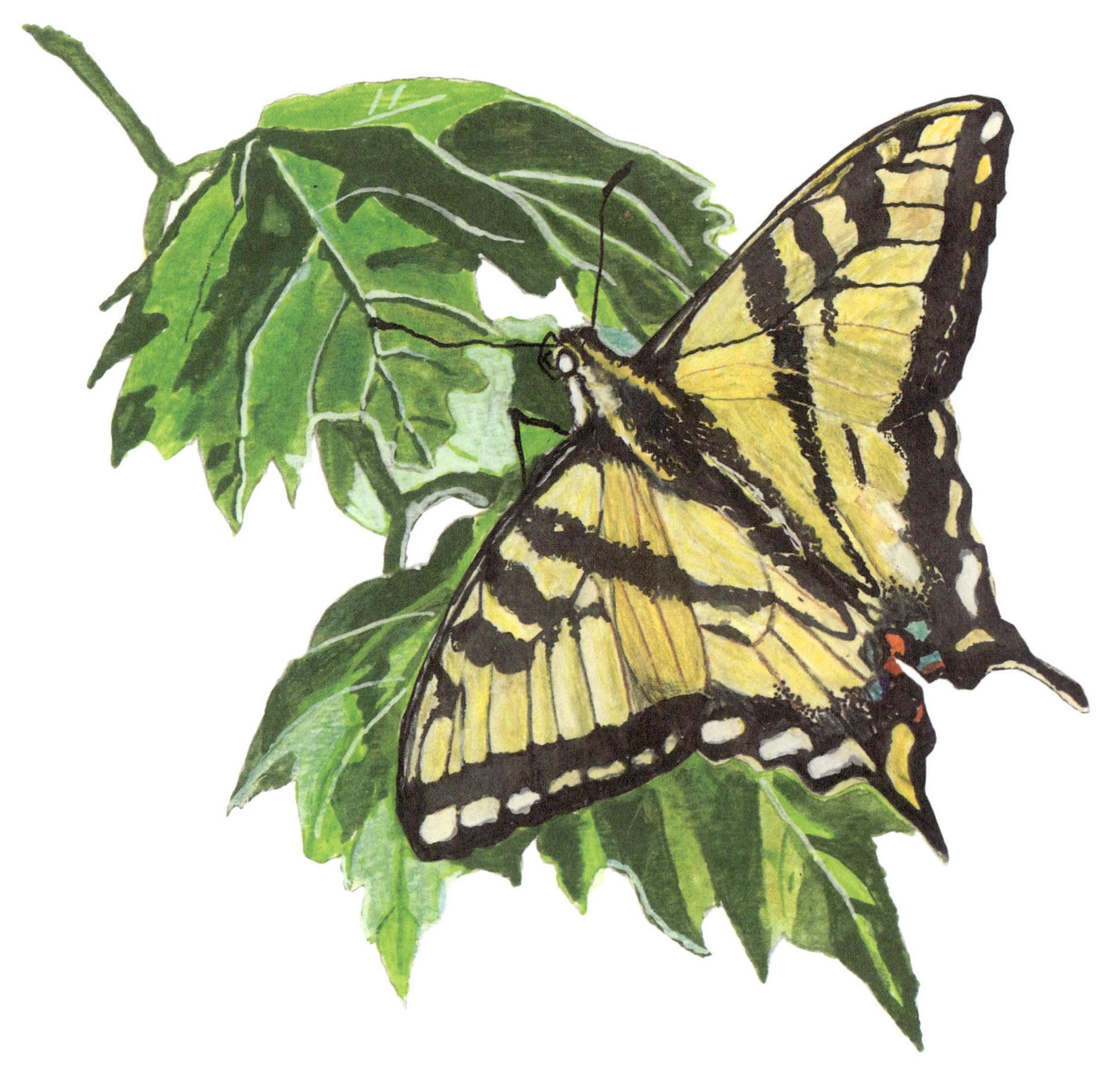

either side. From the point of view of the creature, Market Street might as well be the Kern River, a canyon with dappled sunlight filtering through these beloved trees. All in all, the butterfly is there, folks, because the ecosystem mirrors its natural setting. Because the London plane tree is also one of the butterfly's host trees, we have the entire life cycle playing out over the heads of the downtowners. The large female lays her eggs in the canopy, the males following the females' pheromone trails like road maps to her location. We went and inadvertently created the perfect swallowtail habitat.

With the success of my first butterfly conservation project, the Green Hairstreak corridor in 2007 (see page 255), my mind began to shift to a place where it's more difficult to connect people to nature. It's one thing if patches of unused land can be filled with the right native plants; it's a whole other thing to work with an antiseptic boulevard of concrete and skyscrapers. But I was not alone in this endeavor.

When I first met Amber Hasselbring, she was an installation artist with a project on Angel Island. We'd both worked on the poster committee for Nature in the City's Earth Day celebration in 2009, and as I got to know Amber, my respect for her only grew. She has mad graphic skills and a wonderful sense of color, and it was about this time she became the executive director for Nature in the City. We decided to team up for an idea I had: Tigers on Market Street. Amber had done a mural on the street, and it reminded me that there were many ways of telling a story, especially this story, which most people knew nothing about.

Amber and I received a small grant to look deeper into the butterfly's behavior and what other invertebrates live on the street. Market Street became our field station. We wanted to know just where they were getting their nectar sources, mud-puddling locations (if any), and just how many were there in total. (Sidebar: a bizarre moment when we were stopped behind crime tape in proceeding with our transect. I turned to Amber: "I bet we are the only field researchers that have to wait for a dead body to be removed," I said.)

And we discovered that these leafy streets were a little less than perfect, after all. Much peril exists for any caterpillar falling out of the canopy, no doubt trampled by the commuting feet below. Even as adults, butterflies usually need a good thirty minutes to harden up, to stiffen, after eclosing (emerging) from the pupa. The butterfly is vulnerable at many stages. To support this accidental swallowtail corridor, we had this idea that if we could get some of the big business owners to swap out a portion of their banal landscaping plants for real natives and flowers, then the butterflies just might have a chance down here. One of the more memorable moments of that

Western Tiger Swallowtail on London plane tree

season afield was watching a butterfly at Sansome and Market come down and check out flowers at a cut flower stand. Think about that for a moment. A fleeting moment of adaptation in a non-natural setting. Extraordinary.

We lead many walks to connect as many people as well as we can to the story. A classroom of elementary school children re-created the corridor in miniature with paper butterflies flying over downtown. We even created a small "field guide" of all the creatures, from mice to pigeons to dragonflies, one might be able to see. All the dismissed nature. There was an art contest for ideas with installations that would help reimagine Market Street as the city set off with its Better Market Street campaign. Amber's idea was accepted, and here is an update on the project from her in 2024:

> There is much to count as positives as the Tigers on Market Street story continues to unfold—parks with wild patches throughout that provide safe harbor for migrating and residential birds, dragonflies and more; POPOS (publicly owned private spaces) with creative flair transforming roof decks, plazas indoors and out, living walls, lush gardens into spaces more habitable by nature. Salesforce Park above the Transbay Transit Center is a new, grand example at 5.4 acres, diverse with plantings, stunning architectural views, birds, butterflies and insects. Nature in the City's BART/Muni Canopy Living Roofs project is a dream vision that may or may not come to be. If it could be successful, via combining public and private funds, San Francisco may see a new elevated

parkway 16 feet above the sidewalks just for wildlife in the canopy of the London plane street trees that the butterfly uses to lay her eggs on.

So why is any of this important? The butterflies are there; they weren't trucked in or released. I'm thinking of my younger self again sitting on that wall at lunch—what if I had looked up? I would have gasped at seeing such a large butterfly soar by. "What the heck was that?" Like the swallows of Capistrano, the Tigers on Market Street could be an annual festival, watching for the first of the season to return. It would connect people, a whole lot of people, to an unusual neighbor when and where an encounter with nature is least expected. That makes it important.

The other reason this butterfly is important is for our city. I walked downtown recently, having not been down there in years, since the COVID pandemic locked everything down. Empty businesses and buildings outnumbered the ones in use. The ever-growing homeless encampments took up a great deal of sidewalk. After the Better Market Street campaign pushed through the removal of cars on Market Street, small businesses collapsed and foot traffic dried up. Downtown feels like a ghost town. There has been a pall over the city these last couple years, but if San Francisco has a history of anything, it's rebirth.

No doubt the last thing on any city official's mind is the presence of a big, yellow butterfly. Yet, that same day, a stranger looked up, and pointing to the sky, said, "Look. There's one!" A large female floats by, a jarring sight against the skyscraper's glass walls. But there they still are. The phenomenon continues. From changing my life one afternoon in a backyard years earlier to me trying to change its life for the better now, I hope in the future someone picks up this baton to keep improving the downtown area of San Francisco.

ANISE SWALLOWTAIL

Lucas 1852

Papilio zelicaon

Everything changed for this butterfly with the introduction of sweet fennel (*Foeniculum vulgare*) into the West. Along with a name change (this butterfly used to be called the Western Parsley), it now flies virtually year-round because of this non-native plant. Despite this affinity for fennel by the butterfly, I advise against planting it in your yard. It will take over and you will curse the day you did it. If you want this butterfly in your garden so you can see the beautiful caterpillars, plant parsley or dill, annuals that you replace each season.

This butterfly is a strong hill-topper, so look for them on peaks. The San Francisco Butterfly Count in 2014 had a one-day national high in observances: we found 144 of these. Helped to have 42 hills to check. It is our most common and widespread of our true swallowtails.

Many elementary school lessons use the Anise Swallowtail to teach metamorphosis. They are easily found by teachers in the community. Collect local and release local in order to help these butterflies thrive—and this also avoids using the butterfly kits with Painted Ladies being fed manufactured pellets.

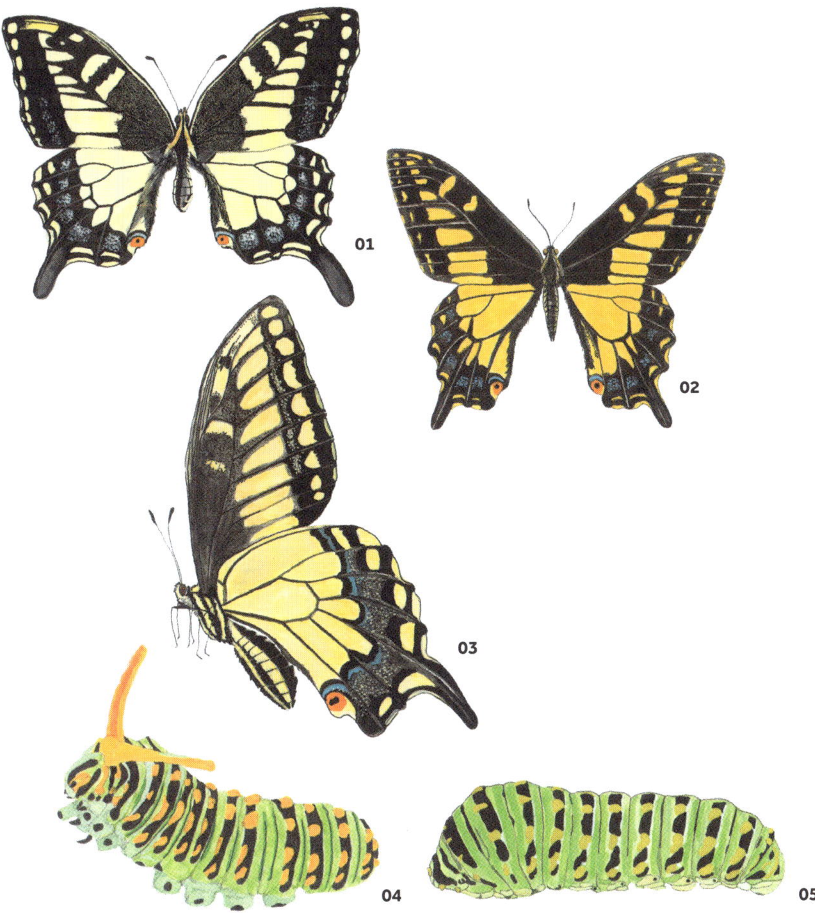

01: dorsal ♀
02: dorsal ♂
03: ventral
04: larva with osmeterium extended
05: mature larva

Habitat: Many, from vacant lots in cities to pristine, natural mountaintops
Host Plants: A variety of plants in Apiaceae, the carrot family. *Lomatium* spp., *Angelica* spp., cow parsnip, poison hemlock, and yampah
Life Phases: The larvae are green with black vertical stripes broken up by orange spots on each segment. Pupa is light grayish. Overwinters as a pupa and emerges in early spring. In adults, aside from a size dimorphism (the female is larger—she is carrying the eggs, after all), sexes look similar with a creamy yellow field or ground color and predominant black shoulders (see **Appendix B, Tableau 4**). Both have developed false heads at the base of their hind wings. Multivoltine (January–December).
Counties They Fly In: All
Great Places to See Them: Twin Peaks and Grandview Park (both San Francisco County) at their summits; Miller/Knox Regional Shoreline, Point Richmond (Contra Costa County)

TWO-TAILED TIGER SWALLOWTAIL

Kirby 1884

Papilio multicaudata

This is our largest butterfly in Central California. I lay down on the edge of a puddle once to try and get the scale of this large insect while it was mud-puddling. With its long face and wings held up high like Pegasus, it had the nobility of a stallion.

Because their host plant has such a limited range in the Greater Bay Area, assured places to see this leviathan are equally narrow in scope.

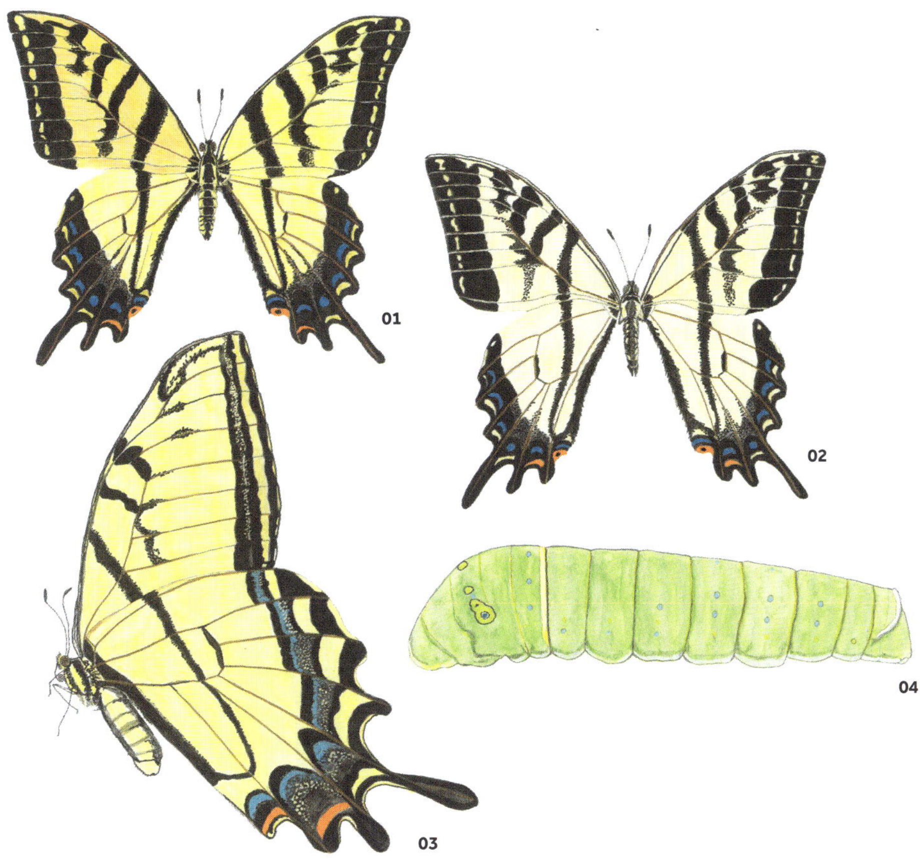

01: dorsal ♀
02: dorsal ♂
03: ventral
04: mature larva

Habitat: Riparian creeks and valleys, chaparral where the host plant is present

Host Plant: Western hoptree (*Ptelea crenulata*)

Life Phases: Chartreuse-colored larvae big as a human finger. Pupa is light green or brown. Female adults have more of an orange hue to their ground color of yellow, while the male is a creamy lemon yellow. Both females and males have two tails on their hind wings—one short, one long. Multivoltine: early spring to late summer.

Counties They Fly In: Alameda, Contra Costa, Napa, Santa Clara, Solano, and Sonoma (most prevalent in Contra Costa and Solano)

Great Places to See Them: Mitchell Canyon Trail, Mount Diablo State Park (Contra Costa County). Further up the trail toward Deer Flat, one can witness it nectaring on cobweb thistle (*Cirsium occidentale*). Along the creek trails at John Muir National Historic Site (Contra Costa County).

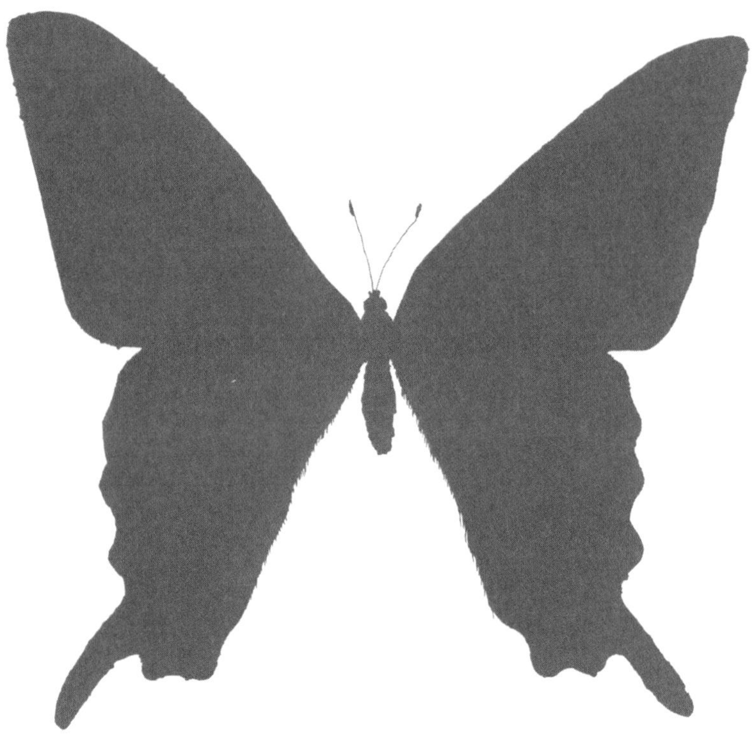

PALE TIGER SWALLOWTAIL

Lucas 1852

Papilio eurymedon

Referred to sometimes in this book the way most Lepidopterists I know refer to it as just **"PALE"** or **"THE PALE"**

Absence is the first thing you notice when one of these pied creatures floats by—a total lack of color. Those are the males. A merit badge to you when you're advanced enough to key out the ever-so-slightly-yellow females on the wing (rough when Western Tigers are about). In 2021, I was surveying Mount Sutro for Golden Gate Bird Alliance when a female landed above me. The last confirmed sighting in San Francisco County was 130 years before, by the legendary lepidopterist F. X. Williams. The landscape in San Francisco is so fractured that it was hard not to conclude it was a vagrant from across the San Mateo County line.

Some people think that all the work has been done on butterflies. It's just not true. There is still a great deal everyone can add to our butterfly stories, especially regarding ranges.

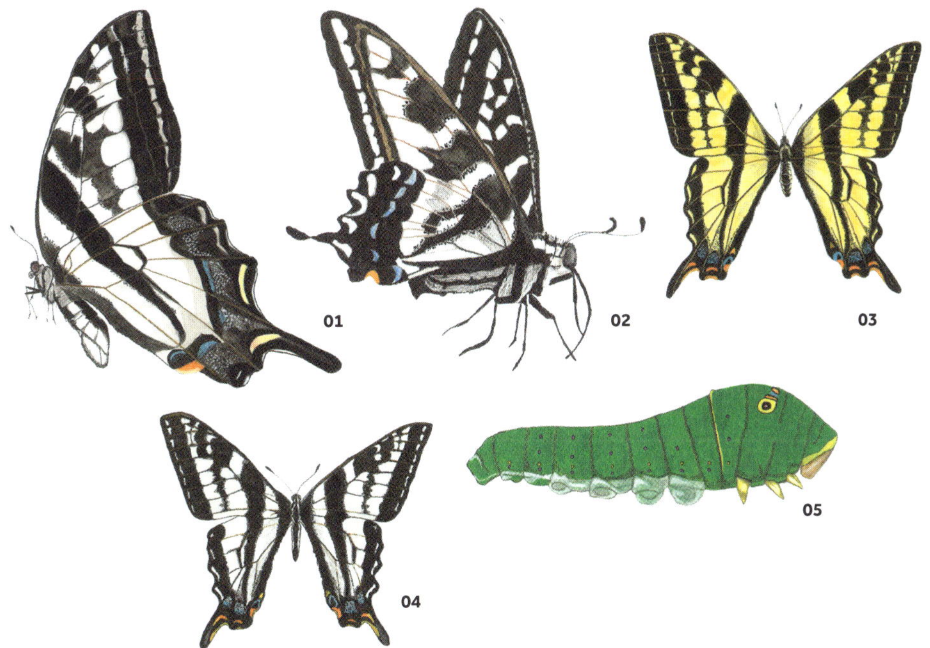

01: ventral
02: in situ
03: dorsal ♀
04: dorsal ♂
05: mature larva

Habitat: Connected communities of coastal scrub. Does not seem to penetrate the urban landscape using a horizontal corridor of tree canopies as the Western Tiger does.

Host Plants: Coffeeberry (*Frangula californica*) with some reports of buckthorns (*Rhamnus* spp.), and California lilac (*Ceanothus* spp.) (Glassberg.2012.30)

Life Phases: Eggs laid on top of leaves. Pupa like that of Western Tiger. Eyespots on larvae are smaller than those on the Western Tiger, and the head of the caterpillar has a purplish hue. In adults, both sexes have bold, black vertical stripes running through both wings—a defining trait.

Counties They Fly In: All except San Francisco . . . well, sort of in San Francisco

Great Places to See Them: A strong hill-topper, the Pale Tiger can be seen at Vollmer Peak in Tilden Regional Park (Contra Costa County); the summit of Mount Tamalpais (Marin County); along the Ridge Trail at San Bruno Mountain County and State Park (San Mateo); Mitchell Canyon Trail, Mount Diablo State Park (Contra Costa County). I imagine one could run into them along the Mount Hamilton range and summit in San Jose (Santa Clara County) too.

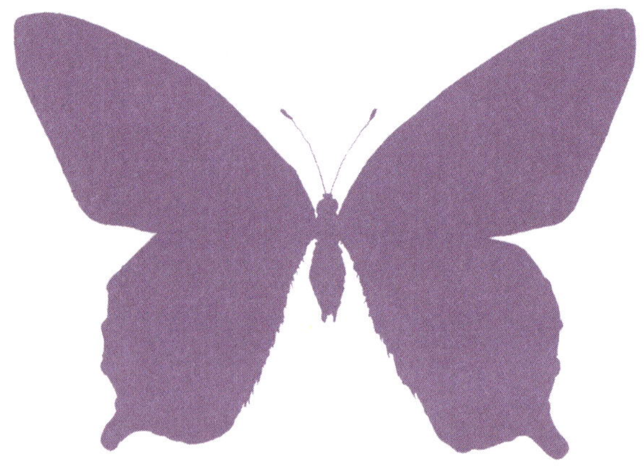

INDRA SWALLOWTAIL

Reakirt 1866

Papilio indra

Get yourself to a piece of serpentine on open hilltops in Sonoma or Napa County. I've observed that this butterfly seems to enjoy blooming thistles just below summits. The first time I saw one, I had to jump a fence to get close. (In general, follow my example, but not in this case—trespassing is bad for your psyche and bad for the sport, obviously.) The only other swallowtail this dark on the wing is a Pipevine. It shares hilltops with the Pale and Anise Swallowtails. All the different Swallowtail species will bombard one another like they're in an unofficial wrestling match, the prize being a mate. The nine subspecies of Indra in the US are divided up by the length of their tails; an endangered subspecies flies along the rim of the Grand Canyon—*P. i. kaibabensis*.

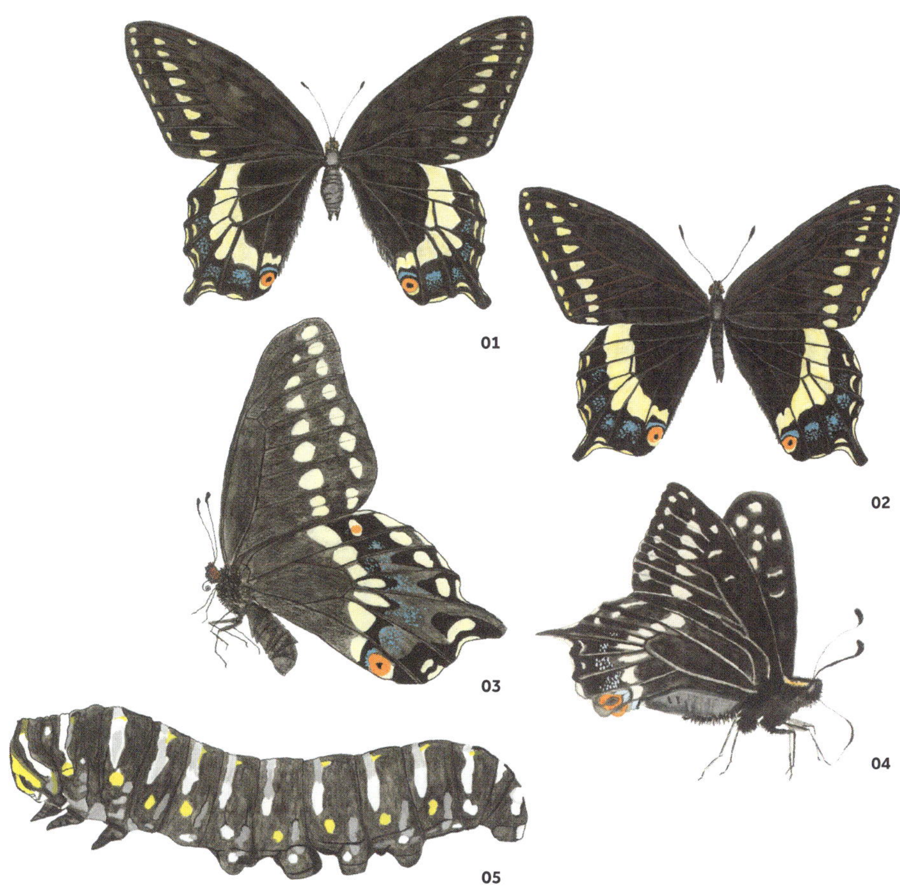

Habitat: Serpentine and its nearby hilltops

Host Plants: In the Greater Bay Area it primarily feeds on biscuitroot (*Lomatium marginatum*) and *Cymopterus terebinthinus* (Steiner.1980.52).

Life Phases: Last instar of larvae is black with crossbands of pink. Pupa comes in a multitude of colors. Adults look similar, with fields of black on their fore and hind wings. (Anise Swallowtails have black primarily only on their forewings.) Univoltine: May–June.

Counties They Fly In: Napa, Solano, and Sonoma

Great Places to See Them: Extremely elusive and a rough one to pinpoint. I saw my first at a hilltop near Clear Lake (Napa County). A few sightings are known west of the town of Venado (Sonoma County). Good luck.

01: dorsal ♀
02: dorsal ♂
03: ventral
04: in situ
05: mature larva

PIPEVINE SWALLOWTAIL

Skinner 1908

Battus philenor

Female adults are dishwater brown and slightly bigger than the males. His gunmetal-blue hind wings shine in flight. This exquisite butterfly rocketed through all of my ten transects in 2007 when I surveyed San Francisco County to record which butterflies were still flying there. Many of my transects weren't anywhere near its host plant, Dutchman's California pipevine, proving that this creature has the ability to find its host over great distances. The native host plant is rare in certain parts of the Bay Area, but the butterfly isn't. Still, I get lots of calls from folks that don't understand why the butterfly hasn't shown up to their gardens—"I planted the host." Nature just isn't that neat, I tell them—it's messy. Be patient because it just might happen.

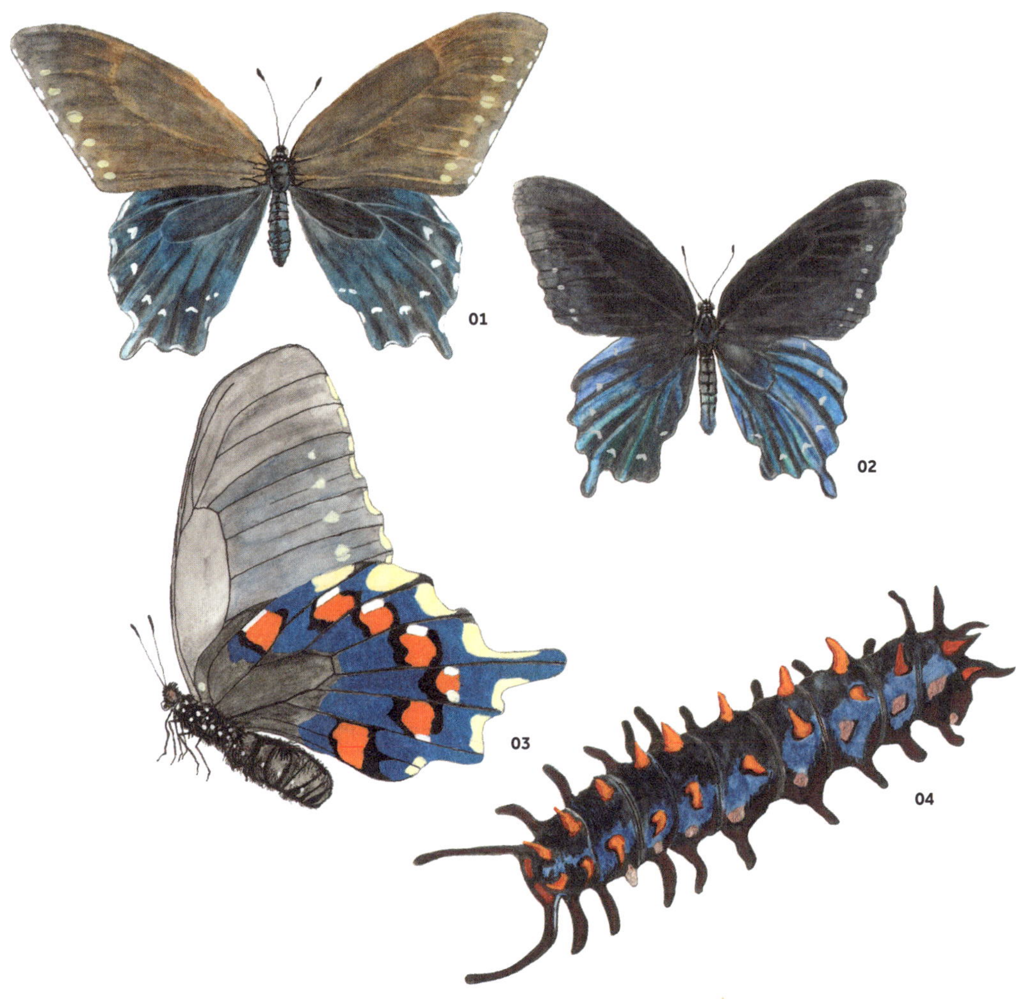

01: dorsal ♀
02: dorsal ♂
03: ventral
04: mature larva

Habitat: Riparian woodland, oak woodland, gardens, and most open spaces
Host Plant: Dutchman's California pipevine (*Aristolochia californica*)
Life Phases: Red eggs in group clusters. Pupa is brown, green, or gray depending on the substrate. The larvae are big, blackish blue with red spikes protruding out of them. (I've always thought they'd make an excellent foe for Godzilla in a Japanese horror film.) Pupae are either mossy green or primarily brown. Overwinter as the chrysalides. The red polka-dotted underside of the hind wing in the adult is definitive in keying this species. Bivoltine: March and again in the summer.
Counties They Fly In: All
Great Places to See Them: Hilltops in the Marin Headlands (Marin County) and San Bruno Mountain State and County Park (San Mateo County); San Francisco Botanical Garden (San Francisco County); and both the Tilden Regional Parks botanical gardens (Alameda County).

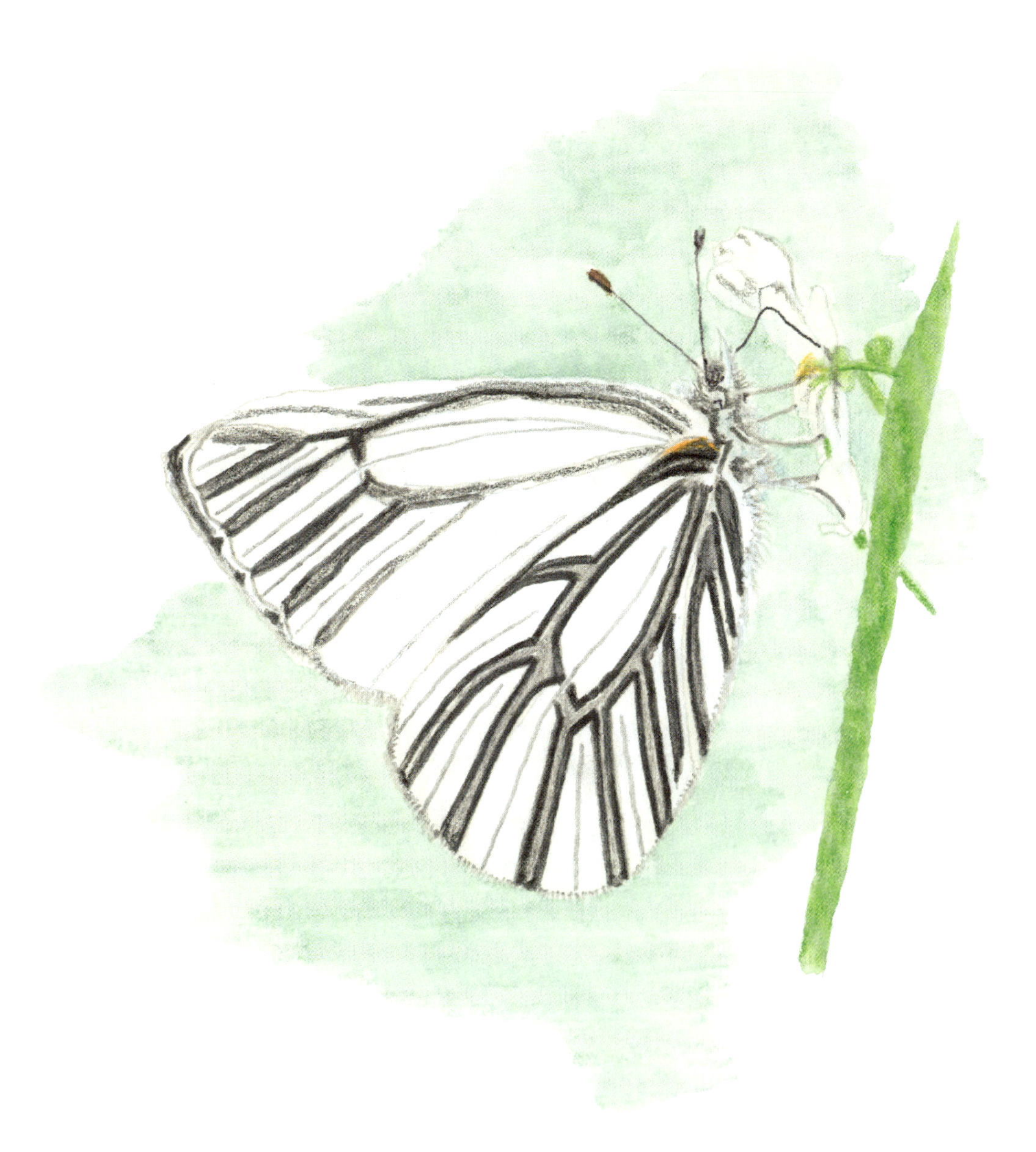

The Whites, Sulphurs, Marbles, and Orangetips (*Pieridae*)

This family contains some of our most abundant butterflies, with a worldwide distribution of two thousand species. King and queen of this mountain here is the much-dismissed Cabbage White (*Pieris rapae*) with its ever-lazy flight and ubiquitous presence. No, they are not moths as many believe, and if one can press through the fact that not all white butterflies wafting by are Cabbage Whites, then there is an astonishing array of colors to behold in this family, from the "bleeding" tips of the orangetips, to the green marbling of the underside of the marbles, from the yellow speeding-by appearance of the sulphurs, to Pine Whites exploding some seasons in such abundance that they look like summertime snow flurries.

Patterns in ultraviolet on the wings are used by the butterfly itself in identifying the opposite sex. No other family of butterflies disperses at such a rate as the Pieridae. Many migrate.

We have twelve by my count in the Greater Bay Area. Markings are very important to learn and can be confusing at first due to the subtlety of the differences. And don't be tripped up by the fact that a yellow butterfly can also be a white butterfly.

Margined White

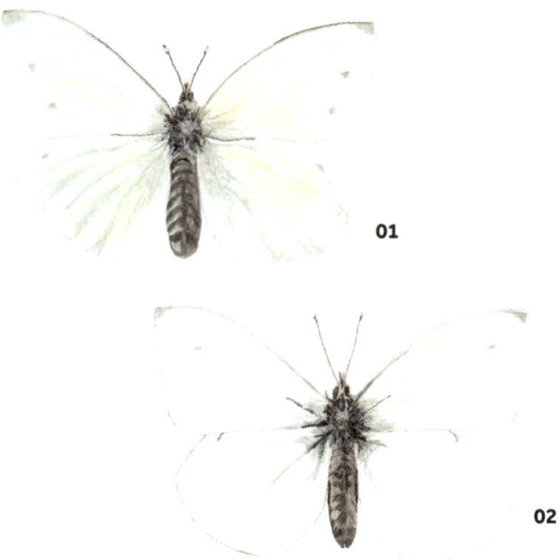

CABBAGE WHITE

Pieris rapae

Linnaeus 1758

Many consider the Cabbage White a "rat bug" to be dismissed. I must be wired wrong, because I still get a dopamine rush every time I see one. It is an exotic butterfly that came to our country through Quebec around 1860. The male has one dot on each of its forewings; the female has two on each. For more identification info, see **Appendix B, Tableau 5**.

The females exhibit a "rejection pose" where they lift their abdomens perpendicular to the thorax and drop their wings low, signaling to the eager paramour, "No thanks" (Scott.1986.217). At UC Davis, retired professor and butterfly expert Dr. Art Shapiro buys a beer for anyone who can validate the earliest sighting of the Cabbage White in the new calendar year.

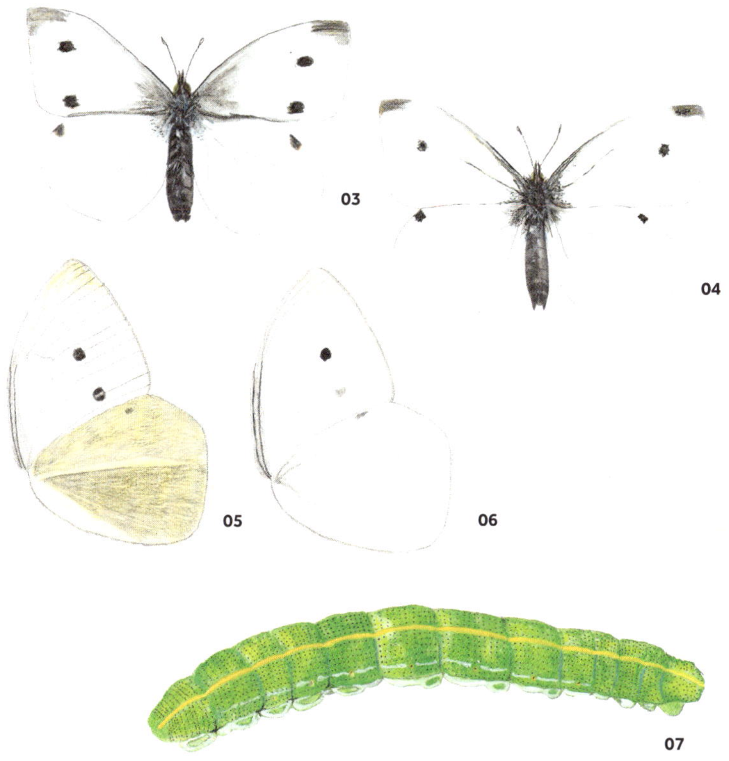

Habitat: Many, from vacant lots to city streets, from agricultural fields to roadside cuts. Not normally found in shady enclaves.

Host Plants: A generalist in the mustard family (Brassicaceae); it even lays its eggs on wasabi (*Eutrema japonicum*).

Life Phases: Little green larvae are considered a pest in many gardens. Pupa brown, green, or gray matching its final location. Early spring flights are not as heavily marked as subsequent generations. Adult females have a soft, lemon yellow to their ventral wings. Multivoltine: almost year-round.

Counties They Fly In: Everywhere humans breathe outside, except in dense forests. (An excellent addendum added here by my friend Paul Johnson, who pointed out that perhaps my generality is a bit extreme. It is a rather rare visitor to the Pinnacles national park.)

Great Places to See Them: One would be hard pressed not to see them in urban areas, but they become rarer in sunny, hot environments. Backyards are a good place to start. At the Point Bonita Lighthouse in the Marin Headlands (Marin County), one can actually see this butterfly hosting on cabbage, as the plant has jumped the fence from the Victorian lighthouse keeper's garden and naturalized down the slope. Another good spot is Alemany Farm (San Francisco County).

01: dorsal ♀, early spring
02: dorsal ♂, early spring
03: dorsal ♀, summer
04: dorsal ♂, summer
05: ventral ♀
06: ventral ♂
07: mature larva

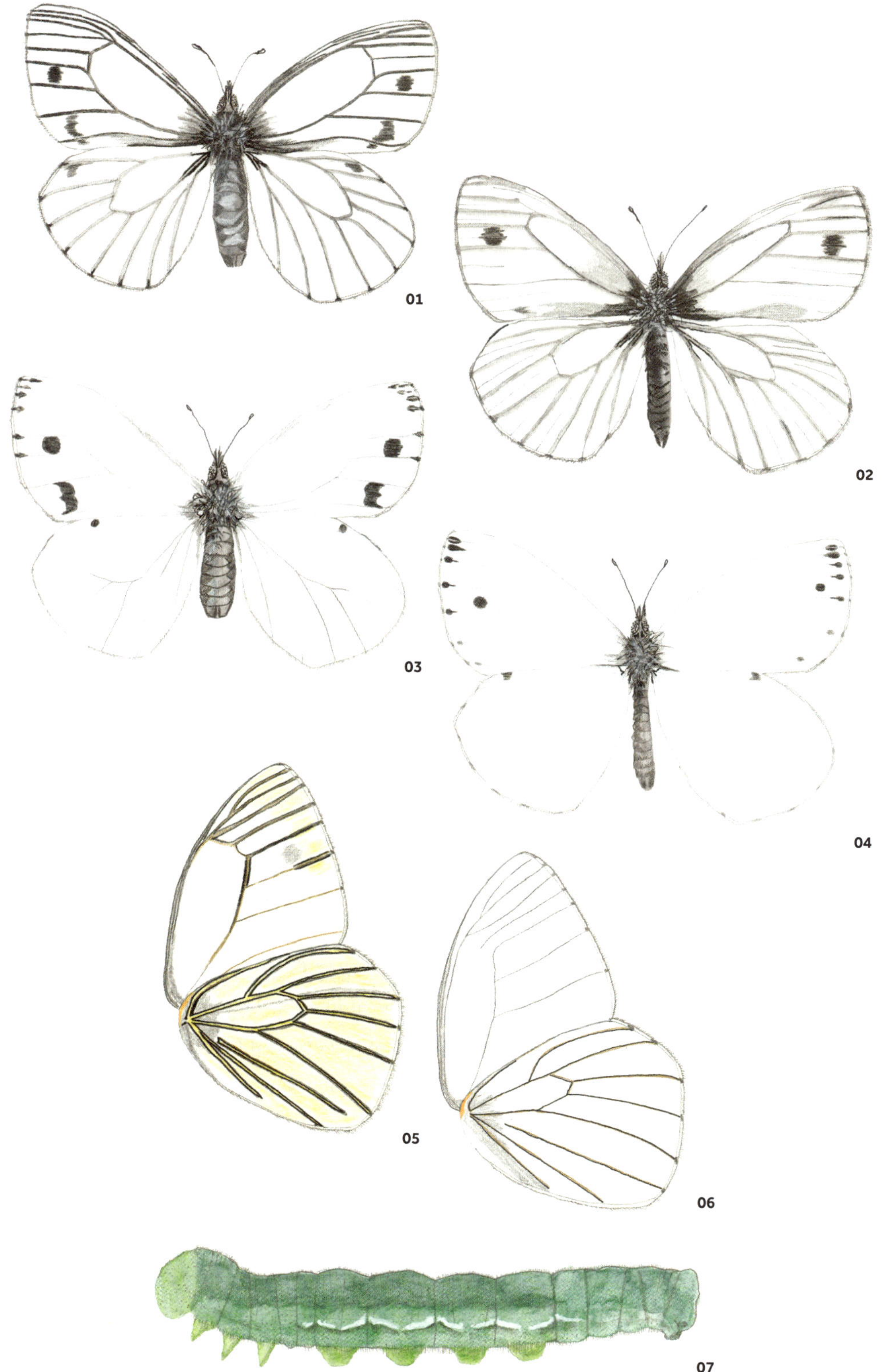
01
02
03
04
05
06
07

MARGINED WHITE

Shudder 1861

Pieris marginalis

I've got a bone to pick with this common name. When I first started out, they were referred to as Veined Whites, because that is what is prominent about them—the veins! Now, due to splitting, the Veined White name only refers to the East Coast species. I'm not even sure what "margined" is referring to . . . lives on the margins of a forest? If you love butterflies, fasten your seatbelts for the never-ending common name swap.

Habitat: Wet, damp forests; roadside clearings and glades. The creature is almost always found in the shade.

Host Plants: Milkmaids (*Cardamine californica*) and other Brassicaceae

Life Phases: This species grows fast: larvae graze for eighteen days and then the adult emerges after eight more (James & Nunnallee.2011.96). Pupa light brown, blue gray, or green. Bivoltine (two flights): the second generation of adults being less heavily veined than the first. The window of the two separate flights can be January–November.

Counties They Fly In: All except San Francisco

Great Places to See Them: Alum Rock Park, San Jose (Santa Clara County); Marin Headlands—strangely not in the shade there (Marin County); Reinhardt Redwood Regional Park (Alameda County); Garrapata State Park, Big Sur (Monterey County)

01: dorsal ♀, early spring
02: dorsal ♂, early spring
03: dorsal ♀, summer
04: dorsal ♂, summer
05: ventral ♀ and ♂, early spring
06: ventral ♀ and ♂, late spring
07: mature larva

CHECKERED WHITE

Pontia protodice

Boisduval & Le Conte 1852

I was surveying out at Candlestick Point in San Francisco County one day. A large group of school kids was playing nearby. I saw a white butterfly on a flower, whiter than a Cabbage White (it's an advanced skill set that comes when you have been doing this long enough—some white butterflies are so white they seem . . . bluish? This is one of them.) It was a Checkered White. I'm so hardcore, I have the county historic record lodged in my brain somewhere. I knew this was an important and rare butterfly and not known from the county. I thought I should collect it as a specimen to give to the California Academy of Sciences for their collection. It's only on these occasions that I use my net in the park, which doesn't really justify breaking the law, I know, but I believe it's a gray area. And I'm willing to chance it. No sooner had I wielded it with one swing than I had thirty little faces looking like the cast of the old Our Gang series up close to me. (Children are magnets to butterfly nets.) "Whatcha got, Mister? Whatcha got?" —they were all so incredibly close all I could think were two words: *pink eye*. I explained to them that this was an important butterfly that needed to go into a museum. One little boy asks me in a pathetic tone, "You're not gonna kill it, are you Mister?" Pause. *Damn it*, I thought. Pause. A sea of Bambi eyes looking up at me. Then I . . . let it go. A round of yippees and yahoos rang out. *Damn it*, I still thought.

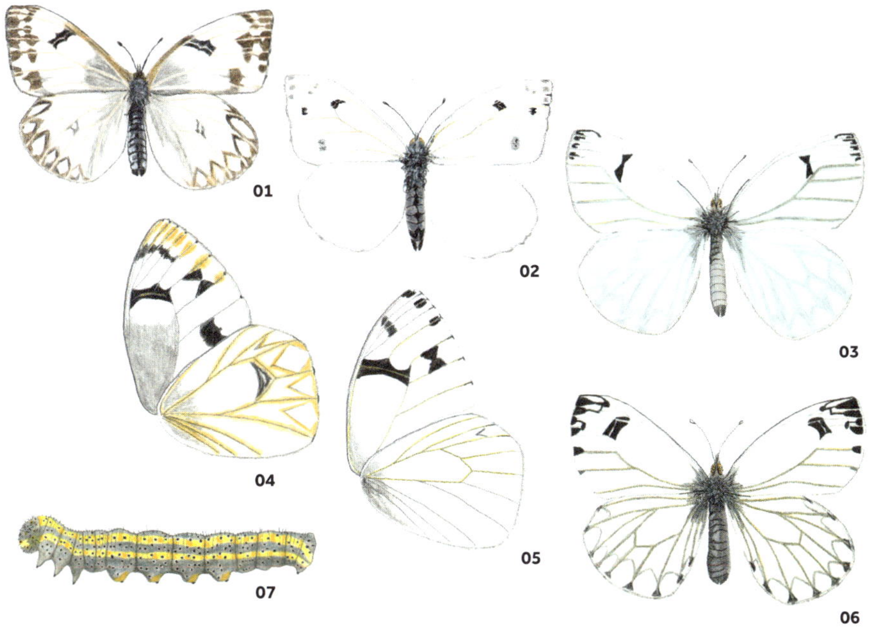

Habitat: Disturbed areas. Think of the edges of agricultural fields, vacant lots, and weedy pathways.

Host Plant: Generalist in Brassicaceae

Life Phases: Mature larvae are gray blue with lateral, middorsal stripes. Pupa blue gray with a smattering of black dots. The adult spring flight is much more heavily checkered than the summer flight. It is easily confused with the Western White (*Pontia occidentalis*), and when I painted the two, I was shocked at the similarities. Even experts are hard pressed to tell them apart in the field. Within the scope of this book, chances are you're 99.9% looking at a Checkered White. Multivoltine, with the fall flight being more noticeable.

Counties They Fly In: All—penetrates San Francisco County at Candlestick Point

Great Places to See Them: They seem to be more inland, away from the coast. Benicia State Recreational Area (Solano County), also below the Benicia–Martinez Bridge along the jogging path (Benicia/Solano County side) up on a levee running parallel to the cyclone fence; Del Puerto Canyon (Santa Clara County); side canyons of San Bruno Mountain County and State Park (San Mateo)

01: dorsal ♀, summer
02: dorsal ♂, summer
03: dorsal ♀, spring
04: dorsal ♂, spring
05: ventral ♀, summer
06: ventral ♂, summer
07: mature larva

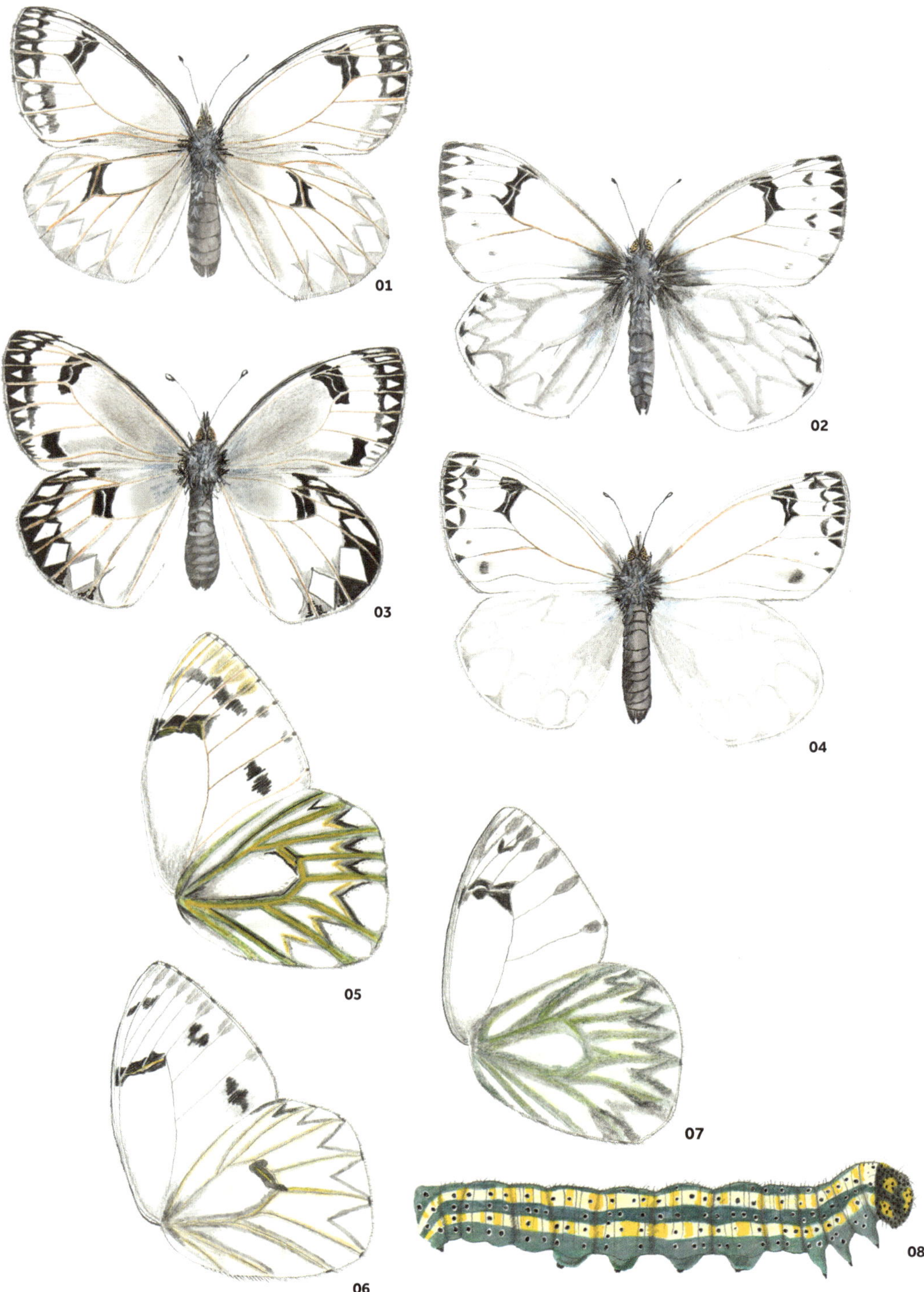
01
02
03
04
05
06
07
08

WESTERN WHITE

Reakirt 1866

Pontia occidentalis

This is not a common butterfly—it may not even fly in the Bay Area at all. So you may be asking yourself at this point, *Why is Liam even including this species here?* Because one of our common butterflies looks strikingly similar to it—the Checkered White (*Pontia protodice*). And if you find yourself in the northern reaches of Napa or Sonoma Counties, a stray is possible. Trust me, even the experts have a tough time keying these two butterflies apart.

01: dorsal ♀, spring
02: dorsal ♂, spring
03: dorsal ♀, summer
04: dorsal ♂, summer
05: ventral ♀, summer
06: ventral ♂, summer
07: ventral, spring
08: mature larva

Habitat: Rocky outcrops primarily at summits
Host Plants: Native mustards (Brassicaceae), tower mustard (*Arabis glabra*)
Life Phases: Larvae similar to the Checkered White. Pupa light gray. Adults fly in a deliberate, straightforward manner sometimes described as "with a purpose." Males are strong hill-toppers (Kaufman & Brock.2003.48). Multivoltine: March–September.
Counties They Fly In: None to note. There have been some reports of it from northern parts of Napa and Sonoma Counties (Shapiro & Manolis.2007.106).
Great Places to See Them: Absolutely no postings of it on iNaturalist in the Greater Bay Area for the last five years

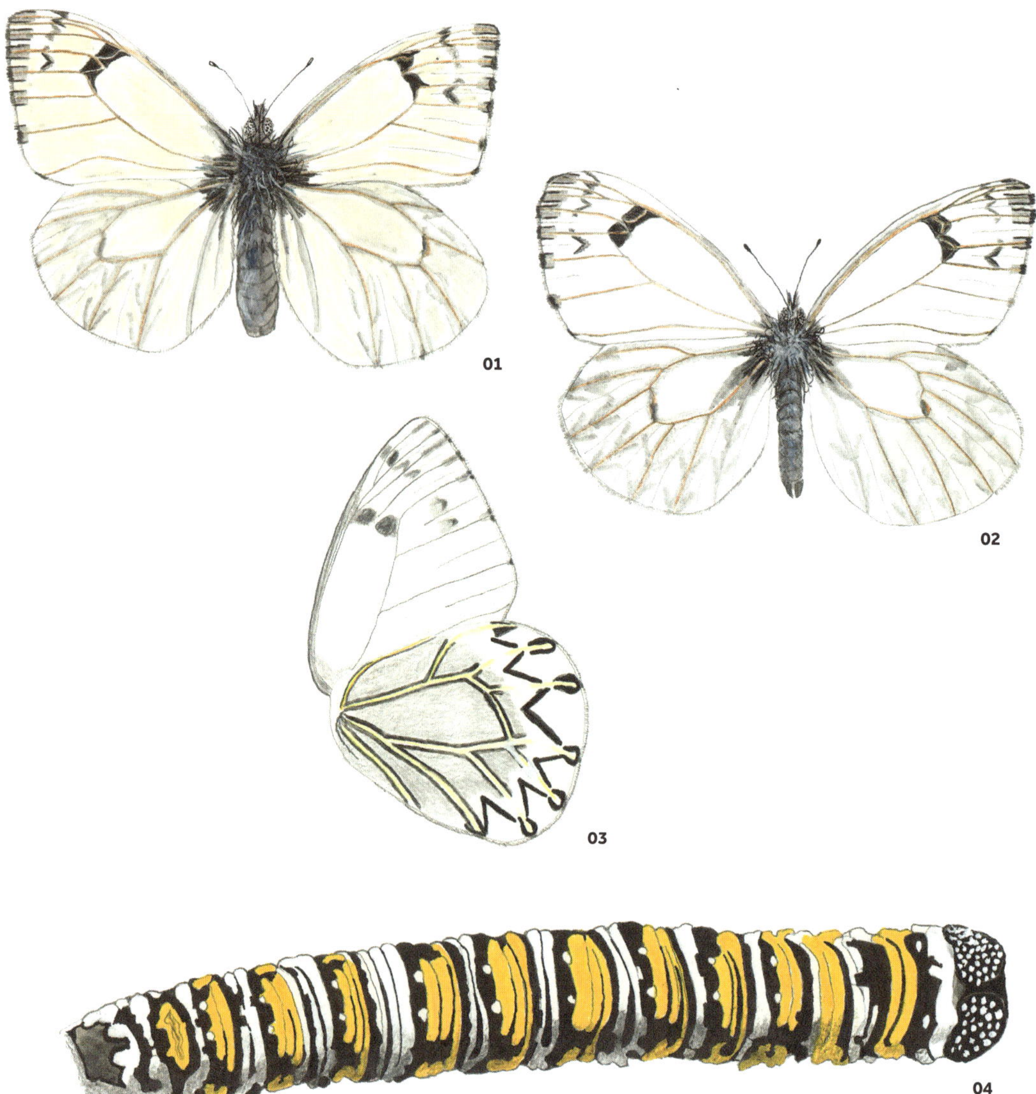
01
02
03
04

SPRING WHITE

Boisduval 1852

Pontia sisymbrii

also known as **CALIFORNIA WHITE**

You'll find this butterfly blasting about open habitats where there is a lot of sun, but interestingly enough, it seems to prefer cooler weather (Emmel & Emmel.1973.17). A true creature of the serpentine ecosystem, the Spring White is also a strong hill-topper.

Habitat: Serpentine outcrops; dry, open rocky slopes

Host Plants: Serpentine jewelflower (*Streptanthus* spp.) and rock cress (*Arabis* spp.)

Life Phases: One of my favorite caterpillars, *P. sisymbrii* looks like Mayan hieroglyphics with smiling yellow frogs staring back at you. Pupa dark brown with black markings. The adults come out very early in the spring. Univoltine: late February–early May.

01: dorsal ♀, yellow form flava
02: dorsal ♂
03: ventral ♀ and ♂
04: mature larva

Counties They Fly In: Northern section of Napa County. Sporadic sightings in Alameda and Santa Clara Counties.

Great Places to See Them: I found my caterpillar toward the summit of Mount Saint Helena (Napa County).

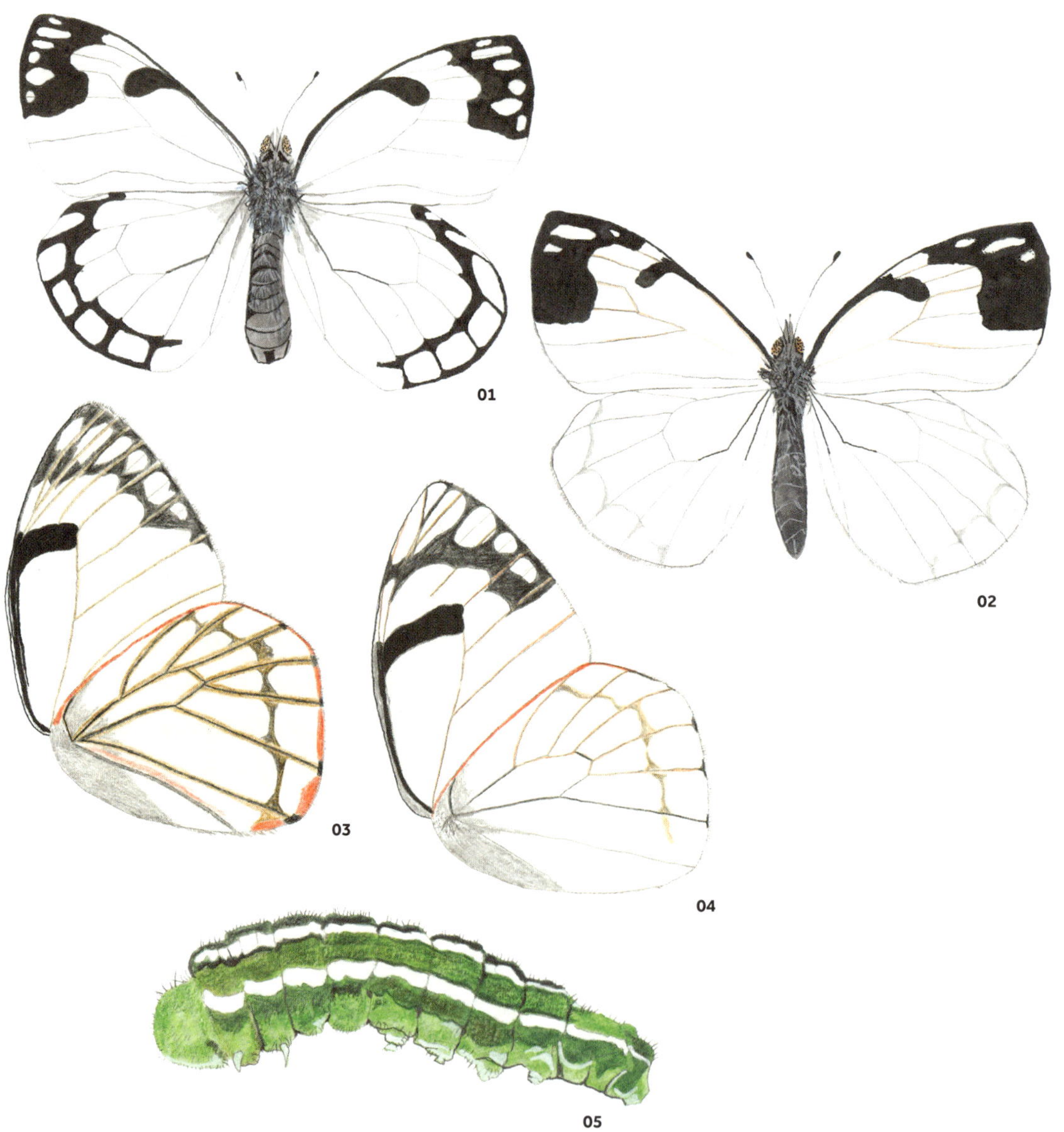
01
02
03
04
05

PINE WHITE

C. & R. Felder 1859

Neophasia menapia

No other white flies or looks like a Pine White. Their wings are elongated and rounded at the apex, and they waft and float more like a swallowtail or Monarch (*Danaus plexippus*). It's the most abundant butterfly in a hemlock or Doug fir forest (Miller & Hammond.2003.67). Many people have reported seeing, on occasion, explosive numbers—by the millions—to the point they thought it was snowing.

Habitats: Pine forests

Host Plants: Larvae eat ponderosa pine (*Pinus ponderosa*) and Douglas fir (*Pseudotsuga menziesii*) needles.

Life Phases: Long, thin, and dark green with horizontal white stripes, the larvae blend right into their food source. The pupae are sexually dimorphic: light green (male) and chocolate brown (female). The female adult has a crazy border of fuchsia on her hind wing, and both sexes have strong, black graphic markings on their forewings. Adults live most of their lives up in the canopy but are known to have the crepuscular behavior of visiting flowers early in the morning and late in the day. Univoltine: June–September.

Counties They Fly In: Sonoma

Great Places to See Them: Get yourself to the intersection of Seaview Road, Hauser Bridge Road, and Kruse Ranch Road (Sonoma County). Walk a hundred yards southeast along the transmission right of way. Also Tom Wysham reports them at Old Skaggs Springs Road 1.2 miles from Skaggs Springs (Sonoma County) in late June (Wysham.pers comm.2022).

01: dorsal ♀
02: dorsal ♂
03: ventral ♀
04: ventral ♂
05: mature larva

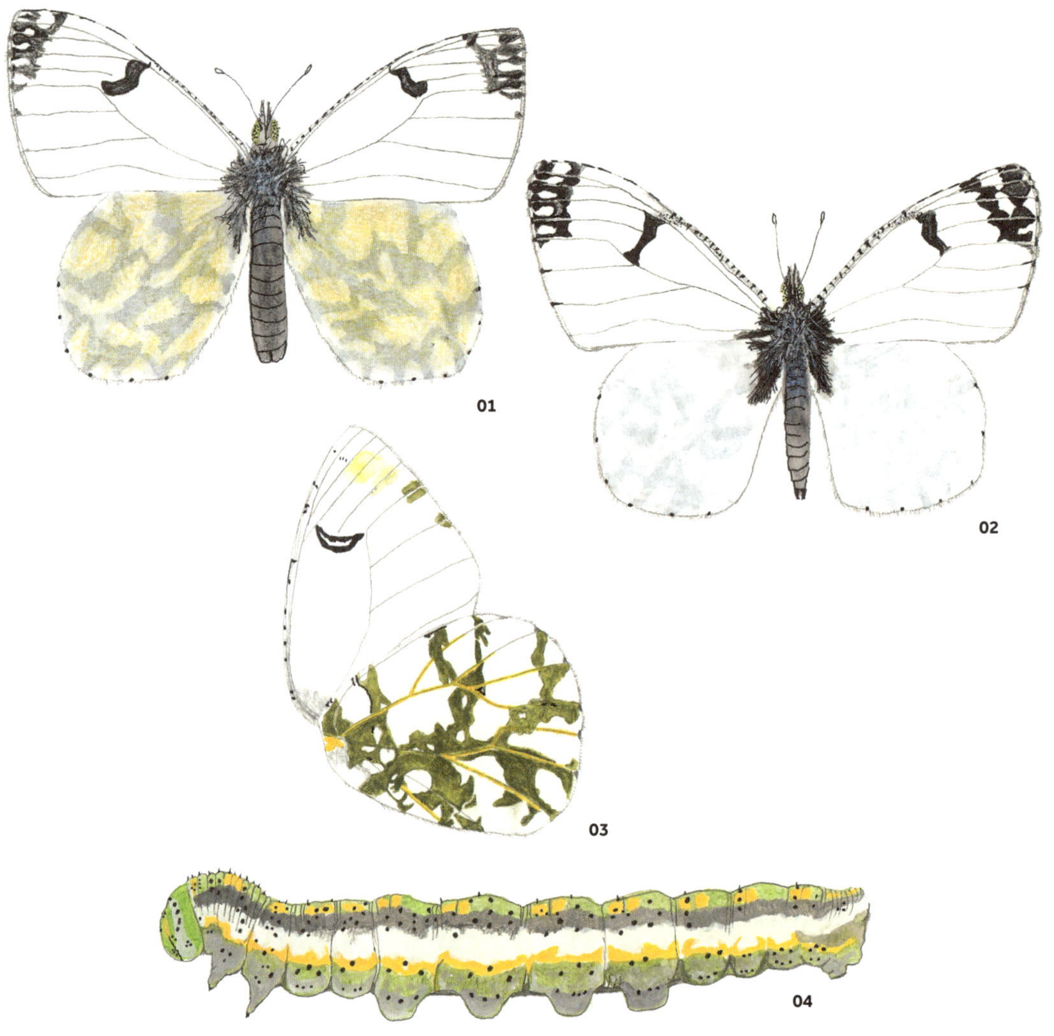
01
02
03
04

LARGE MARBLE

Lucas 1854

Euchloe ausonides

Like the Spring Whites mentioned above, Large Marbles fly with a direct, purposeful trajectory as if they are aiming at a target (Cabbage Whites kind of float aimlessly). Once you master these differences and the ability to key Large Marbles in flight, you will notice you don't even really need the butterfly to land to ID it. The green marbling is visible in flight as well, but when these beauties land, I still sort of gasp. The Large Marble has always been my favorite butterfly for its pattern of marbling, distinct flight, and the surprise when one lands before you. For more identification info, see **Appendix B, Tableau 5**.

Habitat: Meadows, vacant lots full of weeds, vegetable gardens, coastal scrub
Host Plants: Crucifers within the family Brassicaceae, both native and non-native mustards (*Arabis* spp. and *Sisymbrium* spp.).
Life Phases: Egg color goes from blue green to orange (Pyle.2002.142). Larva dark green with a yellow lateral line. Pupa dirty brown with darker markings toward the center. Overwinters as pupa. Adults are marked with slightly different forewing patterns, but both sport exquisite green marbling ventrally. Female is larger because she's carrying the eggs. Bivoltine: March–July.
Counties They Fly In: All
Great Places to See Them: Twin Peaks and Alemany Farm (San Francisco County); Garin Regional Park (Alameda County); Marin Headlands (Marin County). In the headlands, take Alexander Avenue, the first right on the Marin side of the Golden Gate Bridge, then the first left around the bend called Bunker Road, then first right and then a quick left into a small parking lot before the tunnel. They are usually blasting about there in the spring.

01: dorsal ♀
02: dorsal ♂
03: ventral ♀ and ♂
04: mature larva

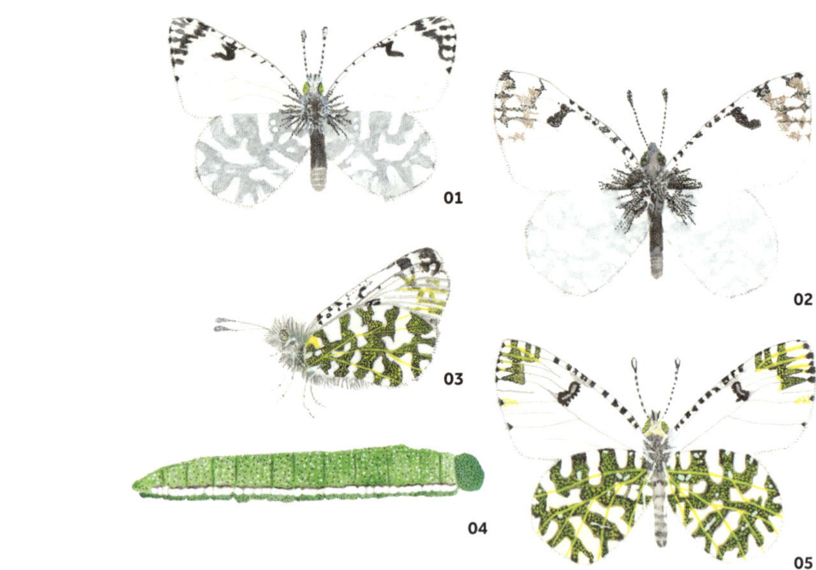

SMALL MARBLE

Edwards 1871

Euchloe hyantis

also known as **CALIFORNIA MARBLE**

Smaller than a Large Marble in flight, Small Marbles have many rapid wing-beats per minute and are strong hill-toppers. If it stops at a flower, try to get a shot or look at the underside marbling—the markings are a great deal heavier than those of the Large Marble, looking almost solid across the hind wing. Search around the host plant for males awaiting females.

Habitat: Serpentine hilltops, mountain ridgelines, rocky outcrops and south-facing canyon walls (Howe.1975.386).

Host Plants: Native and non-native mustards (Brassicaceae), native rock cress (*Arabis* spp.), jewelflowers (*Streptanthus* spp.)

Life Phases: Caterpillar green with lateral yellow or green stripe. Pupa gray or light brown with light brown lines. Univoltine: March–June.

Counties They Fly In: Napa and Sonoma

Great Places to See Them: Around Clear Lake (Napa County); iNaturalist sightings near Healdsburg (Sonoma County) and Oat Hill Mine, near Aetna Springs (Napa County)

01: dorsal ♀
02: dorsal ♂
03: in situ
04: mature larva
05: ventral ♀ and ♂

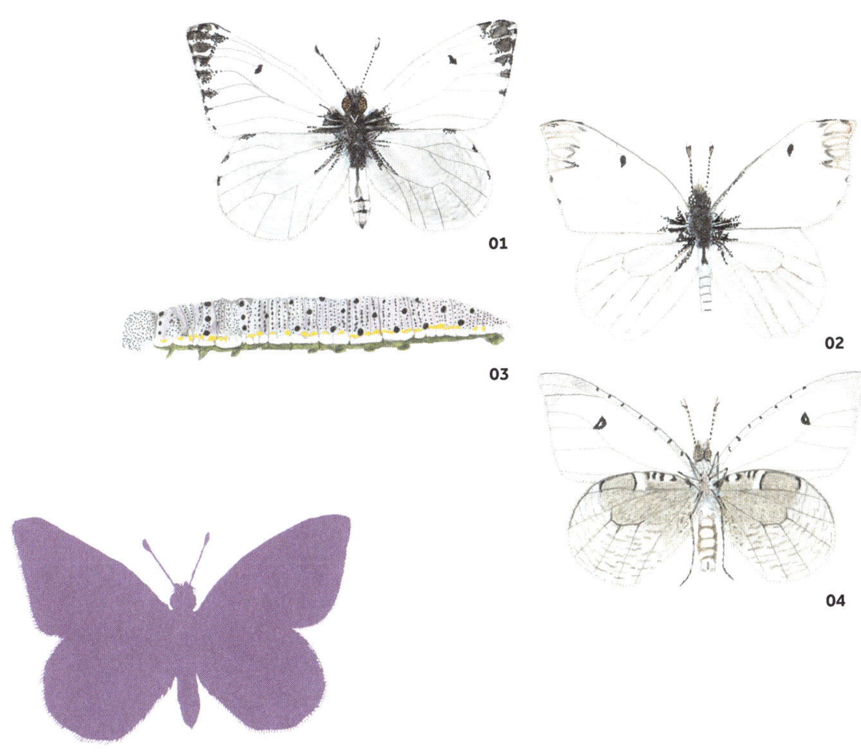

GRAY MARBLE

Lucas 1852

Anthocharis lanceolata

I've got to be honest here: this is a butterfly you can't really set out to find—you happen upon it (which is true for a great deal of butterflies, in fact). It flies direct and purposeful like the other marbles. It's white above, but the underside appears gray or slightly maroonish. Lacks green that is present on the other marbles. Check out the bottom of valleys to encounter patrolling males.

Habitat: Open woodland canyons, rocky cliffs
Host Plants: Many species of rock cress (*Arabis* spp.), jewelflower (*Streptanthus* spp.), hedge mustard (*Sisymbrium officiale*). Native and non-native Brassicaceae.
Life Phases: Larvae are known to eat blossoms and fruit. Pupa black, brown, or gray. Overwinters as pupa. Adults distinctive with the apexes of their wings coming to a falcate point. Univoltine: April–June.
Counties They Fly In: Napa and Sonoma
Great Places to See Them: A good lead comes from iNaturalist: Pine Flat Road near Healdsburg (Sonoma County). Worth checking out!

01: dorsal ♀
02: dorsal ♂
03: mature larva
04: ventral ♀ and ♂

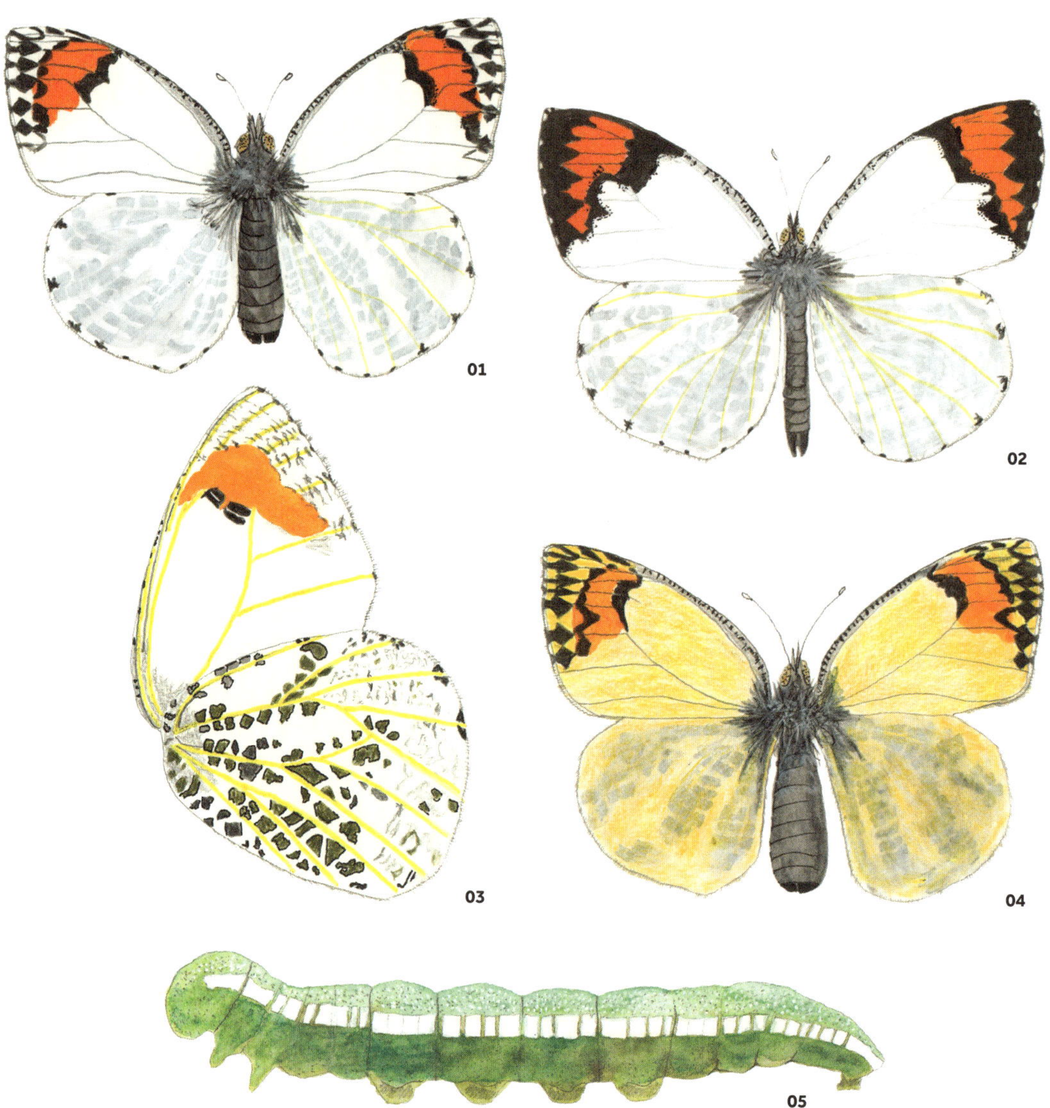
01
02
03
04
05

SARA ORANGETIP

Lucas 1852

Anthocharis sara

I'm not sure where I first read a reference to Sara Orangetips as "bleeding snowflakes," but the description is spot-on indeed. On the wing this butterfly looks like it had its forewing tips dipped in crimson paint. One of these zipping by is a slam dance to the senses. It's also an easy butterfly to learn when you are first starting out because of these orange epaulets—red orange patched wing tips of Sara can key it away from others if you're in an area where several orangetips fly.

Habitat: Streams and creeks, edges of forests; sunny places from dry chaparral to forest glades

Host Plants: Mustards (Brassicaceae)

Life Phases: Pupa is light brown with black markings. Male adults are difficult to approach for photography. They emerge before the females and are in a constant state of motion, rarely going to a flower. They seem to land more when the females come out later in the flight. Female comes in a yellow form. Univoltine: March–May.

Counties They Fly In: All except San Francisco (though I have personally witnessed a stray on the lower flanks of Twin Peaks)

Great Places to See Them: Mitchell Canyon Trail, Mount Diablo State Park (Contra Costa County); Alum Rock Park, San Jose (Santa Clara County)

01: dorsal ♀
02: dorsal ♂
03: ventral ♀ and ♂
04: dorsal ♀, yellow form
05: mature larva

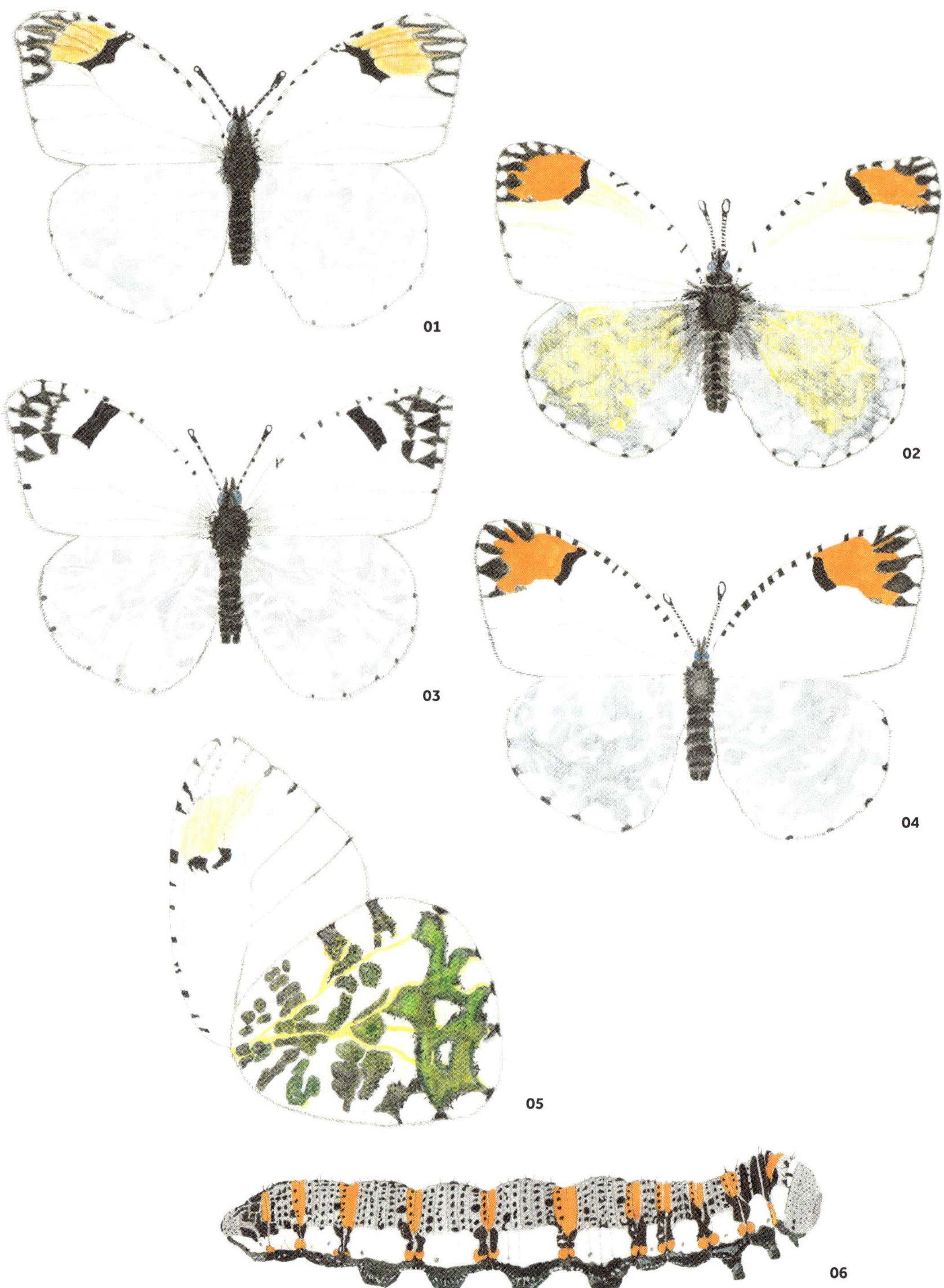
01
02
03
04
05
06

DESERT ORANGETIP

C. Felder & R. Felder 1865

Anthocharis cethura morrisoni

The man who discovered this subspecies flying out there in the arid Panoche Hills was named Bill Swisher. He dedicated much of his advanced years to this creature and was known to have pure joy in his approach (Johnson,P.2021.pers comm). Warning: the Desert Orangetip flies with the Sara Orangetip out there as well. Even female Orange Sulphurs can trip you up here. This is a highly elusive creature, and finding it may take multiple excursions. Explore any of the valleys off New Idria Road on your search. Males patrol hilltops all day in search of females. Females can be seen on the slopes of hills in the afternoon (Glassberg.2012.51).

Habitat: Rocky slopes, chaparral

Host Plants: Various native and non-native mustards

Life Phases: Pupa gray to brown to black. In adults, the ground color (or field) of both males and females varies from cream to yellow (Kaufman & Brock.2003.60). The female's upperside can come in a white form that lacks orange patches on the forewings. The underside hind wing looks like a dense series of mottled green bands (as opposed to the green marbling of the marbles).

Counties They Fly In: San Benito

Great Places to See Them: Griswold Hills BLM Area (San Benito County). Park and cross the small creek to head toward the trailhead that switchbacks up the hill. Summit the hill and look for the butterfly (known to land on small flowers up there). Do not be alarmed by the incessant gun blasts at targets out here in the distance—"recreation" means different things to different people. I recommend folks follow safety practices such as always wearing blaze orange and maintaining awareness of any shooting in the area. Best thing about this Bureau of Land Management place? One can carry a net if one so choses.

01: dorsal ♀
02: dorsal ♂, yellow fin
03: dorsal ♀, form that lacks orange patch
04: dorsal ♂
05: ventral ♀ and ♂
06: mature larva

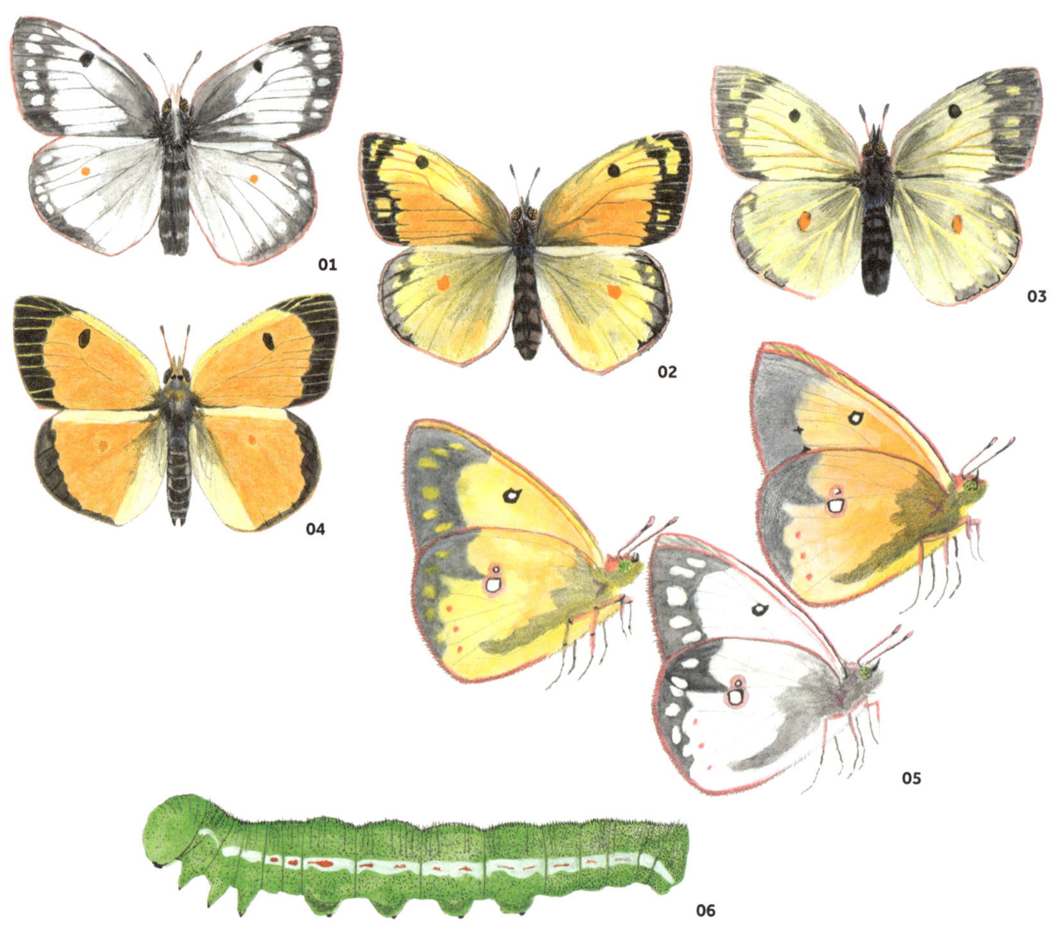

01
02
03
04
05
06

ORANGE SULPHUR

Boisduval 1852

Colias eurytheme

It's a real breakthrough for the novice when a white female can be identified on the wing as a Sulphur—that party trick might take you a few seasons to acquire. The sexual markings on the forewing margin are heavier than on a Cabbage White (see **Appendix B, Tableau 5**). It flies "with a purpose" in a direct, linear manner. This butterfly is so white there's almost a faint blue to it like the Checkered White—at least to my freakish eyeballs.

Habitats: This is the most widespread of the Sulphur butterflies (Kaufman & Brock.2003.60). You'll find it in disturbed habitats, vacant lots, alfalfa fields, coastal scrub when the host is present, dunes.

Host Plants: Legumes, primarily alfalfa (*Medicago sativa*), white clover (*Trifolium repens*), vetches (*Vicia* spp.), deerweed (*Acmispon glaber*), and plants in the pea family (Fabaceae)

Life Phases: Eggs are white and look like a spindle (Dornfeld.1980.47). Pupa green with yellow and black markings. Adults: spring and late fall flight forms (f. ariadne) are smaller than summer forms (f. amphisada). Females come in three color forms with the most striking one being white (f. alba). Multivoltine: throughout the season.

Counties They Fly In: All

Great Places to See Them: An Orange Sulphur, only slightly less omnipresent than the Cabbage White, can show up just about anywhere. Drive I-80 north out of the San Francisco area toward UC Davis (Solano County) and just try not to have this butterfly smash into your windshield—it's an annual pest on alfalfa. Watch them use their native host deerweed at the Lobos Dunes within the Presidio (San Francisco County), or in Point Reyes National Seashore (Marin County).

01: dorsal ♀, form alba
02: dorsal ♀, regular form
03: dorsal ♀, yellow form
04: dorsal ♂, regular form
05: ventral ♀, yellow form; ventral ♀, form alba; ventral ♂, regular form
06: mature larva

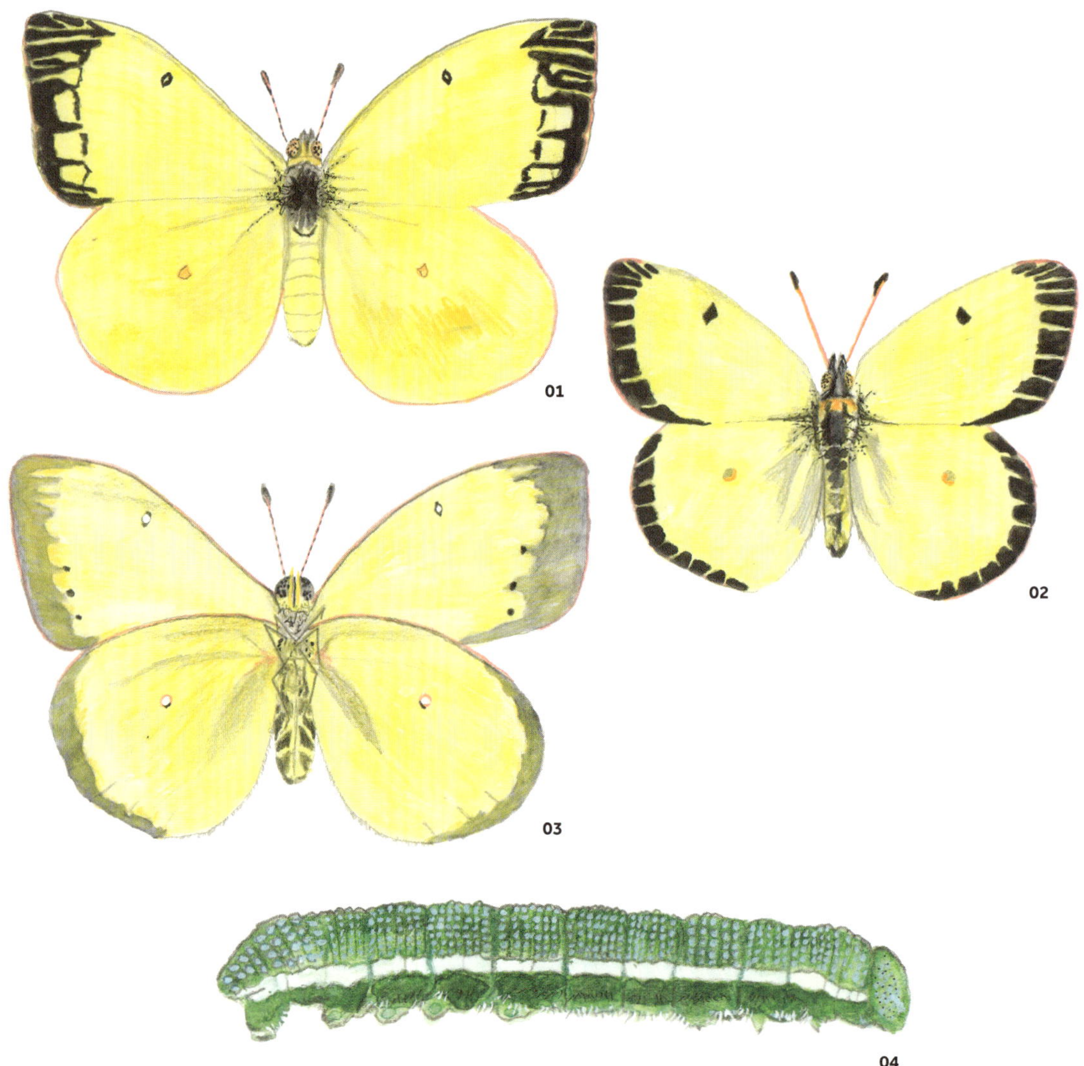
01
02
03
04

HARFORD'S SULPHUR

Hy Edwards 1877

Colias harfordii

This Sulphur is a rather rare visitor to Monterey County even though its range is only from Santa Barbara County south, as on occasional years it pushes its northern range. It needs its host plant to exist, and rattleweed is not widespread in coastal California. The plant is, however, in San Benito County. This is a butterfly to watch for. It's got a different shade of yellow, more like tangy lemon in flight, than the Orange Sulphur, which is a washed-out yellow.

Habitat: Coastal scrub, dry canyons
Host Plant: Rattleweed (*Astragalus douglasii*)
Life Phases: Eggs are laid below the leaf. Overwinter as larvae. Pupa yellow green. In adults: female and males look alike aside from black border intensity—females' borders are more faded or even absent. Both have a thicker black border on the forewing and a thinner one on the hind wing. Each presents a vivid yellow field. Bivoltine (two flights): March–May and again June–August.
Counties They Fly In: Monterey
Great Places to See Them: If you can get into the University of California Research Station Landels-Hill Big Creek Reserve along Highway 1 on the Big Sur coast (Monterey County), do so. Generally closed to the public, but if you volunteer for their butterfly count, you get access to it annually, usually in June. My friend Jim Kruse saw one there.

01: dorsal ♀
02: dorsal ♂
03: ventral ♂
04: mature larva

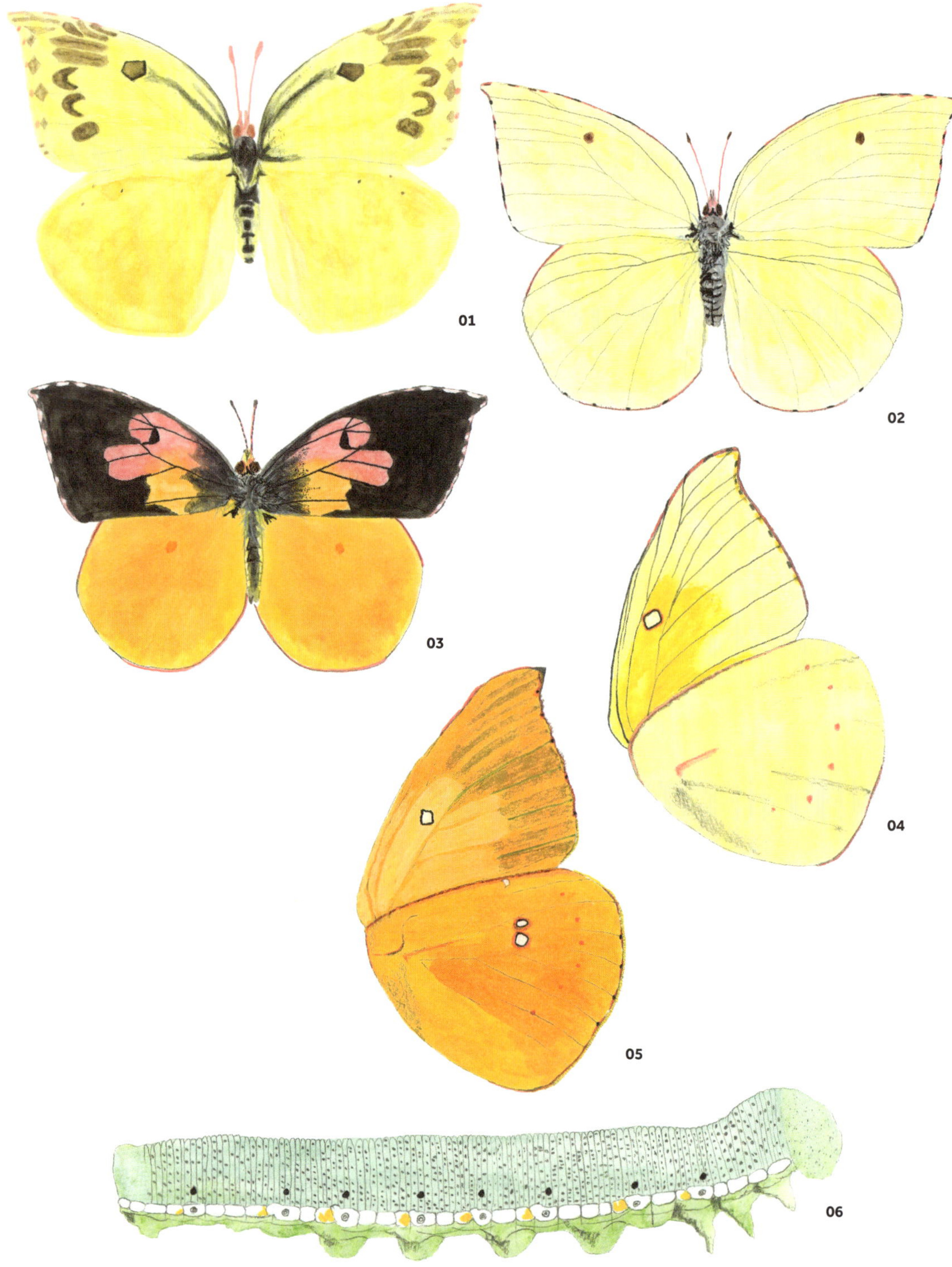
01
02
03
04
05
06

CALIFORNIA DOGFACE

Boisduval 1855

Zerene eurydice
also known as **THE FLYING PANSY**

There is a reason this butterfly is our official California state insect—it's spectacular! A form of it named amorphae, in which the females don the dog's head outline, has been observed at least twice in Pinnacles National Park (San Benito County).

Habitat: Riparian and oak woodlands
Host Plant: False indigo (*Amorpha californica*)
Life Phases: Pupa is light green with lines in the midsection. The adult female is lemon yellow with one black spot on each forewing. The male on each of his hind wings has a yellow-orange field, while his forewings are jet black with the profile of a poodle (even with an eye!) bathed in a reflective mauve color. Bivoltine: April–June and again July–September.
Counties They Fly In: Marin, Monterey, San Benito, and Sonoma
Great Places to See Them: Behind Alpine Dam (Marin County). Walk the Kent Pump Road (see **Best Butterfly Walks** for directions). The butterfly's host begins appearing along the road at about the three-mile mark. Pepperwood Preserve (Sonoma County) has the plant in abundance below an oak forest. (One has to take a class or be invited into this facility. It is not open to the general public. I've taught a class there in the spring for the last ten years.) I've also seen the butterfly about five miles up on Tassajara Road in Carmel Valley (Monterey County). It's the road to Chews Ridge (see **Best Butterfly Walks** for directions).

01: dorsal ♀, form amorphae
02: dorsal ♀
03: dorsal ♂
04: ventral ♀
05: ventral ♂
06: mature larva

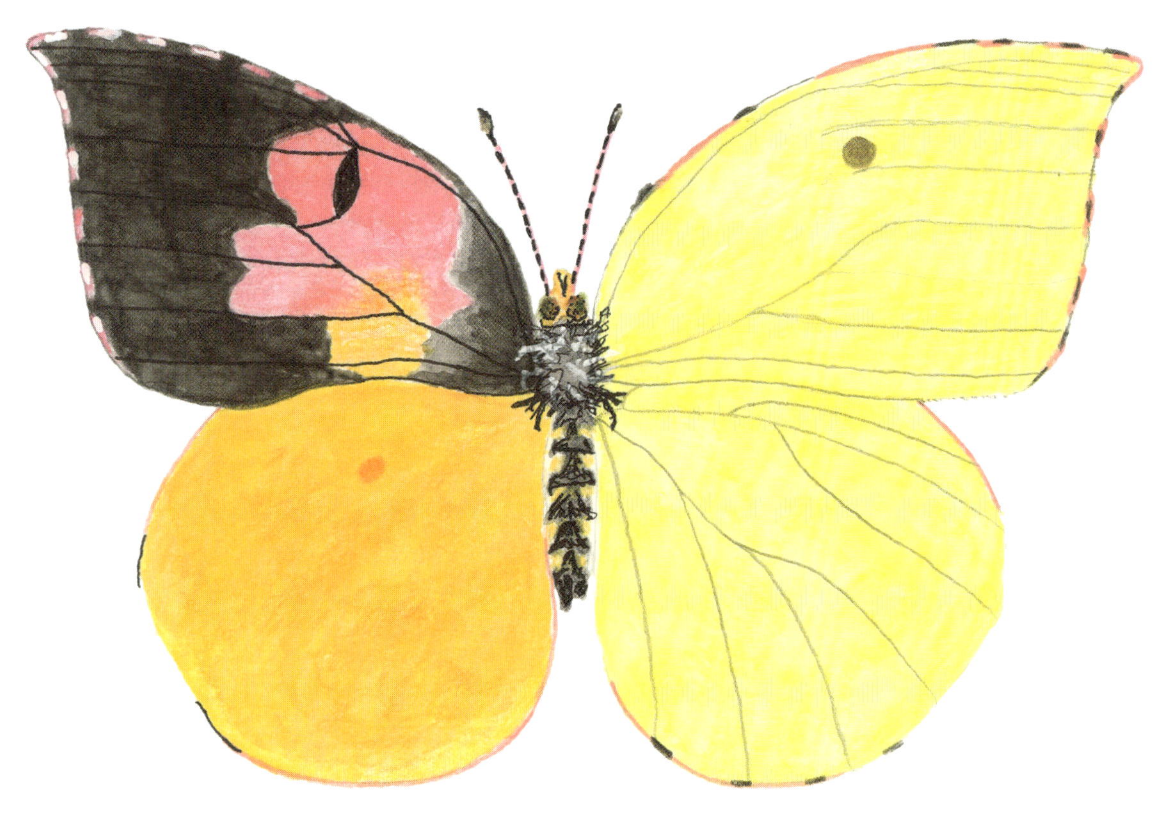

BILATERAL GYNANDROMORPHISM

In the 1930s in the San Bernardino Mountains of Southern California, lepidopterist Janet Riddell collected a series of hybrids in a zone where the Southern Dogface (*Zerene cesonia*) to the south and the California Dogface (*Zerene eurydice*) to the north interbreed. She obtained every imaginable mix of these two, and today her series is housed in the collection at the California Academy of Sciences. One day while swinging her net she struck gold—a bilateral gynandromorph of a California Dogface butterfly.

A gynandromorph is an individual expressing both male and female characteristics. Split evenly down the center of the thorax and abdomen (bilaterally) these creatures possess both sexes' genitalia. The individuals are sterile, incapable of mating. They are considered one of the holy grails to obtain for any university's collection. The phenomenon has been observed mainly in invertebrates but also in birds. A lobster fisherman off the coast of Maine had caught a gynandromorph lobster in one of his traps.

I wish I could express how . . . joyous, utterly humbling, and mind-blowing it is just to know something like this exists out there. A true chimera roams the earth.

Also, a shout-out to Janet Riddell and to all female entomologists and lepidopterists everywhere. Too many times I've gone to national conventions and it is a sea of old white men (like myself now, I guess, if I'm honest). Things are slowly changing, but I like to share stories like hers to encourage little girls that it's cool to be into bugs. And that we need you.

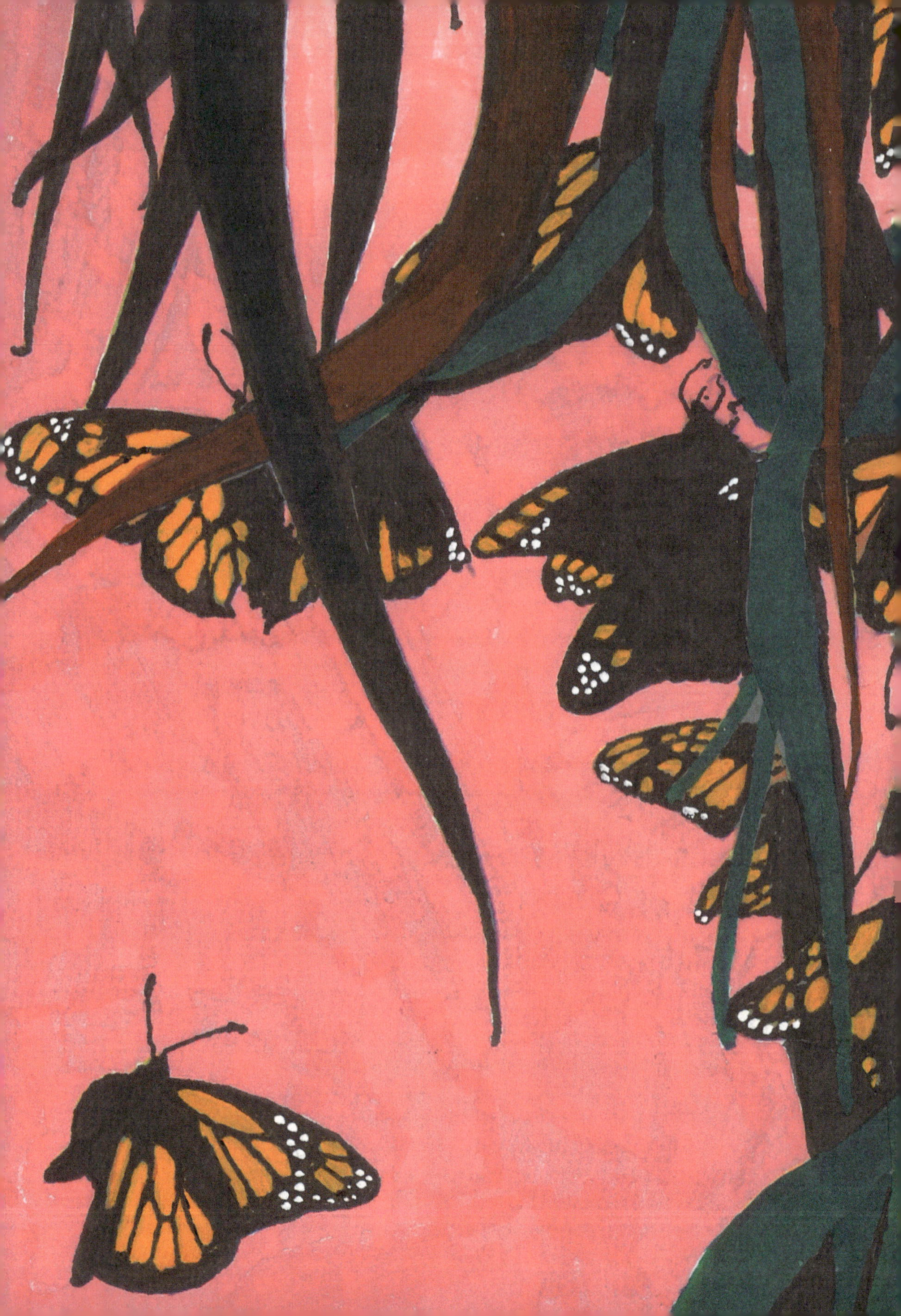

The Brush-foots (*Nymphalidae*)

This is the largest family of butterflies, with several thousand species known worldwide.

They are so named for their two reduced front feet, which are not used for walking any longer but are now delegated to help the female find her host plant—they sort of look like brushes. With her combined senses of smell and touch anchored in the pads of these atrophied appendages, she lands and senses whether the plant is the one she's looking for. If she identifies it as the correct plant, ovipositing occurs.

Our most vivid examples of color and wing shape are found within this family. Many larvae have developed spines, some stinging, some not. The *Vanessa* genus is found circumpolar as its most famous member, the Painted Lady (*Vanessa cardui*) lives on five of the seven continents. With stripes and epaulets like those on a naval uniform, the Admirals (tribe Limenitidini) take their name from the bands on their wings. In general, these are our medium-sized butterflies. They consist of several subfamilies in North America, five of which are pertinent to this book: Danainae (milkweed butterflies), Satyrinae (wood nymphs, satyrs, and arctics), Heliconinae (fritillaries and longwings), Limenitinae (admirals and sisters), and Nymphalinae (true nymphs and checkerspots). Within the Heliconinae, we have several greater fritillaries and one lesser fritillary, the Pacific Fritillary (*Boloria epithore*); and one longwing, confusingly called the Gulf Fritillary (*Dione vanillae*).

This group can be rough to learn with so many of them a variation on a brown, black, orange, and red palette. Start with the common ones in your backyard and branch out from there.

Sunset with Monarchs

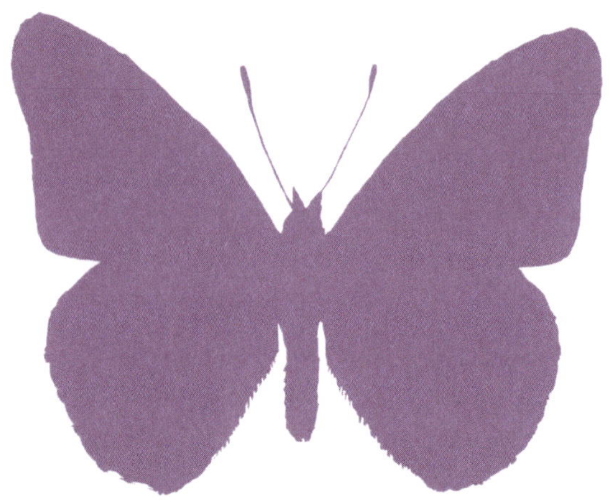

LORQUIN'S ADMIRAL

Limenitis lorquini

Boisduval 1852

Pierre Lorquin was a Frenchman who came to California in the 1850s to find gold. Luckily for us he brought his butterfly net, collecting many new butterflies to science. Thirty-four species in this book made their way, along with others he collected, to Paris, where his countryman J. B. Boisduval named all of the species in Lorquin's California collection. His collections in San Francisco alone resulted in thirteen type specimens (first of their kind), including the Orange Sulphur (*Colias eurytheme*), Mylitta Crescent (*Phyciodes mylitta*), and Edith's Checkerspot (*Euphydryas editha*). Lorquin never did find gold, but his name is at the top of the pantheon of California lepidopterists.

Lorquin did the majority of his collecting in San Francisco, making the absence of this namesake butterfly in the city all the more ironic. (Though one in San Francisco is noted in John Steiner's thesis [Steiner.1980.121], decades of my own fieldwork in the city have come up with none. It is reported rather frequently now on the San Francisco Butterfly Count just over the county line into San Mateo.)

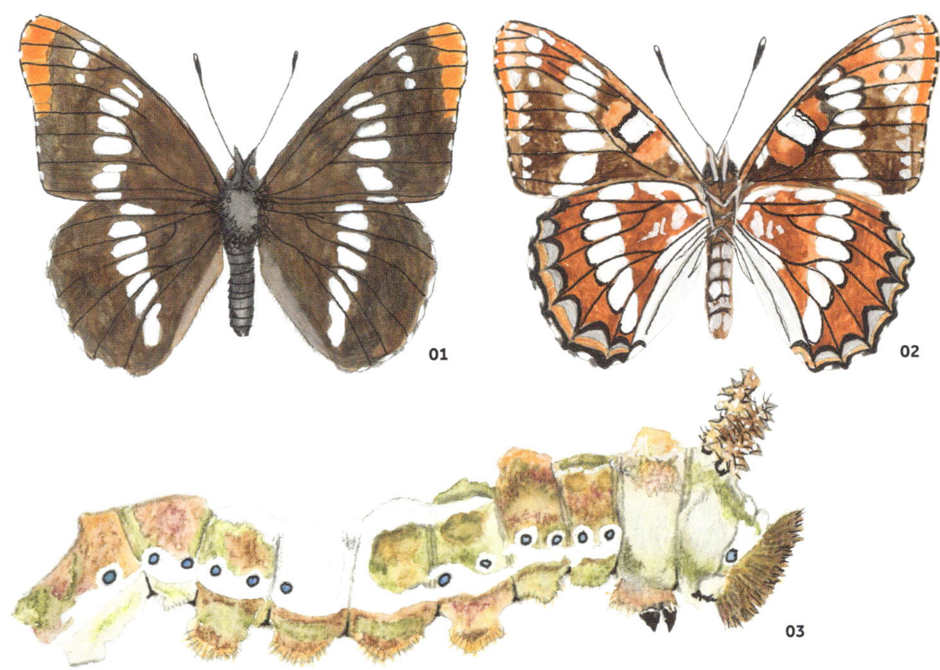

01: dorsal ♀ and ♂
02: ventral ♀ and ♂
03: mature larva

Habitat: Riparian corridors, streams, and creeks heavily lined with willow
Host Plants: Willows (*Salix*), specifically *S. lasiolepis* in our area, and cottonwoods (*Populus*)
Life Phases: Overwinters as second larva. Pupa purplish brown dorsally, white ventrally, and green on its sides. Larvae look a lot like bird poop in its final instar (or a rhinoceros in drag). Multivoltine: April–December.
Counties They Fly In: All except San Francisco
Great Places to See Them: Mitchell Canyon Trail, Mount Diablo State Park (Contra Costa County); Owl Canyon, San Bruno Mountain State and County Park (San Mateo County). I've had one at the parking lot of the visitor's center at the northern end of the Golden Gate Bridge (Marin County).

CALIFORNIA SISTER

Adelpha californica

Geyer 1837

Though females have been recorded in San Francisco during their fall dispersal, no established population exists there. While it looks similar to Lorquin's Admiral (*Limenitis lorquini*), there are clues that differentiate them once you're paying close attention. Flight: Sister—"flap, flap, glide, flap, flap, glide." Lorquin's—"flap, flap, flap, flap, flap, flap, flap." Both usually return to a fixed point or perch about shoulder high; Sister may remain higher up in the oak canopy. Environment*:* Sister is a denizen of an oak forest. Lorquin's is a member of the riparian/willow community. If the two habitats mix, one may see both. Markings: at the apex (tip of the wing), the orange spot or patch (referred to as an epaulet) on a Sister is clearly delineated from the rest of the forewing pattern. On Lorquin's it bleeds into the white, vertical bands. The bands on a Sister terminate somewhat at the epaulet. The bands on Lorquin's go all the way up to the wing margin.

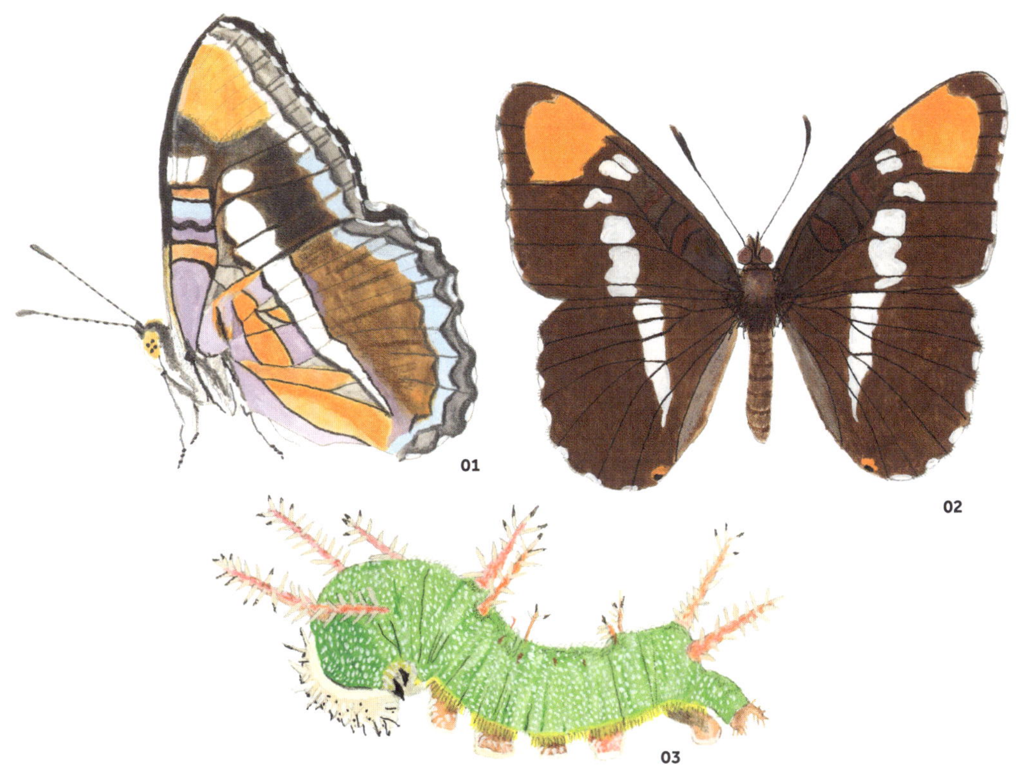

01: ventral ♀ and ♂
02: dorsal ♀ and ♂
03: mature larva

Habitat: Oak woodlands, riparian woodland with oaks

Host Plants: Oaks—several species within *Quercus*: coast live oak (*Q. agrifolia*), blue oak (*Q. douglasii*), California black oak (*Q. kelloggii*), canyon live oak (*Q. chrysolepis*)

Life Phases: Five instars with the final one hibernating. Last instar (larva) is a crazy-looking thing with long spikes and an overall chartreuse color. Pupa tan. Bivoltine: April–November.

Counties They Fly In: All

Great Places to See Them: Mitchell Canyon Trail, Mount Diablo State Park (Contra Costa County) up on buckeye tree blossoms; Chews Ridge, Carmel Valley (Monterey County); Kent Pumphouse Road, behind Alpine Dam (Marin County). Really, anywhere oaks are present in large numbers.

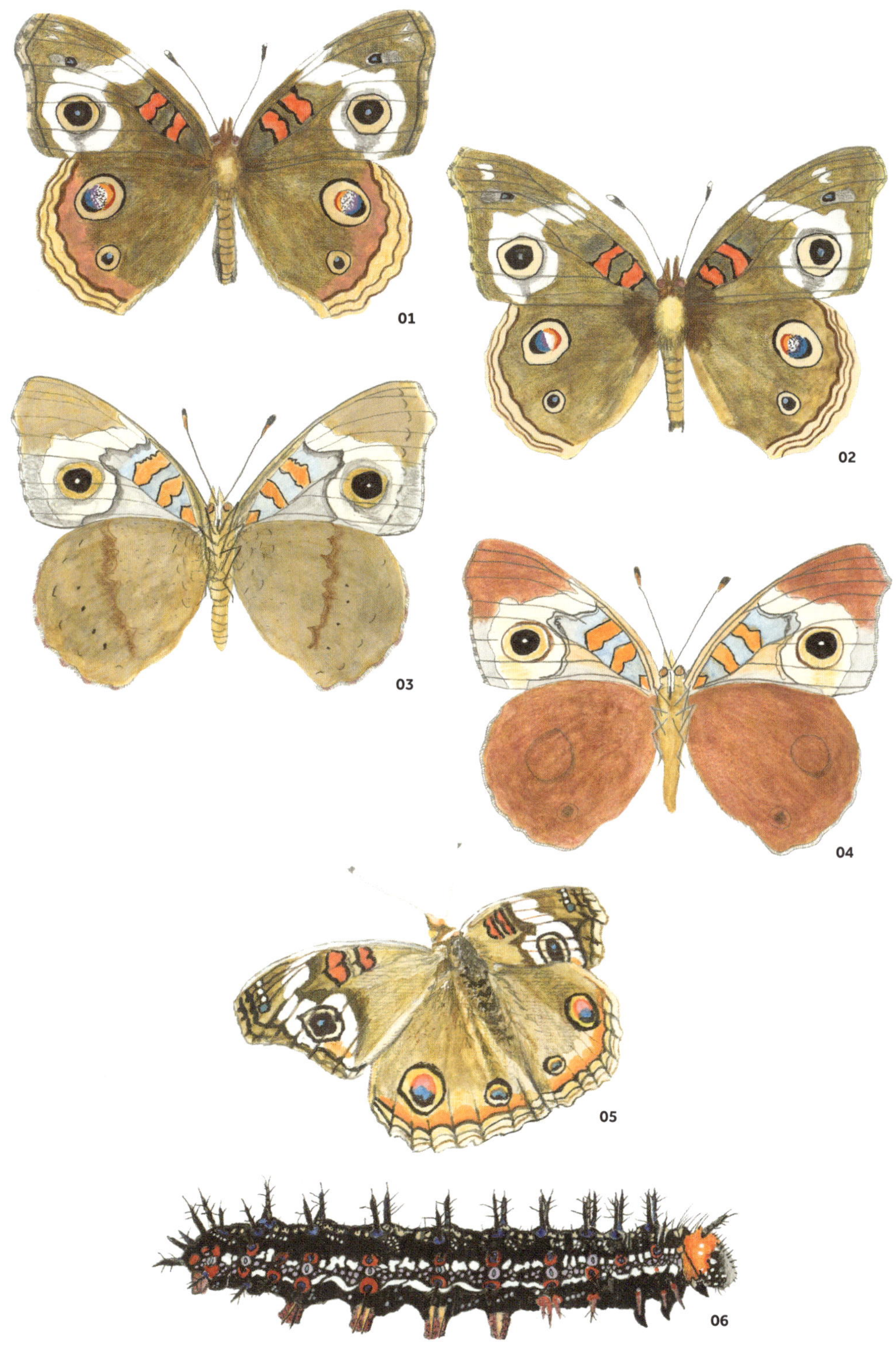

01
02
03
04
05
06

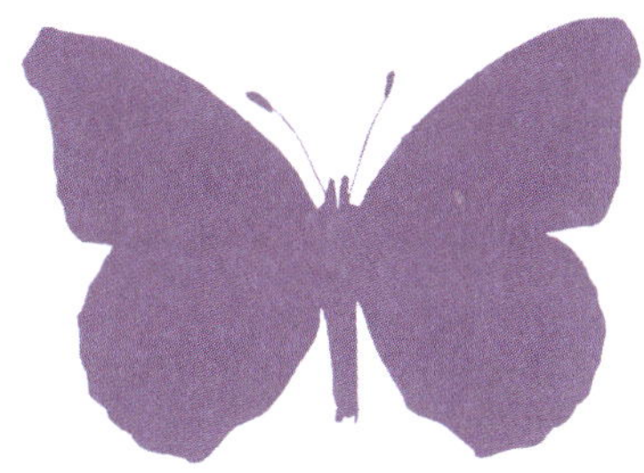

GRAY BUCKEYE

Austin & Emmel 1998

Junonia grisea

Gray Buckeyes defend their territories with such ferocity that they're enormously fun to watch. They dart out at any moving object, whether a bird, a dragonfly, or any other butterfly in hopes of finding a mate or fending off other suitors. Toss a rock gently over their heads to see what I mean: they'll jump up at it. Even two males will check one another out over and over again if they are neighbors along the path, shooting straight up into the air till the pheromones reveal they're the same sex, and then they split apart (if one were the opposite sex, they'd go off and mate).

This butterfly went through a recent renaming and reworking that even I am still getting tripped up on. In many of your field guides it will be referred to as the Common Buckeye (*Junonia coenia*). The descriptor "common" still saddled on the Common Checkered Skipper (*Burnsius communis*) seems pejorative in the labeling of either creature.

Habitat: Grasslands, edges of coastal scrub, oak woodlands
Host Plants: Plantain (*Plantago* spp.), bee plant (*Scrophularia* spp.), verbena (*Vervain* spp.)
Life Phases: The mature larvae are dark with longitudinal stripes of pale yellow and rows of branching spines. Pupa is extremely light brown with two protrusions at the top. Overwinters as a caterpillar. First adults to appear in the New Year are small (form rosa) with a slight rose-colored hue to them. Males and females look the same. Bivoltine: January–December.
Counties They Fly In: All
Great Places to See Them: Any county or state park with dirt trails (I don't really see them on cement) and lots of grassland. It's one of those butterflies that kinda comes to you on a hike. Males are territorial and work out who gets what area along a path. It's an easy one for the novice to learn because no other butterfly looks like it with its multicolored ocelli and tan field.

01: dorsal ♀
02: dorsal ♂
03: ventral ♀ and ♂, summer form
04: ventral ♀ and ♂, fall form
05: in situ
06: mature larva

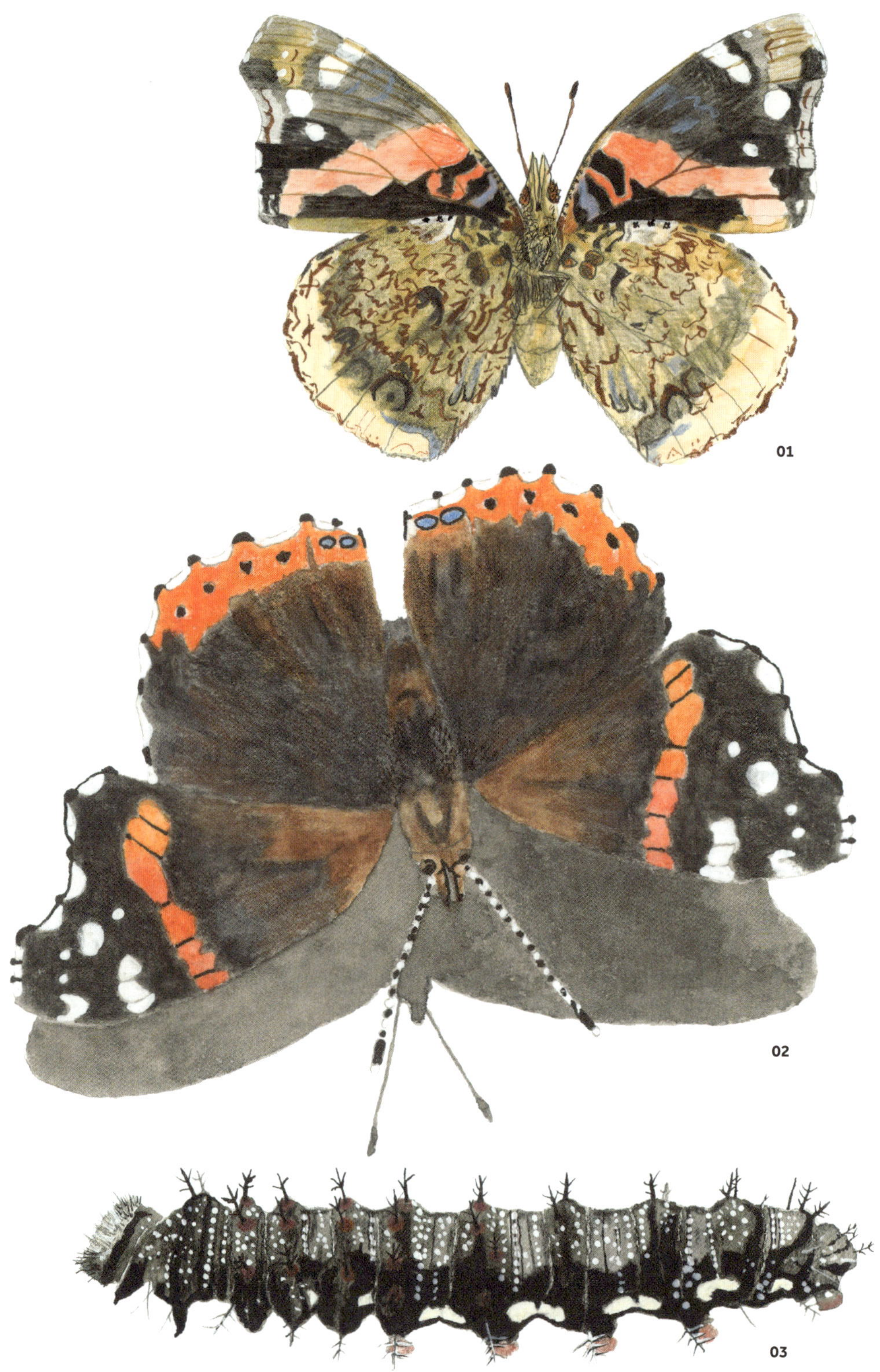

01
02
03

RED ADMIRAL

Linnaeus 1758

Vanessa atalanta

also known as **THE RED ADMIRABLE**

A strong and erratic flier, this butterfly is so agile it moves away from you without any apparent downward thrust. One doesn't even really see it fly away; it's more of a blur. I love that when it lands, it turns quickly upside down with its tail end pointing up, making it appear larger.

Habitat: From coastal scrub to city streets, wet meadows to vacant lots
Host Plants: Stinging nettle (*Urtica californica*), dwarf nettle (*Urtica urens*), western pellitory (*Parietaria judaica*)
Life Phases: Older larvae bite through the main vein of a leaf, making it droop, then they silken up the edges and live within the tent (Scott.1986.280). Pupa with gold spots. Overwinters as an adult. Migratory. A strong hill-topper. Multivoltine: on the wing all twelve months.
Counties They Fly In: All
Great Places to See Them: One of the most common and widespread butterflies, the Red Admiral once landed on a parking meter I was approaching in the Mission District of San Francisco. Highly urbanized, it seems to stick to city sidewalks where the non-native host, western pellitory, abounds. I bet it's visited every garden of every person reading this. Easily found hill-topping any summit, usually jousting around with the other *Vanessas*.

01: ventral ♀ and ♂
02: dorsal ♀ and ♂
03: mature larva

PAINTED LADY

Linnaeus 1758

Vanessa cardui

also known as **THE COSMOPOLITAN**

Found on all continents except Australia and Antarctica, this butterfly is known to use the higher atmosphere for travel. Like the California Tortoise-shell (*Nymphalis californica*), the Painted Lady has explosive movements. One is hard pressed not to see them during one of these events, which occur every four to five years. Every flower in the garden can be festooned with them during these times. It's quite a show, with individuals blasting about every few seconds. Our first indicators whether it will be a Painted Lady Year for the Greater Bay Area are reports of numbers building up in Southern California around Anza-Borrego Desert State Park. They explode off the plants down there by what seems like the millions and immediately head north. I once did an oil painting of the front of my rental car after visiting the park—dead Painted Ladies all over the grill.

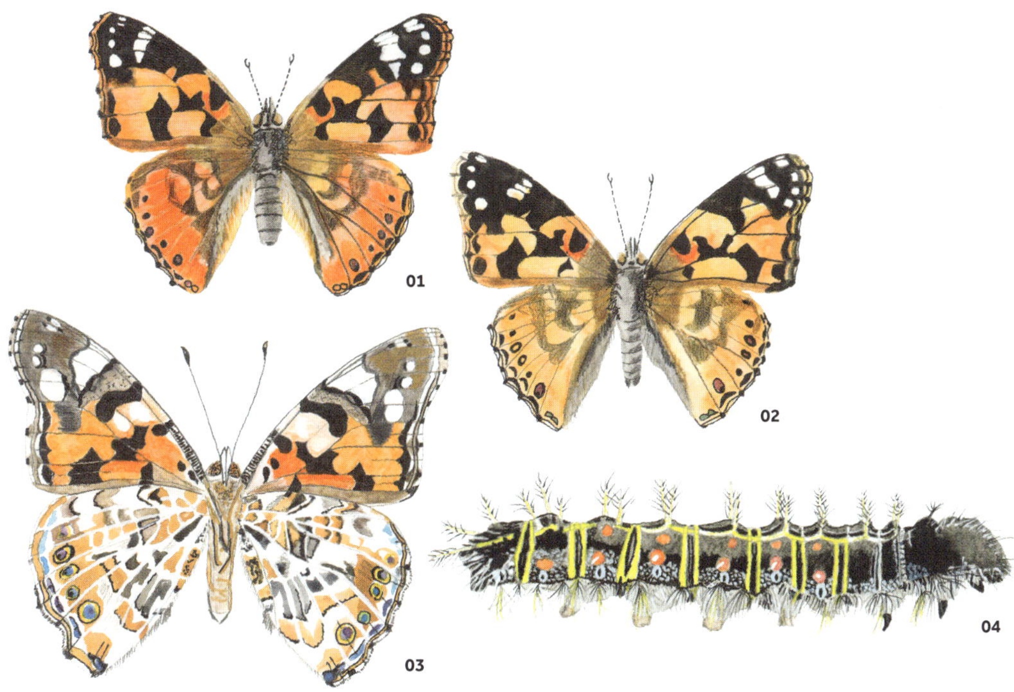

01: dorsal ♀
02: dorsal ♂
03: ventral
04: mature larva

Habitat: Found in all habitats: meadows, deserts, coastal scrub, vacant lots in cities

Host Plants: The greatest generalist after the Gray Hairstreak (*Strymon melinus*) when it comes to host plants. One hundred known hosts have been recorded, including thistles (*Cirsium*) and lupines (*Lupinus*) to name just two. Best place to find caterpillars is on *Cirsium*.

Life Phases: Larvae are lilac to yellow green (Opler.1999.326). Pupa purplish brown. All the *Vanessa* spp. can be confusing for the novice to key apart (see **Appendix B, Tableau 6**). Multivoltine: Febuary–November. A strongly migratory species worldwide.

Counties They Fly In: All

Great Places to See Them: A strong hill-topper, so it would be advisable to visit the summits of some of our higher peaks. Also seen at lower hilltops. Mount Saint Helena (Napa County); Chews Ridge (Monterey County); Mount Tamalpais (Marin County); or at Mount Hamilton (Santa Clara County).

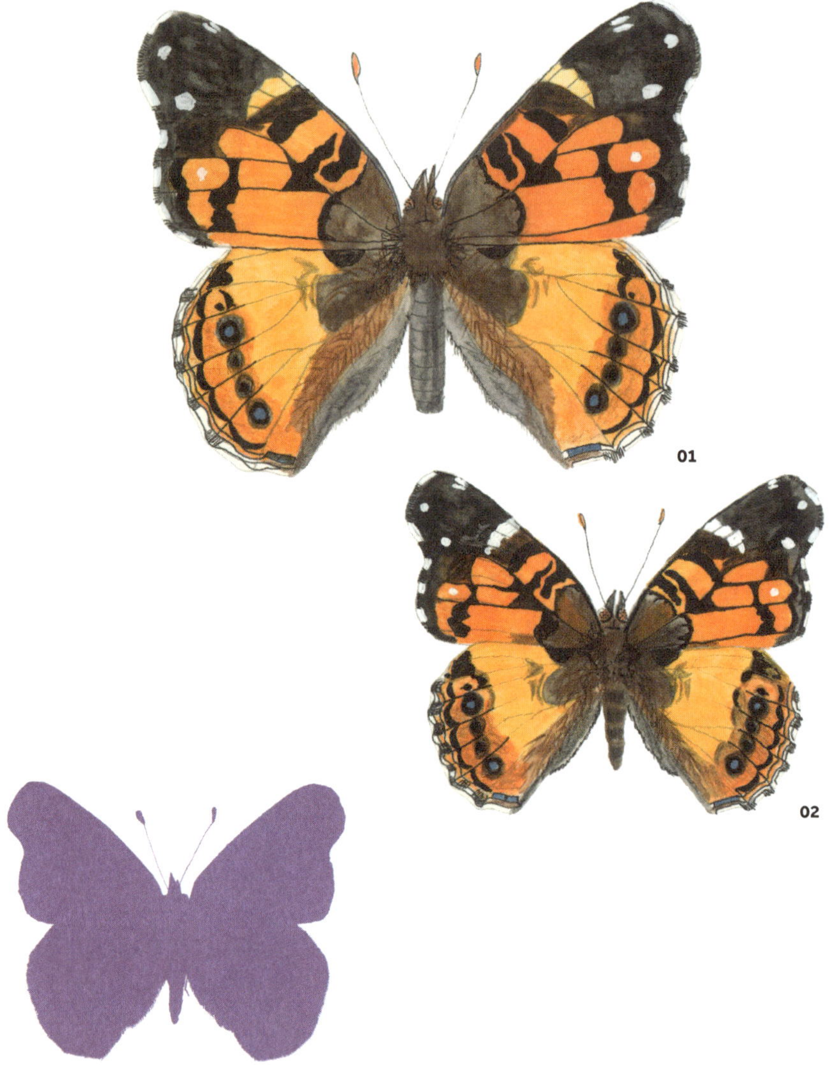

AMERICAN PAINTED LADY

Drury 1773

Vanessa virginiensis

also known as **HUNTER'S BUTTERFLY**

The American Painted Lady (commonly referred to in the field as an "American Lady") is heralded by many as having the Most Beautiful Underside of any US butterfly. The large eyes and cobweb pattern ventrally makes it truly an eye-catching wonder. It has an obvious size dimorphism as in all the ladies: the female is bigger to carry the egg load. Despite this size difference, it's still pretty Advanced Butterflying if you can sex this butterfly in the field. Takes a while to figure it out.

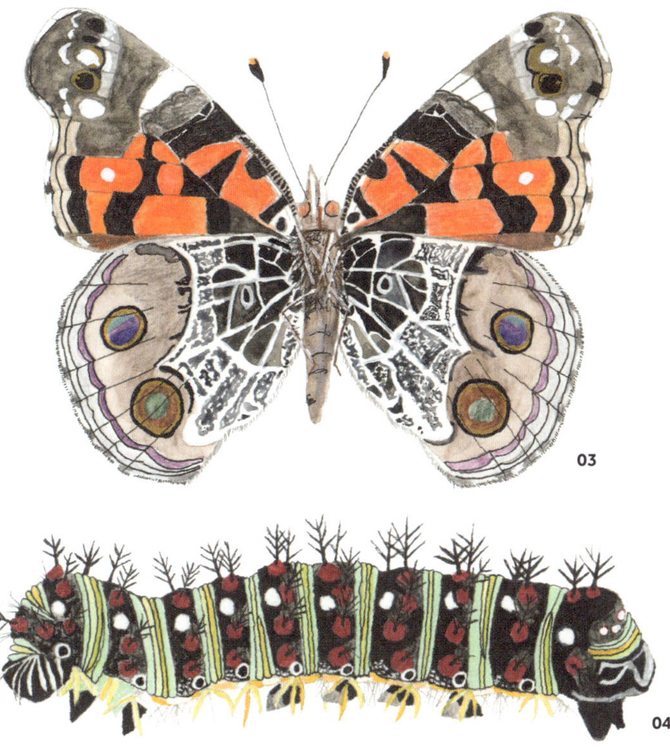

Habitat: Coastal scrub, chaparral, riparian woodland, grasslands, city streets, vacant lots

Host Plants: Pearly everlasting (*Anaphalis* spp.), cudweed (*Gnaphalium* spp.), rabbit tobacco (*Pseudognaphalium* spp.)

Life Phases: Larvae live at the top of the host plant and build their nests near the top flowers. Pupa black with yellow stripes. Males and female adults look the same except for a slight orange shade to the female's forewing discal cell. The male's discal cell is white. Overwinters as an adult. Multivoltine (many flights): March–December.

Counties They Fly In: All

Great Places to See Them. Like the other ladies, the American is a strong hill-topper, so visit peaks around the Greater Bay Area. It seems to be the first *Vanessa* out of hibernation; I see them annually flying in late March when I begin monitoring Mission Blues in the Marin Headlands (Marin County). Can show up to any garden or any habitat throughout the year, and has a broad distribution, flying from the Pacific to the Atlantic in North America.

01: dorsal ♀
02: dorsal ♂
03: ventral
04: mature larva

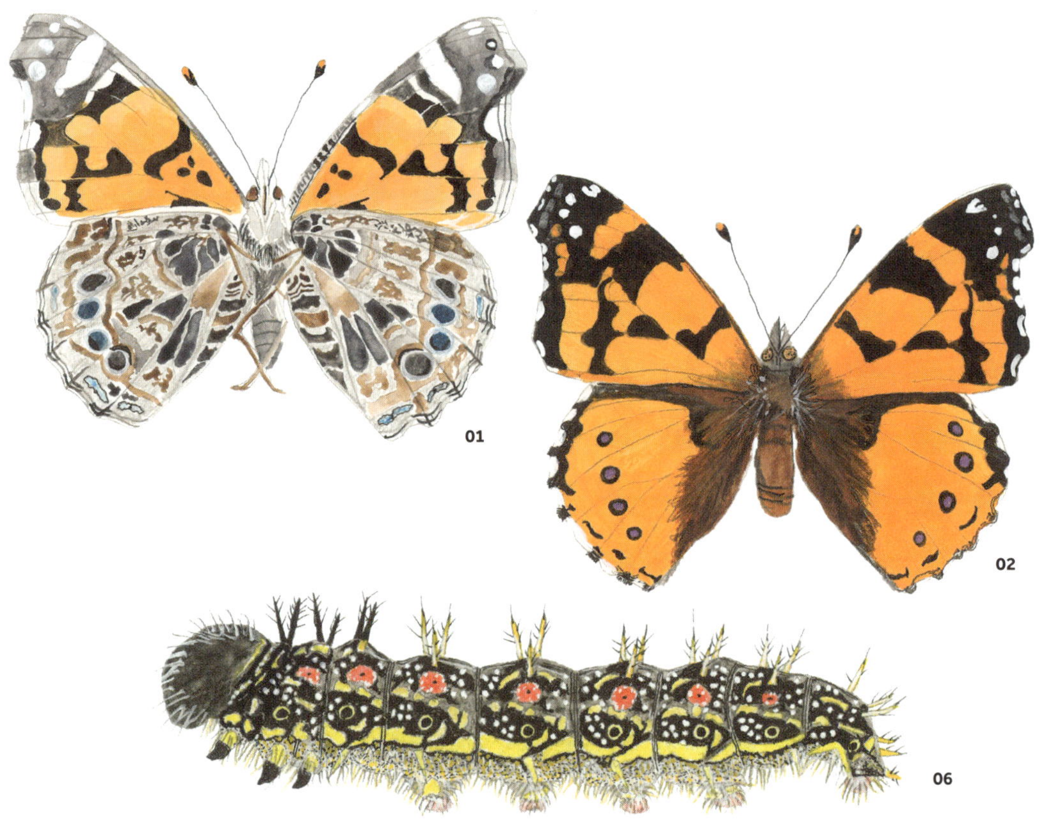
01
02
06

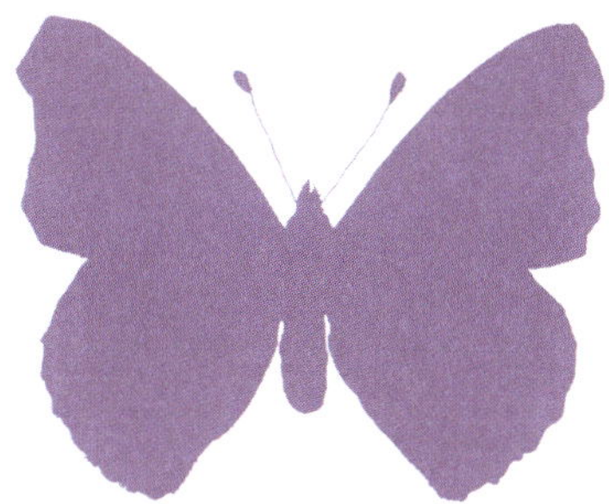

WEST COAST PAINTED LADY

Felder 1971

Vanessa annabella

also known as **THE WEST COAST LADY**

There is a notable push, or migration, seen annually toward the east with this species. Once flew only in the spring using checkerbloom as its host, but since the non-native cheeseweed moved into every vacant lot, the butterfly flies and breeds year-round in the Greater Bay Area. And then there are those years where you can't find it anywhere.

Habitat: Favors open, sunny environments. I see a great deal of them in vacant lots and hill peaks due to all of the non-native mallows.

Host Plants: checkerbloom (*Sildacea malvae fictra*), Malvaceae (*Malva* spp., i.e., cheeseweed, hollyhock)

Life Phases: The larvae live solitary in the leaf shelter (Howe.1975.206). Pupa is creamy and greenish. Overwinters as an adult. Multivoltine: on the wing year-round.

Counties They Fly In: All

Great Places to See Them: No one set place. On top of vacant lots, meadows, and city streets, *V. annabella* is a strong hill-topper and can usually be found jockeying for mating rights with the other four *Vanessa* atop hills. The parking lot around Heron's Head Park (San Francisco County) seems to have a profusion of them in June.

01: ventral ♀ and ♂
02: dorsal ♀ and ♂
03: mature larva

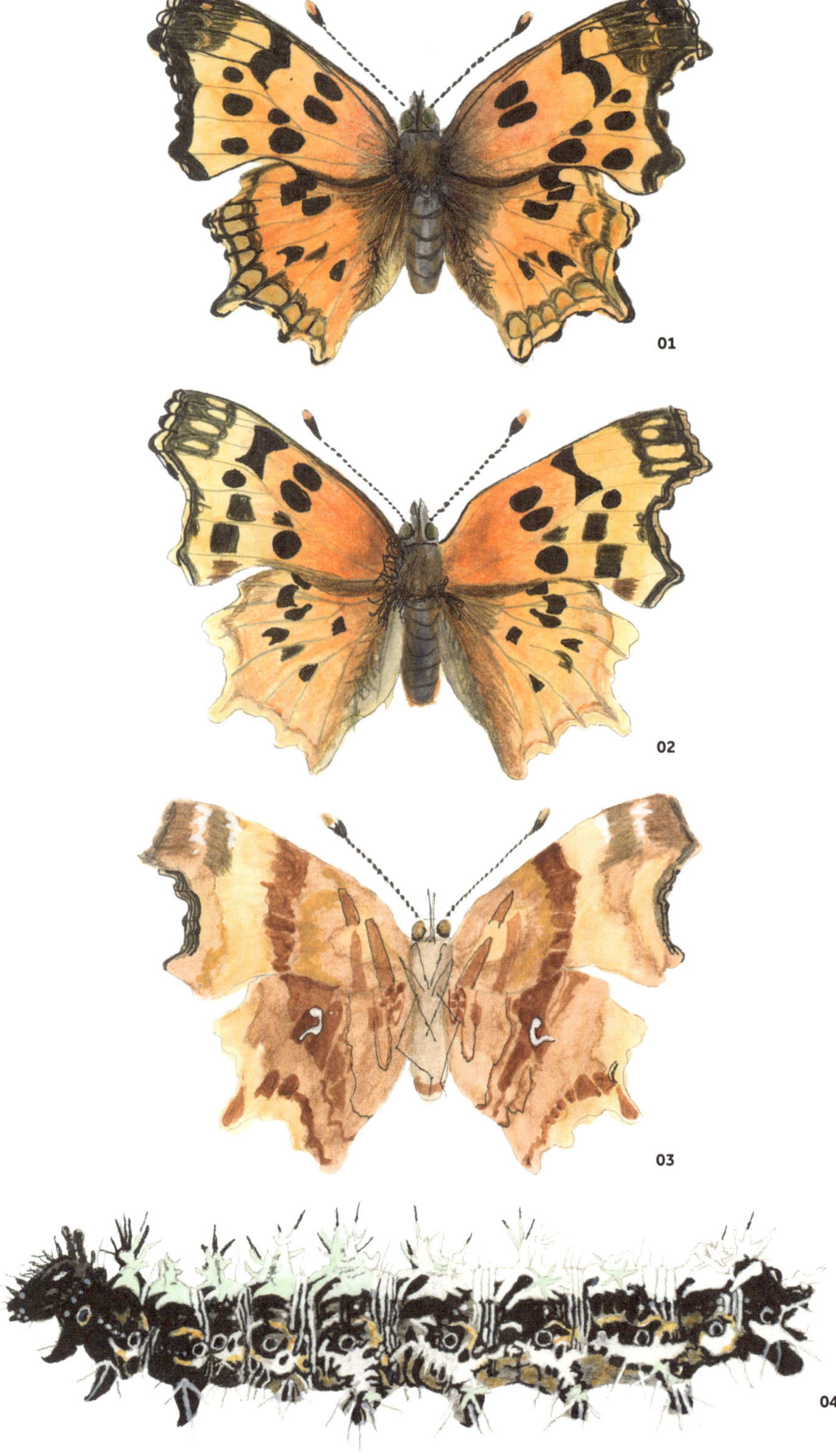
01
02
03
04

SATYR COMMA

Edwards 1869

Polygonia satyrus

There are two forms of this butterfly, one of which emerges in the spring, and the second in the summer. Their seasons can overlap, so there are times both forms are flying, and I remember getting tripped up when I first started: I thought that I was seeing two different sexes. I still find it slightly confusing. Be forewarned that if you go looking for larvae of this species, its host, stinging nettle, is well named: it feels like someone has taken fiberglass and rubbed it on you. Painful. Party trick: if the leaf tent is cut downwards on itself, it's a Satyr Comma. If the sides of the leaf are pulled up over the cut vein, it's a Red Admiral (*Vanessa atalanta*).

Habitat: Glade openings in forests, along streams and creeks containing their host plants

Host Plant: Stinging nettle (*Urtica dioica*)

Life Phases: Spiny caterpillar with bright, saddle-like patterns on both sides and back. Larvae make leaf tents out of the host. Pupa light brown. Summer form of adults is lighter with fewer markings on hind wings than the spring form, which is more heavily marked. Wings are tan and browns below. Bivoltine: early spring with a break around July or August, then second flight through the winter. Second flight overwinters as an adult.

Counties They Fly In: All

Great Places to See Them: Bear Gulch Trail out of the Bear Valley visitor's center/Point Reyes National Seashore (Sonoma County); Reinhardt Redwood Regional Park, Oakland (Alameda County); Lobos Dunes (in the shade of the side going toward the ocean), Presidio (San Francisco County); Garrapata State Park (Monterey County) within the redwoods

01: dorsal ♀ and ♂, spring
02: dorsal ♀ and ♂, summer
03: ventral ♀ and ♂
04: mature larva

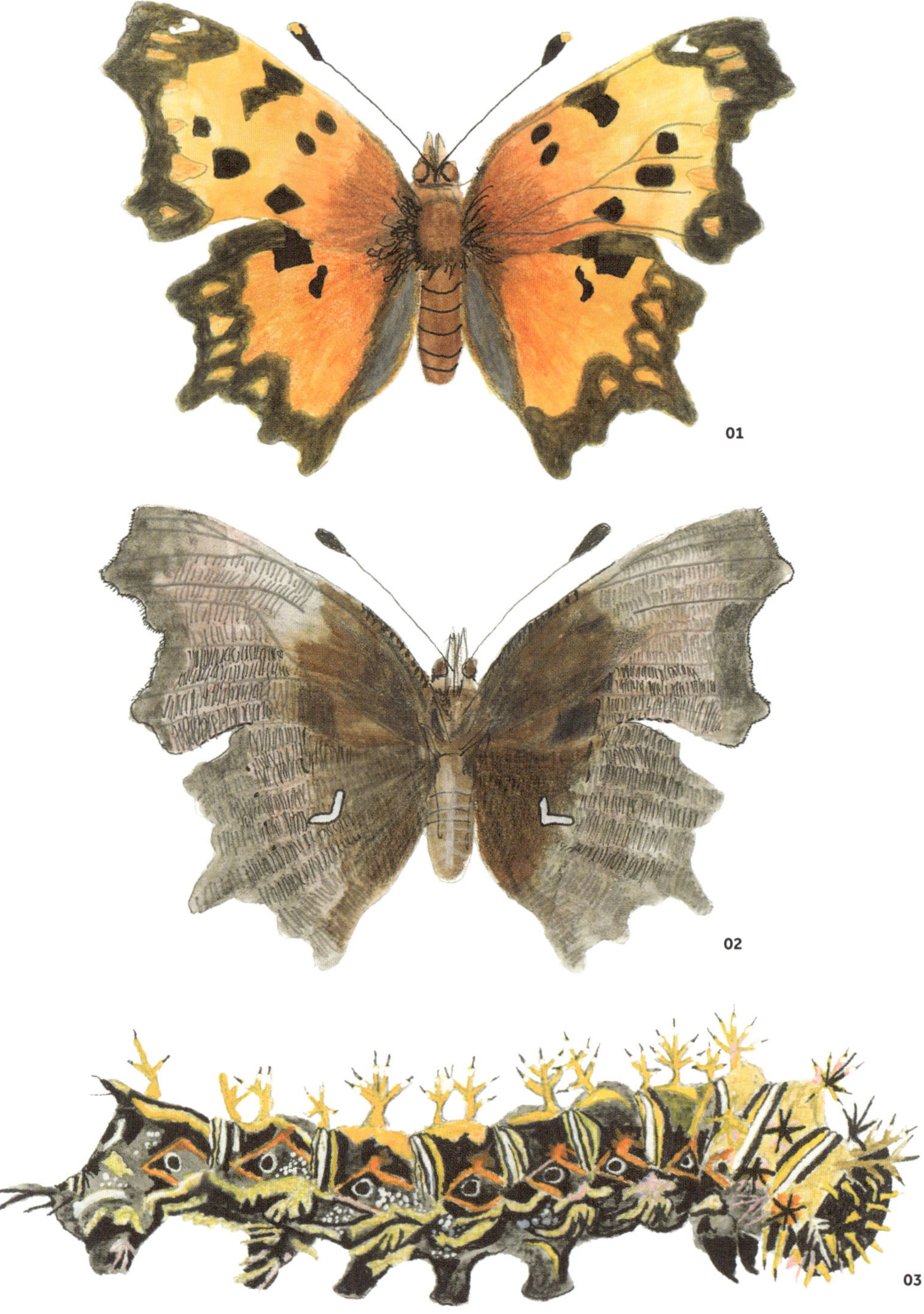
01
02
03

OREAS COMMA

Edwards 1869

Polygonia oreas

In the extraordinary pollinator garden of Barbara Deutsch in Point Reyes Station, I witnessed this butterfly do a fascinating thing. At a dead bull thistle blossom, *Polygonia oreas* had its proboscis extended into the blossom; I watched it spit down the nose-like appendage and then suck it back up, extracting vital salts and minerals from the dead seed head. Amazing. I knew butterflies could do this, but had never witnessed it. This Comma is slightly smaller than our others. The ripped and tattered-looking outer margins gave this genus the name Anglewings for decades. It's only in recent times that the field has shifted to calling them Commas, in reference to the white, punctuation-like marks seen ventrally.

Habitat: Oak and riparian woodland, redwood forests, shaded canyons, and stream beds

Host Plant: Gooseberry (*Ribes divaricatum*)

Life Phases: Caterpillars are a complex pattern of yellow and brown, dark V marks, rings, and blotches. Adults have no spot on the wings, which are dark gray/brown below toward the body. Univoltine: but complicated—June or July, overwinters as an adult, then active again April–June of following year.

Counties They Fly In: All except San Benito and San Francisco

Great Places to See Them: A highly elusive creature. Garland Ranch Regional Park along the Carmel River, Carmel Valley (Monterey County); Point Reyes Station—along where the black iron bridge crosses the creek (Marin County).

01: dorsal ♀ and ♂
02: ventral ♀ and ♂
03: mature larva

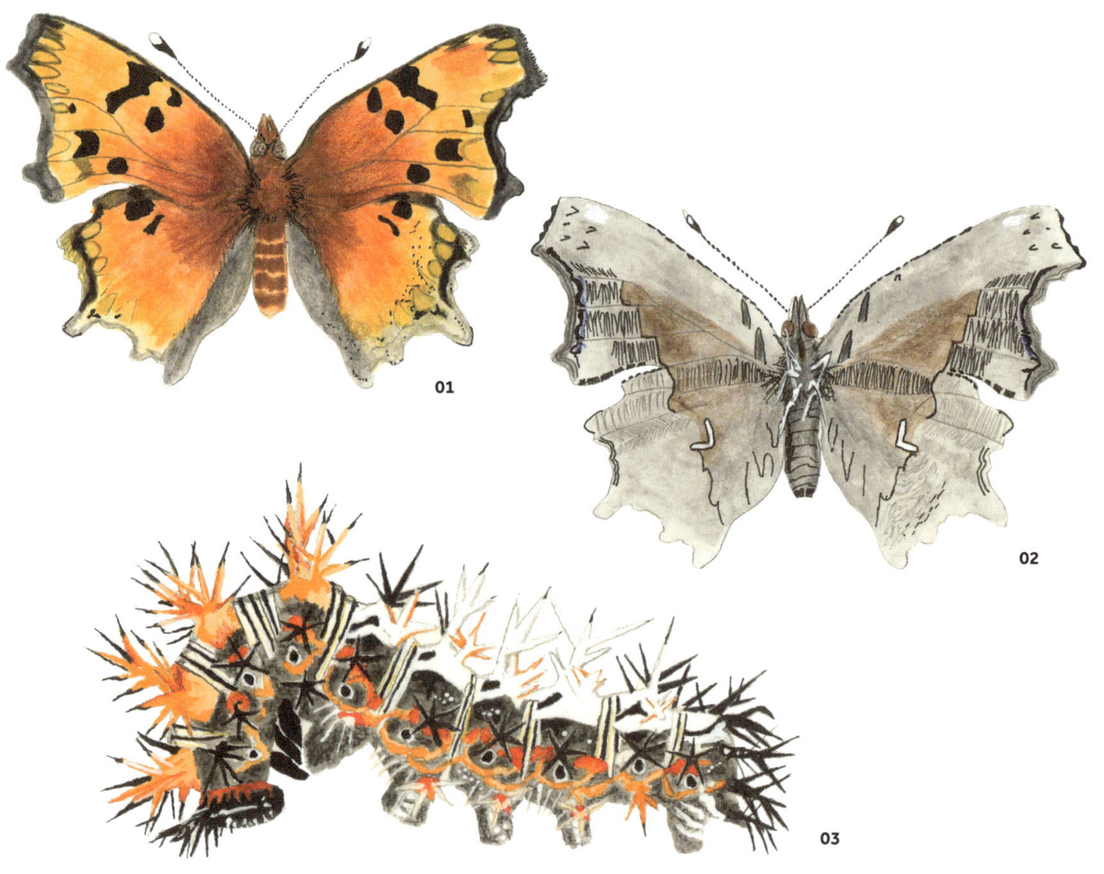

03

ZEPHYR COMMA

Edwards 1867

Polygonia zephyrus

The fifth molt (or instar) before it pupates is a spectacular phase for this butterfly—it has the appearance of a small dragon. In fact, the discovery of the sheer beauty of this genus was one of the highlights in the painting phase of this book. The Zephyr was recently split apart from the Hoary Comma (*Polygonia gracilis zephyrus*).

Habitat: Moist woodlands, usually above 3,000 feet
Host Plant: Currant (*Ribes* spp.)
Life Phases: The black caterpillar is armed with rows of spiky spines, some looking like arrowheads. Light brown pupa with gold markings with sporadic dark brown marking. In adults, the very pale gray underside should help key it away from others if you happen upon it. Zephyr can also be keyed away from other Commas by the lone black spot on the upper forewings' rear margins—all others have two dots. WARNING: This difference is not always obvious, as the second spot on Satyr can be somewhat faded, and even some Oreas exhibit a small, remnant spot there. Bivoltine: with the overwintering phase occurring as adults.
Counties They Fly In: Marin, Mendocino, Monterey, Napa, Santa Cruz, Solano, and Sonoma
Great Places to See Them: According to Steiner's thesis, it is listed in all counties above, but in full disclosure, I have never personally seen it. The nearest sighting on iNaturalist is Onion Springs Camp in Fresno County. It's difficult for even advanced lepidopterists to key in the field, and I'm leery of even including it here.

01: dorsal ♀ and ♂
02: ventral ♀ and ♂
03: mature larva

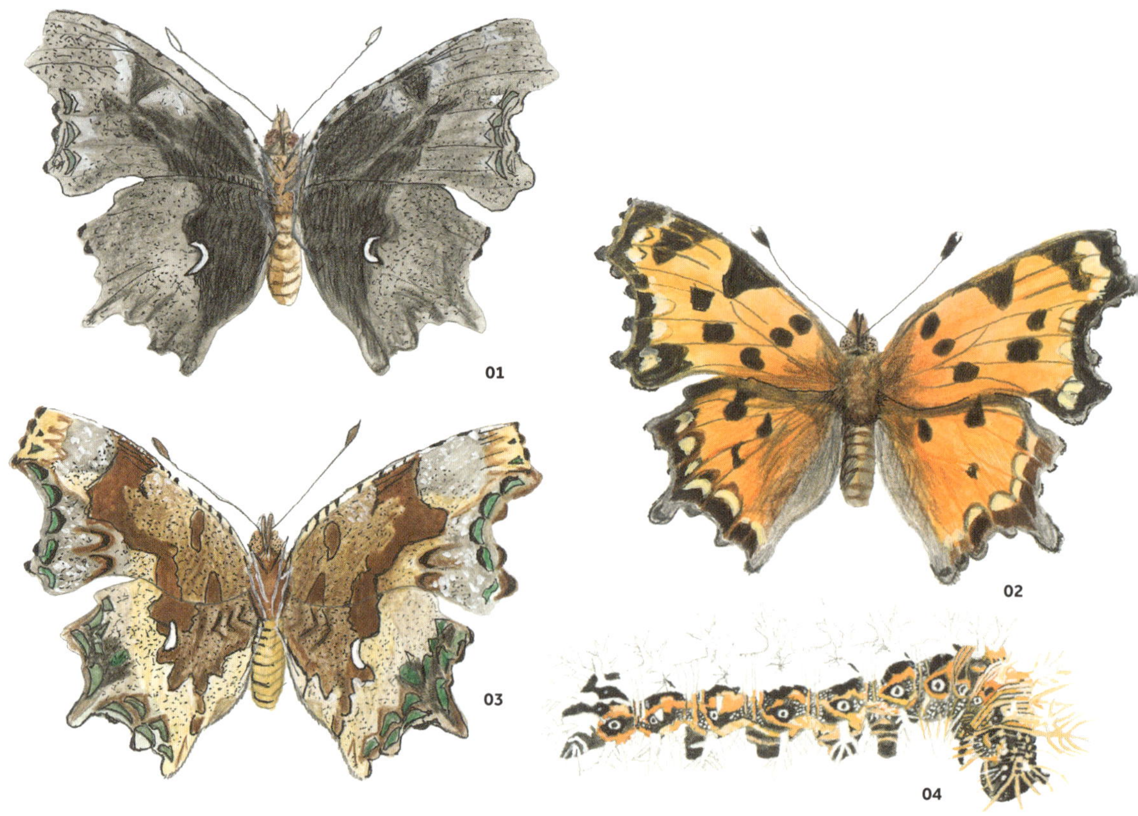
01
02
03
04

RUSTIC COMMA

Edwards 1862

Polygonia faunus rusticus

also known as **GREEN COMMA**

Rather rare butterfly to see in the Greater Bay Area. It's important to note the different undersides in males and females. The males are territorial in their clearings in the woods. Adults feed on nectar, rotting flesh, animal feces, and mud. A single individual lives many months, from fall till next spring. Wings are incredibly jagged.

Habitat: Riparian woodland, redwood and evergreen forests

Host Plants: Willows (*Salix* spp.)

Life Phases: Only a few eggs are laid by the female (Howe.1975.198). The adult female is gray below and the adult male has green, metallic spots. This is the only *Polygonia* with emerald green spots below.

Counties They Fly In: Contra Costa, Marin, Napa, Santa Cruz, and Sonoma

Great Places to See Them: Point Reyes National Seashore (Marin County); Henry Cowell Redwoods State Park, Scotts Valley (Santa Cruz County)

01: dorsal ♀ and ♂
02: ventral ♀
03: ventral ♂
04: mature larva

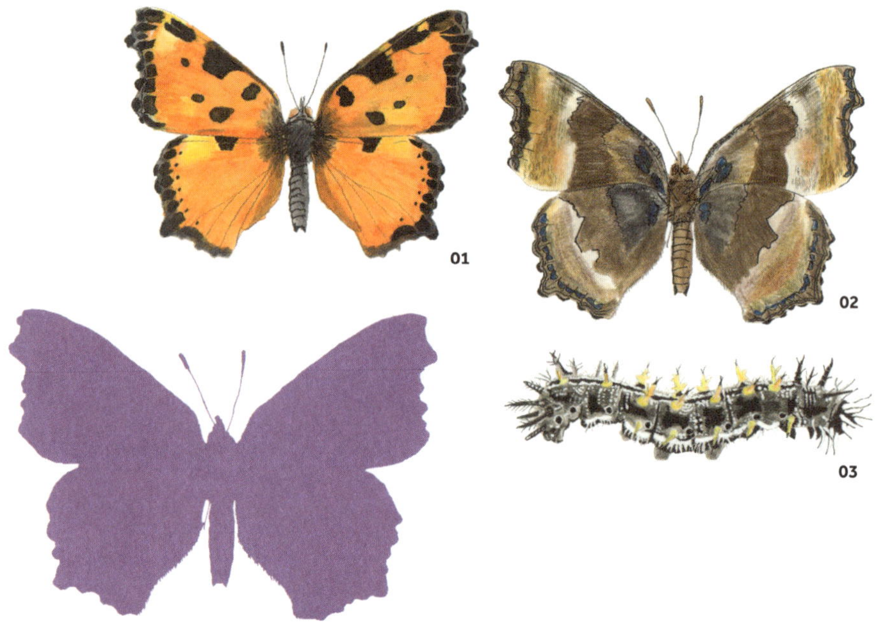
01 02 03

CALIFORNIA TORTOISESHELL

Boisduval 1852

Nymphalis californica

This species has wide swings in its presence and absence—some years it's everywhere; some years you don't see them at all. One time during the Big Creek Butterfly Count in Monterey County I was with Paul Johnson, and the butterflies were emerging from the *Ceanothus* by the hundreds every few minutes! "Counting" was impossible. "Make it stop, make it stop!" I cried, as if we'd wandered into a dangerous swarm—of butterflies.

Habitat: Coastal scrub communities where the host plant is present
Host Plants: An array of California lilac (*Ceanothus* spp.)
Life Phases: Larvae stay together in silken shelters while feeding, then separate after the fourth instar (Steiner.1980.121). Pupa gray, orange, or black with white spots and orange tips. The migrating adults move north and east in May and June and south and west in September and October. Multivoltine: seem to emerge after consecutive days of strong heat (see **On Migration**).
Counties They Fly In: All
Great Places to See Them: Summits of mountains: Diablo (Contra Costa County); Tamalpais (Marin County); Chews Ridge (Monterey County). They'll find you, shooting east in explosive emergences, called movements, that usually occur after consecutive days of heat. When they are not migrating they appear sporadically along many trails.

01: dorsal
02: ventral
03: mature larva

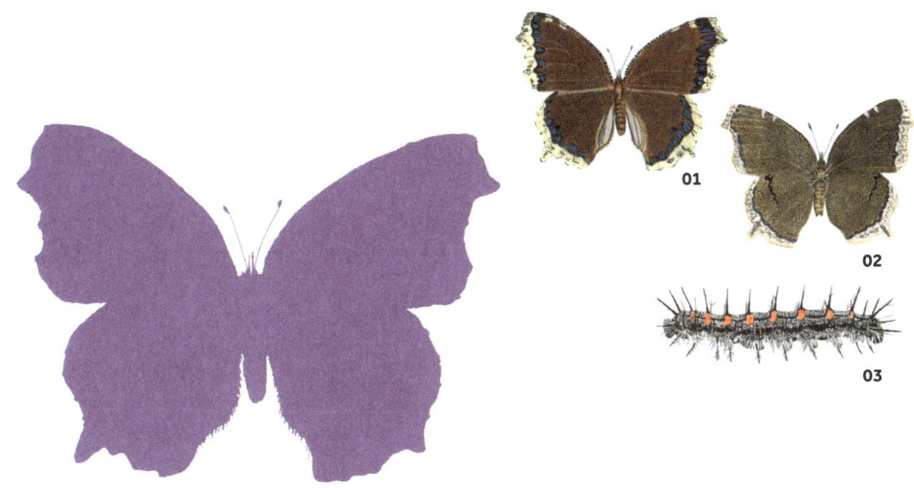

MOURNING CLOAK

Linnaeus 1758

Nymphalis antiopa

also known as **THE CAMBERWELL BEAUTY**

Party trick: if you see a Mourning Cloak in January or February with the prominent, white-fringed border all gnarled and ratty and thin, this is an individual that overwintered and has now re-emerged in late winter. If the fringed border is intact and clean and pristine, this is an individual that has freshly emerged from its chrysalis. The name comes from the town of Camberwell, a suburb of London now.

Habitat: Meadows and forests but primarily riparian zones with willow

Host Plants: I know it to be mainly on willow (*Salix*), but elm (*Ulmus*) and *Celtis* are also reported (Howe.1975.209).

Life Phases: Eggs are laid as a mass. Larvae live socially, feeding and sleeping together. Pupa is light brown with black spots and orange tips. Overwinters as an adult. One individual, including all its stages of life, can live up to ten months, thereby winning, hands down, the Longest Lived Butterfly species award (even the overwintering Monarch generation only lives four to five months). Bivoltine: on the wing year-round, with early spring flight most visible.

Counties They Fly In: All

Great Places to See Them: Somewhat elusive, though widely distributed—this large butterfly might emerge to check you out just about anywhere. Alum Rock Park, San Jose (Santa Clara County); Glen Park (San Francisco County). Check any community park with willow-lined creeks. I've seen them late in the afternoon sitting on the shady part of the Mitchell Canyon Trail (Mount Diablo State Park) as I returned from a day hike, hopping up easily as I approached.

01: dorsal
02: ventral
03: mature larva

VARIABLE CHECKERSPOT

Doubleday 1847

Euphydryas chalcedona

The Variable Checkerspot is quite a handsome butterfly, decorated as it is with a checkerboard pattern. It is not a strong flier, so look for them literally walking down the same trails you are. Like *E. editha*, males use a sphragis after mating to cap the female's entryway and block the entry of another male's sperm packet (a true chastity belt). A form of the Northern Checkerspot (*Chlosyne palla*) looks a great deal like this butterfly. The adults are, aptly, extremely variable in appearance, and sometimes aberrant individuals look more black or white than normal. I led efforts with Presidio Trust wildlife ecologist Jonathan Young to reintroduce this species to the Presidio in 2017. Before our intervention, the butterfly hadn't been seen there since the mid-seventies. It now flies in abundance.

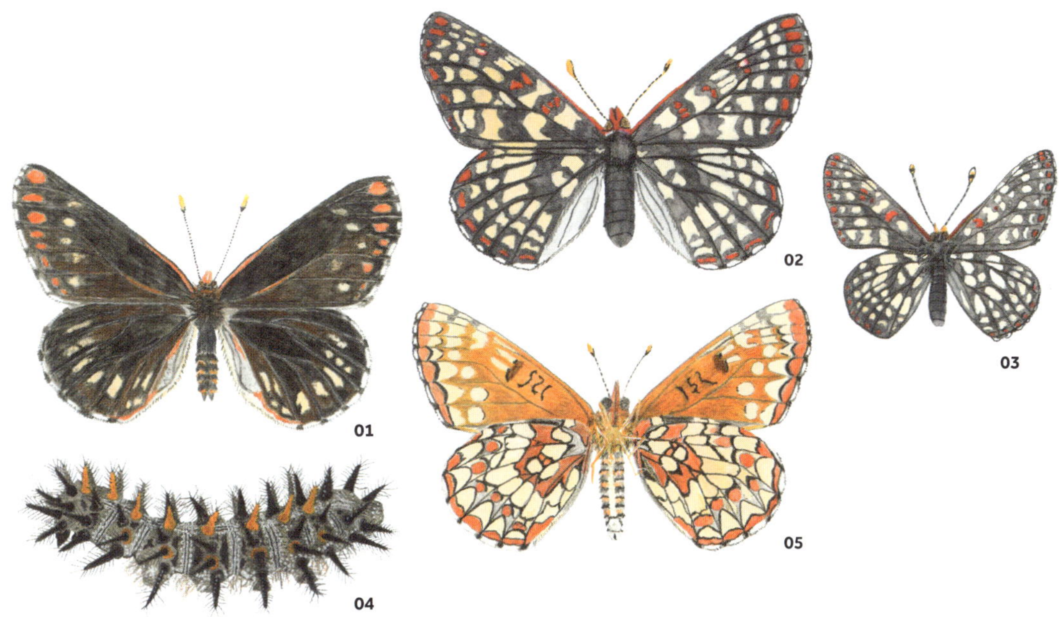

Habitat: Coastal scrub, chaparral

Host Plants: Bush monkeyflower (*Diplacus* spp.), bee plant (*Scrophularia* spp.), paintbrush (*Castilleja* spp.)

Life Phases: Eggs are laid in clusters. Pupa white with black spots and orange bumps. White spots run down the sides of the abdomen (*E. editha* does not have these) of the adults. Orange antennae (*E. editha*'s are black and orange). Univoltine: May–June.

Counties They Fly In: All

Great Places to See Them: Quite widespread when flying. Trails in East Bay Municipal Utility District property (Contra Costa County); San Bruno Mountain, San Bruno Mountain County and State Park (San Mateo County); the Presidio (San Francisco County); and Pinnacles National Park (San Benito County)

An Aberrant Form

01: aberrant form
02: dorsal ♀
03: dorsal ♂
04: mature larva
05: ventral ♀ and ♂

When an individual butterfly has an abnormal appearance due to internal genetics, physical ailments, or environmental impact at the development stage. It just looks noticeably different than how it normally looks. Some of these aberrant forms are so well known they are named.

BAY CHECKERSPOT

Sternitzky 1937

Euphydryas editha bayensis

STATUS: THREATENED

This is a subspecies of Edith's Checkerspot butterflies listed as threatened at the federal level. It might be the most studied of our butterflies as it was used as a population model by Professor Paul Ehrlich, who observed it at Jasper Ridge Biological Preserve, owned by Stanford University. The butterfly went extinct there in 1998. Ehrlich co-authored with Anne Ehrlich the famous book *The Population Bomb* in 1968, about the potential for population growth outpacing food supply. Less well known is his 2004 book with Ilkka Hanski, *On the Wings of Checkerspots: A Model System for Population Biology*.

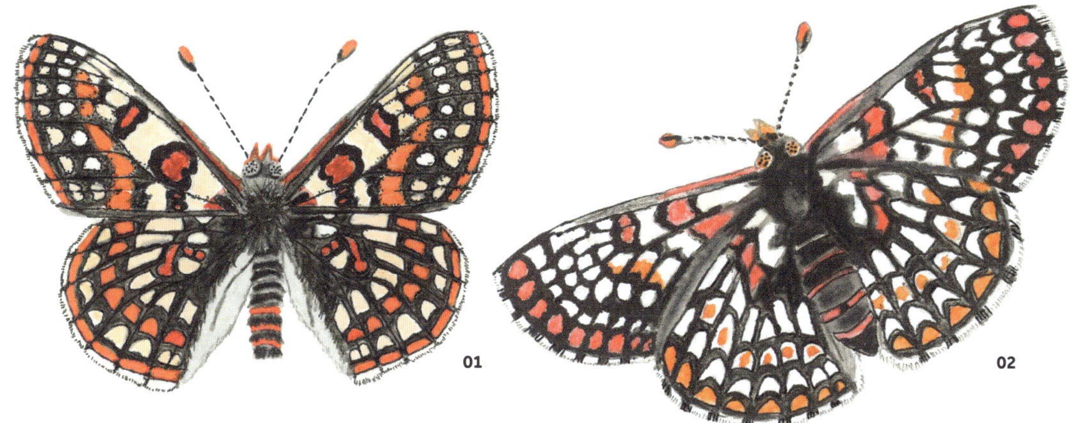

Habitat: Open grasslands
Host Plants: Dwarf plantain (*Plantago erecta*). In seasons when rain is abundant they've been known to move to owl's clover (*Castilleja densiflora*) and English plantain (*Plantago lanceolata*).
Life Phases: Eggs are laid in masses. Larvae sleep together on a communal web. Larvae, after each molting, disperse from the bigger larval crowd (Steiner.1980.142). See Edith's for pupa description. Overwinters in third instar right out on the open ground till the following spring. Univoltine: March–April.
Counties They Fly In: Santa Clara and San Mateo
Great Places to See Them: There was a population of these in the hills of Oakland (Alameda County) until a housing development destroyed their habitat in the 1980s and extirpated the population that had been hanging on. Best place has always been Coyote Ridge Open Space in San Jose (Santa Clara County), where the Máyyan 'Ooyákma trail off of Malech Road provides access to see the butterfly. Hike to the higher Bay Area Ridge Trail, as this threatened species is a notorious hill-topper. They have been reintroduced by Stuart Weiss's team from Creekside Science to San Bruno Mountain State and County Park (San Mateo County)—from the summit radio tower parking lot at the ridge trailhead, look out along the trail and eyeball the highest point. That's where they are flying. Interestingly this population is using the non-native plant English plantain (*Plantago lanceolata*) as its host up there.

01: dorsal ♂
02: dorsal ♀

EDITH'S CHECKERSPOT

Boisduval 1852

Euphydryas editha

Including other subspecies: Edith's Checkerspot (*E. e. editha*), LuEsther's Checkerspot (*E. e. luestherae*), Baron's Checkerspot (*E. e. baroni*)

A strong hill-topper, Edith's is also known as the Ridge Checkerspot. It's known to travel between ridgelines and valleys below in search of flowers. The adults are also known to live only for a week. This butterfly can vary in appearance like the larger Variable Checkerspot (*Euphydryas chalcedona*), so identification can be slightly confusing.

The eastern and southern reaches of our region have a chaparral-inhabiting subspecies of Edith's Checkerspot called LuEsther's Checkerspot (*E.e. luestherae*). This butterfly is closely tied to its host plant, warrior's plume (*Pedicularis densiflora*). In northern Napa and Sonoma Counties, Baron's Checkerspot (*E. e. baroni*) lives in a variety of habitats, and its caterpillars may feed on a variety of plants.

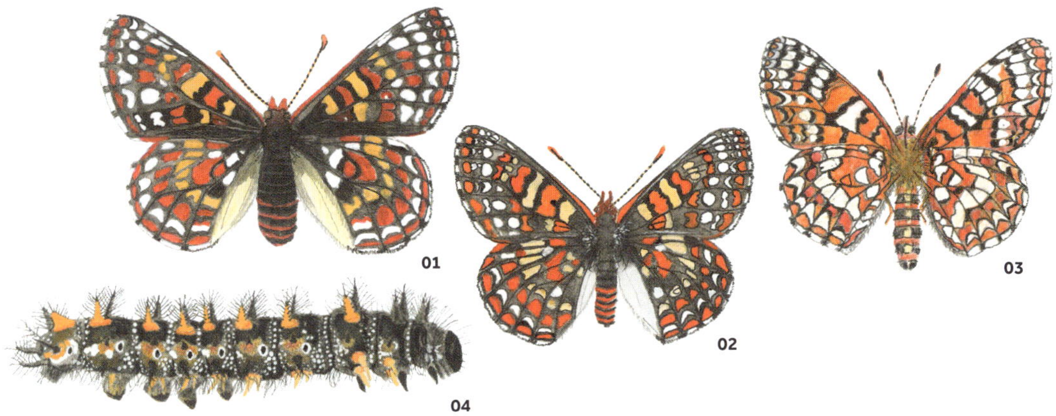

01: dorsal ♀
02: dorsal ♂
03: ventral
04: mature larva

Habitat: *E. e. luestherae*: chaparral, *E. e. editha*: grassland, *E. e. baroni*: various
Host Plants: The genus *Euphydryas* tends to prefer one or a few species of plants in a particular area (Scott, 1986, 296). *E. e. luestherae*: warrior's plume (*Pedicularis densiflora*); *E. e. editha* and *E. e. baroni*: various plants including warrior's plume, paintbrushes (*Castilleja*), and plaintain (*Plantago*).
Life Phases: Larvae live communally in gregarious clusters of silk nests on the underside of leaves (James & Nunnallee.2011.286). Pupa looks much like Variable Checkerspot, some are both white and gray with black spots and orange bumps. Adult's forewings are more rounded and overall redder than the Variable Checkerspot; (*Euphydryas chalcedona*). When the two butterflies are laid side by side, Edith's is much smaller. Males produce a sphragis after mating to block the female's ability to mate with other males. Univoltine: May–June.
Counties They Fly In: *E. e. luestherae*: Contra Costa; *E. e. editha*: San Benito, San Mateo, and Monterey; *E. e. baroni*: Napa and Sonoma
Great Places to See Them: Quite widespread when flying. Trails in East Bay MUD property (Contra Costa County); San Bruno Mountain, San Bruno County and State Park (San Mateo County); the Presidio (San Francisco County); and Pinnacles National Park (San Benito County). On Mount Diablo's Mitchell Canyon Trail (Contra Costa County), hike until you begin to notice the gray-green serpentine rocks and soil and the wildflowers that go with it. Edith's likes to lay on the trail or road. I normally find LuEsther's here in the late spring (May/June). Pure Edith's (*Euphydryas editha editha*) are seen at Pinnacles National Park (San Benito County).

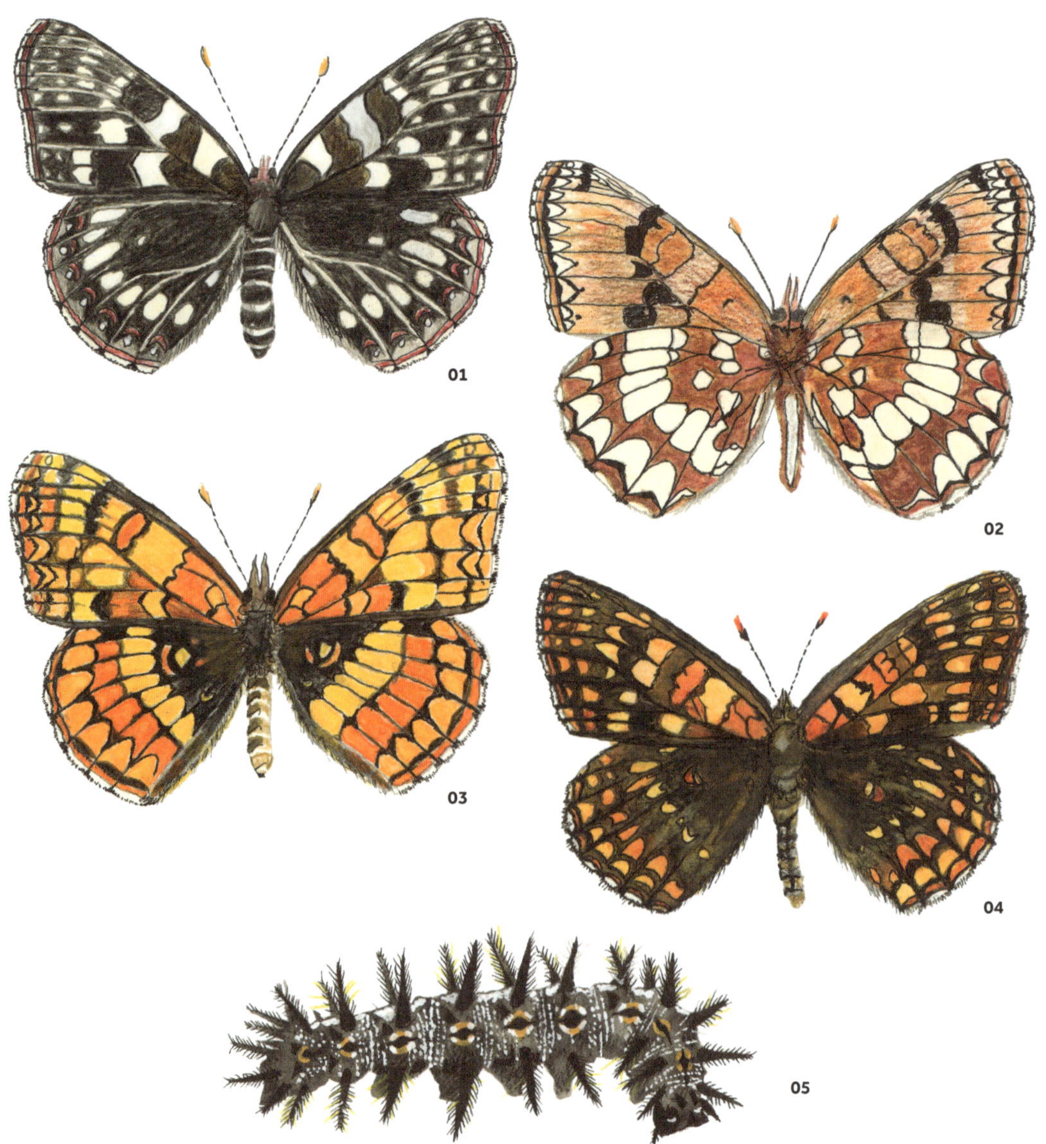
01
02
03
04
05

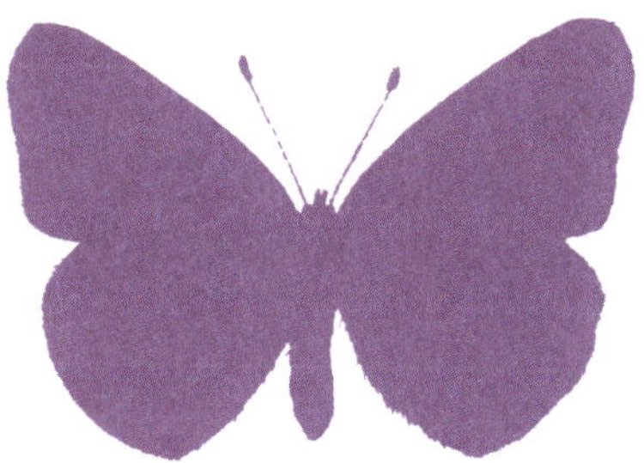

NORTHERN CHECKERSPOT

Boisduval 1852

Chlosyne palla

The lower side of this butterfly has areas of white between the patterns, which offers a definitive way of keying it away from Gabb's Checkerspot, which has more of a creamy pearling in said areas. A form of the adult female is hypothesized to be a Batesian mimic to Variable Checkerspots (*Euphydras chalcedona*) (Steiner.1980.139).

Habitat: Meadows and canyons, trail paths
Host Plants: Paintbrush (*Castilleja* spp.) and native asters (Asteraceae)
Life Phases: Overwinters as larva (Scott.1980.306). Pupa light brown with darker brown markings and glossy bumps. Adults can look like Gabb's Checkerspot, but the two species are separated by geography and never fly together. Univoltine: May–July.
Counties They Fly In: All except San Francisco. In San Benito and Monterey Counties it is replaced by Gabb's Checkerspot (*Chlosyne gabbii*).
Great Places to See Them: It can be quite abundant along the lower Mitchell Canyon Trail in Mount Diablo State Park (Contra Costa County), where it wrestles with Mylitta Crescent and Variable Checkerspots along the path. Another great spot is a small meadow directly east of Volmer Peak between the paved trail and the ridge to the east (Contra Costa County). Carson Falls Trail (Marin County) in the meadow below the waterfall.

01: dorsal ♀, dark form
02: ventral ♀ and ♂
03: dorsal ♂
04: dorsal ♂, including Gabb's
05: mature larva

GABB'S CHECKERSPOT

Behr 1863

Chlosyne gabbii

I'm going out on a slight limb here, folks, because I have in my ear Jerry Powell (to whom this book is dedicated) barking "No such thing as Gabb's. Just an inland form of Northern Checkerspot." Trained by a lumper I was. But because Gabb's is included in so many other field guides and is found in a great deal of the scope of this book, I've decided to put it in mine. Sorry, Jerry. Dorsal-side identification is tricky since it looks awfully close to the Northern Checkerspot, though the colors seem more faded. It's better to use location and an underside photograph (no easy task as the butterfly lies with its wings flat to the ground) for your identification to be valid. Deductive reasoning helps here too: if I'm in one of those three counties and I have something that looks like a Northern, I have Gabb's.

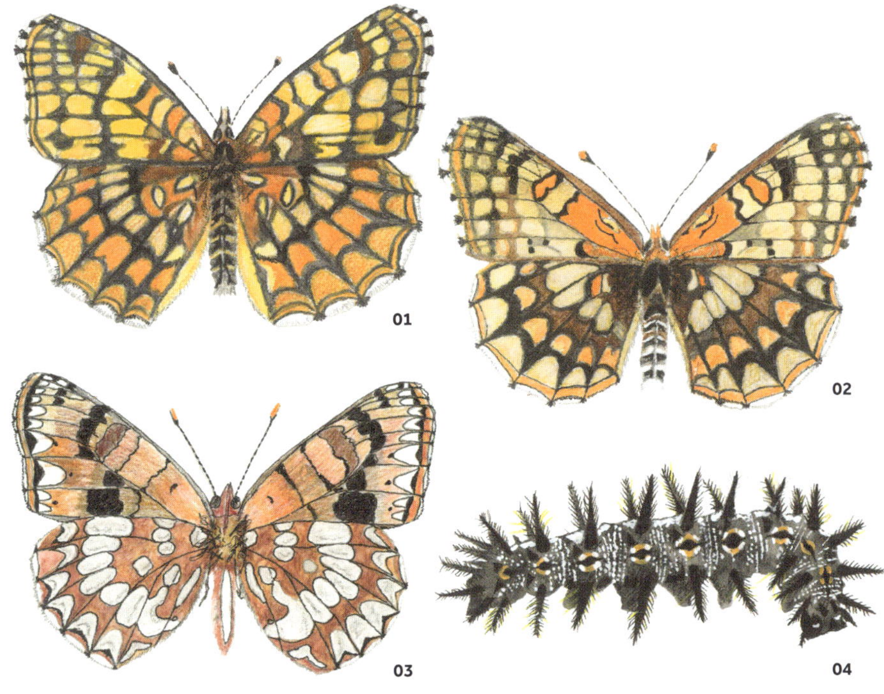

01: dorsal ♀
02: dorsal ♂
03: ventral ♀ and ♂
04: mature larva

Habitat: Canyons, gulches, willow and oak riparian
Host Plants: Paintbrush (*Castilleja* spp.), hazardia (*Hazardia squarrosa*), California aster (*Corethrogyne filaginifolia*)
Life Phases: After the third instar, larvae hibernate. Young stay together in gregarious feedings. Pupa gray covered in black splotches. Adults checkerboarded in brown and orange-red bands above. Females are lighter. Gabb's are identified by the underside: creamy, pearly white spots (areas), whereas Northerns have flat white spots with no sheen. Univoltine: May–June.
Counties They Fly In: Monterey, San Benito, and possibly Santa Cruz
Great Places to See Them: For the last few years I've been assigned the Old Pinnacles Trail during the annual Pinnacles Butterfly Count. Take the trail a few miles in (you may even see them along the way) to a junction where the trail jogs right and up and over the caves—don't take that. Proceed just beyond this intersection toward the actual cave entrance and you will see a small meadow along the creek (usually dry). Like clockwork, I've had Gabb's there annually. I've also seen them at French Camp—Landels-Hill Big Creek Reserve (entrance by invitation) (Monterey County) and Chews Ridge (Monterey County) (see **Best Butterfly Walks**).

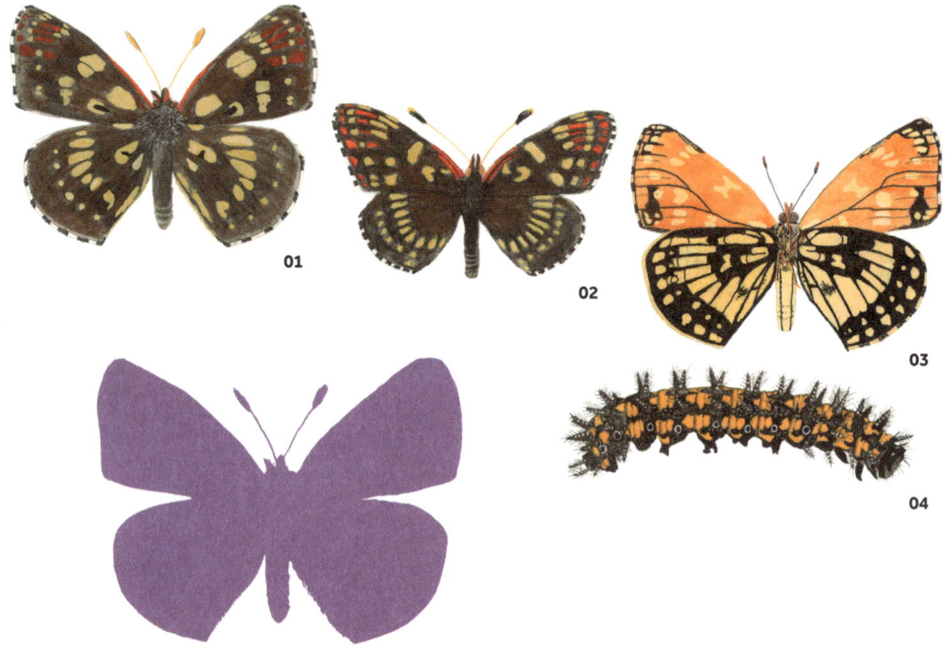

LEANIRA CHECKERSPOT

Felder 1860

Thessalina leanira

Leanira has a unique fluttering flight, like the Sonoran Blue. This beauty is immediately recognizable due to its striking spiderweb pattern from below. No other checkerspot has this pattern. Males both perch and patrol ridgelines.

Habitat: Serpentine, coastal scrub, chaparral, canyons, washes
Host Plants: Paintbrush (*Castilleja* spp.)
Life Phases: Eggs are laid in clusters. Larvae gregariously hang together till the third instar, upon which they separate and hibernate (Steiner.1980.140). Pupa white with orange between black markings. Univoltine: May–July.
Counties They Fly In: All except San Francisco (though I've only seen it there sporadically over the years)
Great Places to See Them: Sierra Azul Open Space Preserve (Santa Clara County); Sunol Regional Wilderness (Alameda County); Mitchell Canyon Trail, Mount Diablo State Park (Contra Costa County). At the fourth switchback up the trail, look for the butterfly's host plant. Males will be patrolling along the trail in search of females. If you have it in you and you haven't seen the butterfly yet, proceed up the steep trail to the summit of Eagle's Peak. Leanira is a strong hill-topper, and you should see many over the slopes.

01: dorsal ♀
02: dorsal ♂
03: ventral ♀ and ♂
04: mature larva

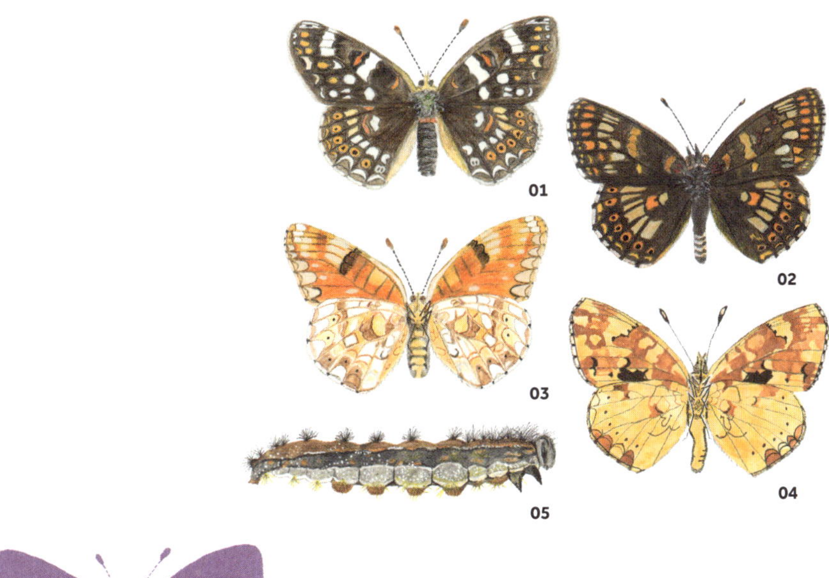

01: dorsal ♀
02: dorsal ♂
03: ventral ♀
04: ventral ♂
05: mature larva

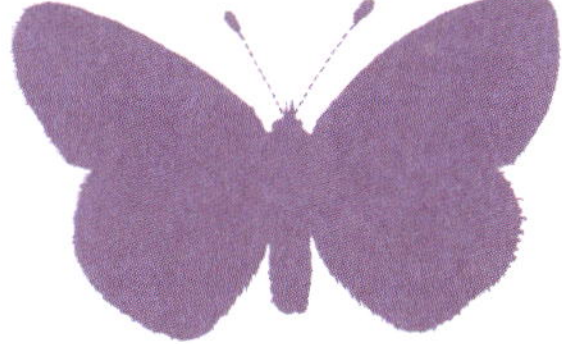

FIELD CRESCENT

Boisduval 1852

Phyciodes pulchella

When you are starting out with this mania, it's easy to get tripped up by sexual dimorphism (sexes looking different) and think, *Oh that's a different species.* This was the species that finally gave me that lightbulb moment. I identified the dark male form initially, but it took a while to figure out that the whiter, pale-orange female was the same species. Gotta learn both sexes, folks, to really know your butterflies.

Habitat: A denizen of many natural open spaces, coastal scrub, marsh, riparian
Host Plant: Pacific aster (*Aster chilensis*)
Life Phases: Pupa have dark and light areas of brown. Adult males very dark above rows of orange spots. The field is black or brown. Female lighter, with rows of orange and cream spots. Multivoltine: February–November.
Counties They Fly In: All
Great Places to See Them: The Coastal Trail in the Presidio (San Francisco County); Alemany Farm (San Francisco County); Mount Diablo State Park (Contra Costa County). I don't think it's an exaggeration to say perhaps every state park mentioned within the scope of this book—the creature is quite ubiquitous.

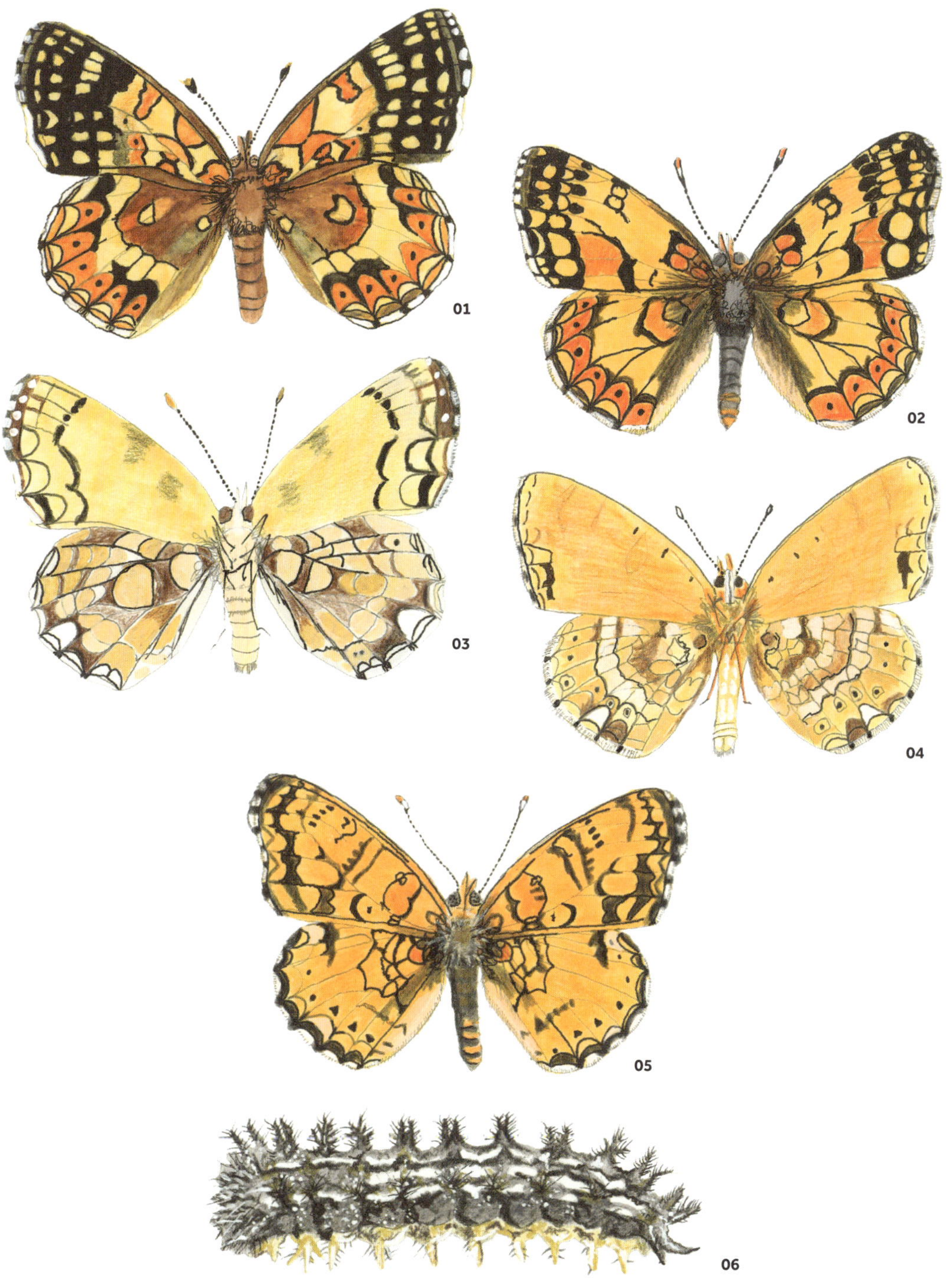
01
02
03
04
05
06

MYLITTA CRESCENT

Edwards 1863

Phyciodes mylitta

Want to hear something wacky? This butterfly is considered a true divining rod. That is, one sees this species both along an actual creek or stream and over slabs of cement in a city where waterways once occurred. I've seen them over Market Street (the city's main boulevard) in San Francisco, where Mission Creek historically ran. How amazing is that?

Habitats: Creeks, meadows, disturbed weedy areas, water concourses
Host Plants: A variety of thistles (*Cirsium* spp.)—weedy, non-native thistles (such as Italian and bull)
Life Phases: Larvae communally stay together and overwinter in their third instar. Pupae have dark and light areas of browns and grays. Adults can be recognized by the white crescents on the underside hind wing. But since one rarely sees the underside (because the butterfly sits with its wings open—dorsal basking) and since this butterfly from below looks hauntingly like a Field Crescent (*Phyciodes pulchella*), a better way to learn it is from above. Males are smaller than the females and he is uniformly bright orange, like the fruit. The female pattern from above looks somewhat blurred and she is more cantaloupe. Then to confuse things even more, nature has given us a dark form and a light form of each. Multivoltine: year-round.
Counties They Fly In: All
Great Places to See Them: It's one of those butterflies that shows up when you can't find any others in the field. If I said the Field Crescent was everywhere, then this one is *everywhere, everywhere*. Usually you'll see patrolling males on the path or trail you are walking on, especially near creeks and water concourses. I find them in vacant lots all the time.

01: dorsal ♀
02: dorsal ♂, dark form
03: ventral ♀
04: ventral ♂
05: ventral ♂, light form
06: mature larva

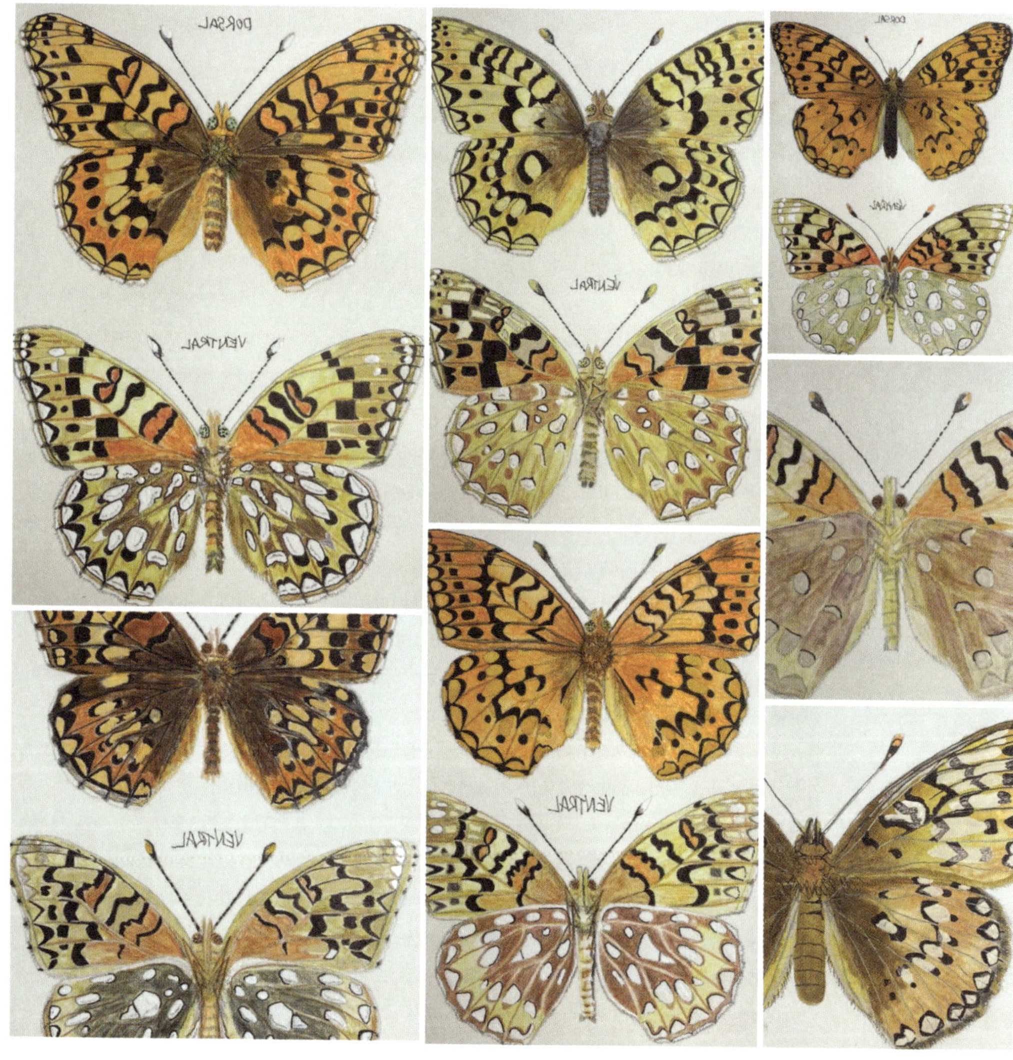

ON SILVERSPOTS, LUMPERS, AND SPLITTERS

Prepare yourself to pull your hair out over these next species, as they all look incredibly similar. And just to add frustration for the novice (and for the old timers—myself included), the genus *Argynnis* is rather new. They're referred to as *Speyeria* in older guides. *Silverspots* is a catch-all term used for all these Greater Fritillaries to come. Some are very clear—the Unsilvered Fritillary (*Argynnis adiaste*) has no silver spots below. But when two subspecies or species fly together (Crown Frits seem to fly with Callippe Frits a great deal), things can get complicated, making individual identification maddening at best. Folks, I'm still working at this, and I've been at it for decades, so go easy on yourselves.

A splitter (an enthusiast who embraces the concept of subspecies) can go into their own backyard and make all those plants they have subspecies because it's based more on location and not unequivocal science. Splitters are most often thought of as those who split single species into multiple species, often raising subspecies to species level. Here's what Coast Ridge Ecology had to say on the matter of Silverspots in their 2009 publication "Distribution Study for Lilian's Fritillary": "The Callippe Silverspot is a subspecies of the genus *Speyeria* [now Argynnis] of which there are 16 species and over 100 subspecies. Subtle wing pattern characteristics such as size, shape and color are used to differentiate subspecies though substantial variation occurs within populations. The Callippe Silverspot (also called the

Silverspots

Callippe Fritillary) can be difficult to differentiate between other callippe subspecies such as Lilian's Silverspot in the northern range and Comstock's Silverspot to the southeast" (63). It's a hall of mirrors, folks.

On the other hand, a lumper (an enthusiast who doesn't adhere to the concept of subspecies) will shout, especially if dealing with Callippe (*Argynnis callippe*), "It's all the same damn bug!" when asked, "Where does *Argynnis callippe callippe* end and *Argynnis callippe lilianna* begin in Sonoma County?" A lumper would argue that there is no answer to that, that they are all the same species just in different forms, like humans are all the same species but come in different forms. People were hoping that DNA sequencing was going to be the big breakthrough and solve all of this, but it doesn't seem to get to the subspecies level. This back-and-forth has gone on since the beginning of the binomial system.

All I can say regarding the Fritillaries (aka the Silverspots) is, do not get too settled in your way of keying them out. Really look at the paintings for detail against the photos you've taken and consider strongly where you've taken the photo when deciding on the most likely identification. "Really close" might be the best we can do.

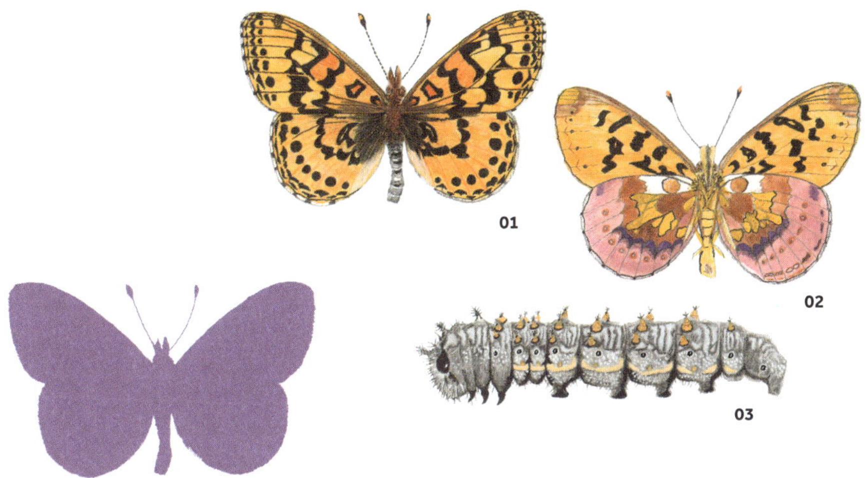

PACIFIC FRITILLARY

Fabricius 1775

Boloria epithore

also known as **THE WESTERN MEADOW FRITILLARY**

The subgroup of the Lesser Fritillaries (within the subfamily Heliconiinae—fritillaries and longwings) in the Bay Area are divided into two tribes: Heliconiinae (which the Gulf Fritillary belongs to) and Argynnini (which includes *Boloria* and *Argynnis*), which includes one of the most beautiful butterflies I've ever painted. The underside was complicated to do—it looks like some sort of mask looking back at you. This butterfly flies very low to the ground. A relict holdover from when the peninsula was colder, this butterfly is also found in the High Sierra.

Habitat: Meadows and moist edge zones near redwood and evergreen forests where the host grows in the understory canopy
Host Plant: Western heart's ease (*Viola ocellata*)
Life Phases: Larvae overwinter at fourth instar (Steiner.1980.136). Pupa is pale brown with white mottling. Univoltine: April—July. Adults are small, about the size of a half dollar, with rounded wings.
Counties They Fly In: San Mateo, Santa Clara, and Santa Cruz
Great Places to See Them: The old, long air-vehicle landing site on Forest Road two hundred yards south of the east end of the Annapolis Road bridge over the south fork of the Gualala River (Sonoma County). Peters Creek Trail, Long Ridge Open Space Preserve, Midpeninsula Regional Open Space District (San Mateo County); Russian Gulch State Park (San Mateo County). Other parks in San Mateo County it's reported in: Purisima Creek Redwoods Open Space Preserve, Half Moon Bay; Palomar Park, La Honda.

01: dorsal ♀ and ♂
02: ventral ♀ and ♂
03: mature larva

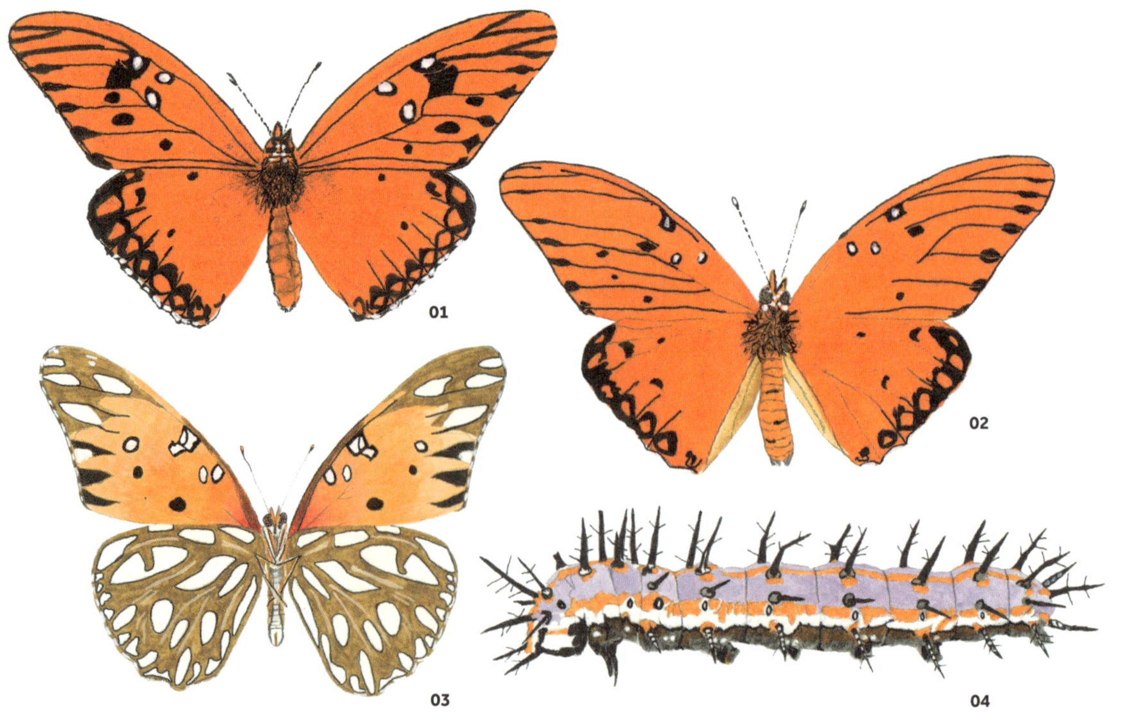
01
02
03
04

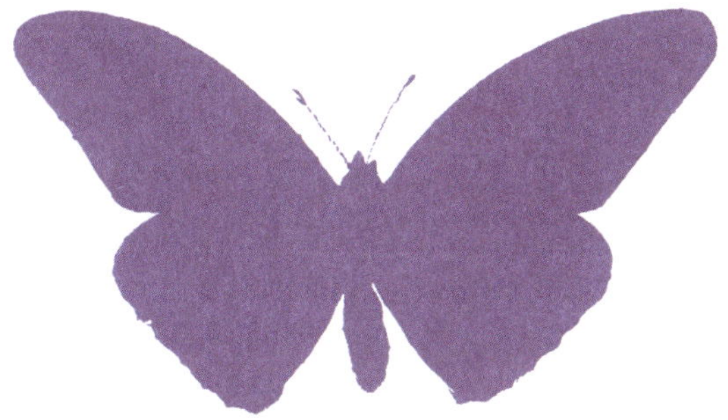

GULF FRITILLARY

Linneaus 1758

Dione vanillae

Passionvine, the host plant, is not native to California. When this tropical plant began going into Southern California nurseries and working its way north as a garden ornamental, up traveled this Mexican butterfly, too. As of the writing of this book, it is nearing the Oregon border, and that probably has much to do with global warming: the butterfly larvae cannot survive a frost—my conjecture is that as frosts are becoming less and less severe, more and more survive. Yes, they have silver spots and are called Fritillaries, but they do not belong to the true Fritillary group and are part of the longwing set.

Habitat: Highly urbanized species—backyard gardens in suburbia, parks, disturbed areas of agricultural fields. Just about anywhere its non-native host plant exists.

Host Plant: Passionvine (*Passiflora* spp.)

Life Phases: Larvae, orange and purple with spikes, eat at night and rest in a J shape (Scott.1986.338). Pupa greenish brown, looks like a dead leaf full of warts. Multivoltine: year-round.

Counties They Fly In: All

Great Places to See Them: Corner of 17th Street and Market Street (San Francisco County). Up 17th Street is a small parklet dedicated to the gay victims of the Holocaust. This silver-studded butterfly bolts about and nectars here, especially on the red valerian (*Centranthus ruber*) flowers. Common on the neighborhood streets of Berkeley (Alameda County) as well as Tilden and UC Berkeley botanical gardens in the same city.

01: dorsal ♀
02: dorsal ♂
03: ventral ♀ and ♂
04: mature larva

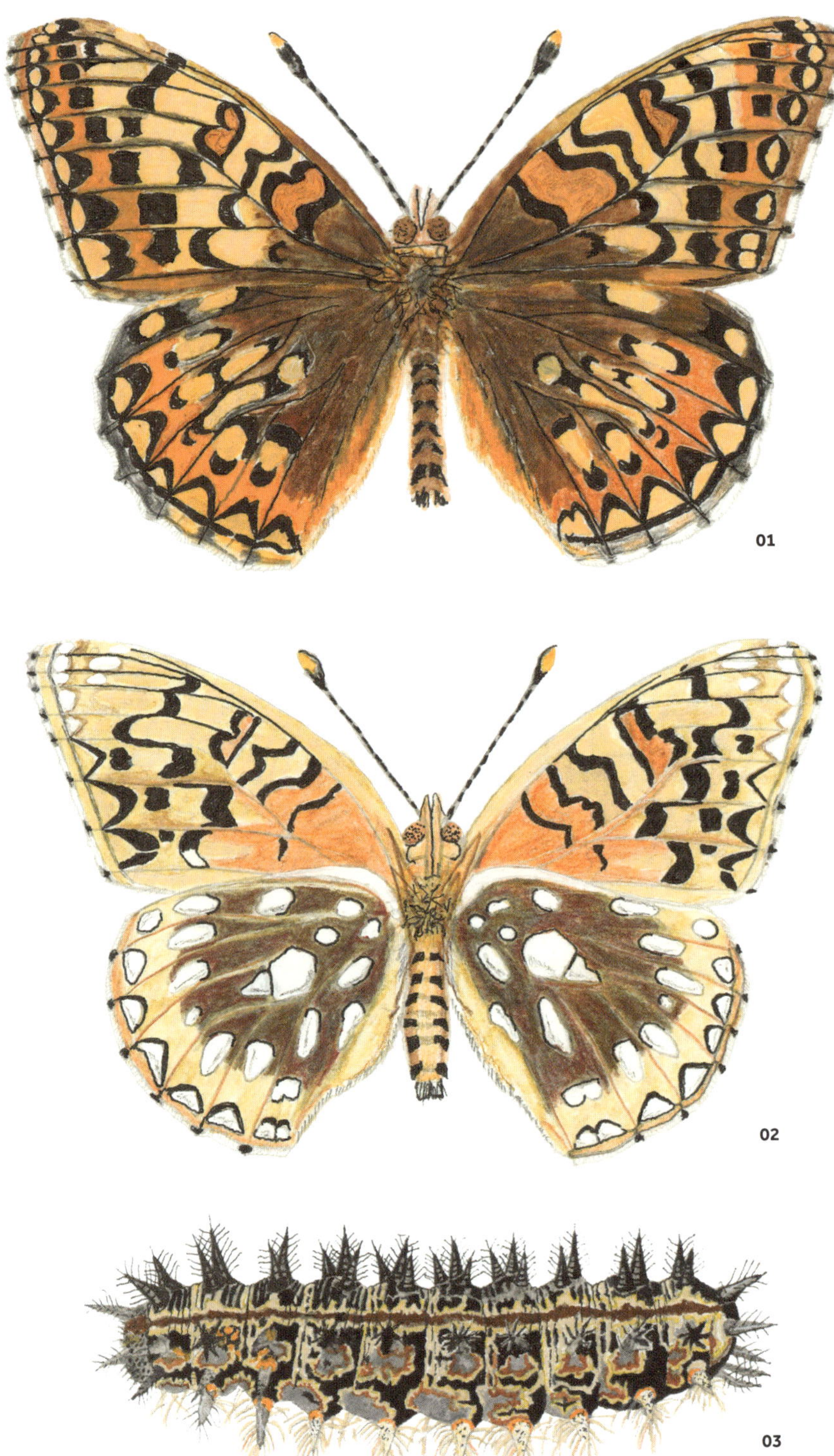
01
02
03

CALLIPPE SILVERSPOT

Boisduval 1852

Argynnis callippe callippe

STATUS: ENDANGERED

A subspecies of the Callippe Fritillary. Because it evolved within the fog belt of the Bay Area, forewings here are much darker than in other Callippes. It's a federal crime to harass an endangered species under the Lacey Act. If you see one, don't chase it for a photo—just enjoy its glorious presence.

Habitats: Open grasslands and chaparral, coastal scrub
Host Plant: Johnny jump-ups (*Viola pedunculata*)
Life Phases: Eggs are laid near the dried-up *Viola* of the season prior (Scott.1986.329). This genus *Argynnis* has six molts for their larvae as opposed to the great majority of butterflies in this book, which have five. The final molt is gray brown with black markings, broken stripes, and rows of spines (James & Nunnallee.2011.238). Pupa white with various dark markings. This is a large, buff-colored butterfly with silver spots ventrally. Univoltine: June-July.
Counties They Fly In: Alameda, San Mateo, and Solano
Great Places to See Them: Ridge Trail, San Bruno Mountain State and County Park (San Mateo County). Walk a good three miles out to the last high-voltage tower to see them nectaring on the bull thistle and the naturalized pincushion flower. You might catch a glimpse of them along the hike out on the hilltops, as these places are known locales for mating. Males are strong hill-toppers. They have also been seen at Blue Rock Springs Park, Vallejo (Solano County).

01: dorsal ♀ and ♂
02: ventral ♀ and ♂
03: mature larva

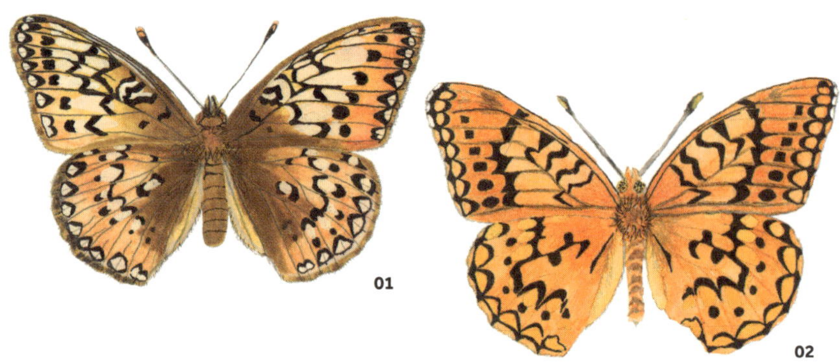

CALLIPPE FRITILLARY

Boisduval 1852

Argynnis callippe

also known as **CALLIPPE SILVERSPOT**

including other subspecies: Comstock's Fritillary (*Argynnis callippe comstocki*) and Lilian's Fritillary (*Argynnis callippe liliana*)

Go easy on yourself. This is a very difficult subdivided species even for the experts. I think the way to go is good photography of the underside markings and a process of elimination by location. First ask yourself: *Where am I seeing this?* If you are not on San Bruno Mountain (San Mateo County), Blue Rock Springs Park (Solano County), or open space near Sonoma Raceway (Sonoma County), then you are probably seeing Comstock's. All aforementioned places are enclaves of *Argynnis callippe callippe*, the endangered subspecies. These subspecies have been known to hybridize, which makes identification even more challenging.

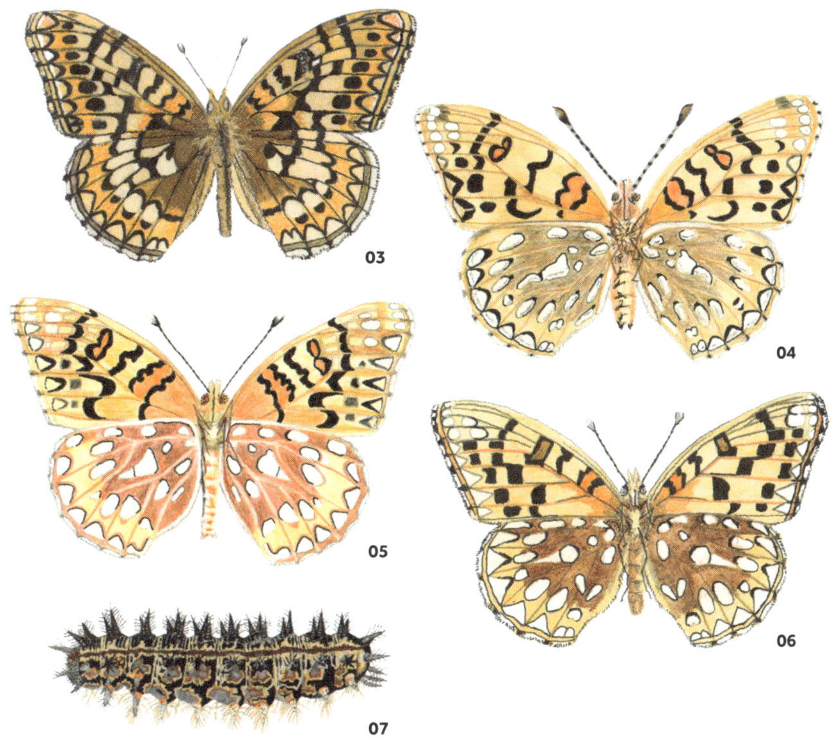

03

04

05

06

07

01: dorsal ♀, Comstock's
02: dorsal ♂, Lilian's
03: dorsal ♂, Comstock's
04: ventral ♀, Comstock's
05: ventral ♂, Lilian's
06: ventral ♂, Comstock's
07: mature larva

Habitat: Open grasslands and coastal scrub.

Host Plant: Johnny jump-ups (*Viola pedunculata*)

Life Phases: Females lay their eggs in and around the dormant host plant (how they find this dormant a foot below the surface is still a mystery). Caterpillars are similar (which shows you just how closely related these are). It is reported the larvae feed nocturnally (Garth & Tilden.1986.74). Pupa white with various dark markings. Comstock's adults are tawny and buff colored. Lilian's is more orange dorsally or reddish on the upper forewings, and the lower two-thirds of the ventral underwing is dark brown. Univoltine (April–July).

Counties They Fly In: Comstock's: Alameda, Santa Clara, San Benito, and Monterey. Lilian's: Sonoma, Napa, and Mendocino.

Great Places to See Them: Comstock's can be seen quite easily in the Pinnacles National Park (San Benito County) and have been reported in Mount Hamilton (Santa Clara County). Most reports for Lilian's are along the county line between Napa and Sonoma. Los Pasados Forest (Napa County). They can also be seen in Los Padres National Forest (Monterey County). Other reports of Mendocino County.

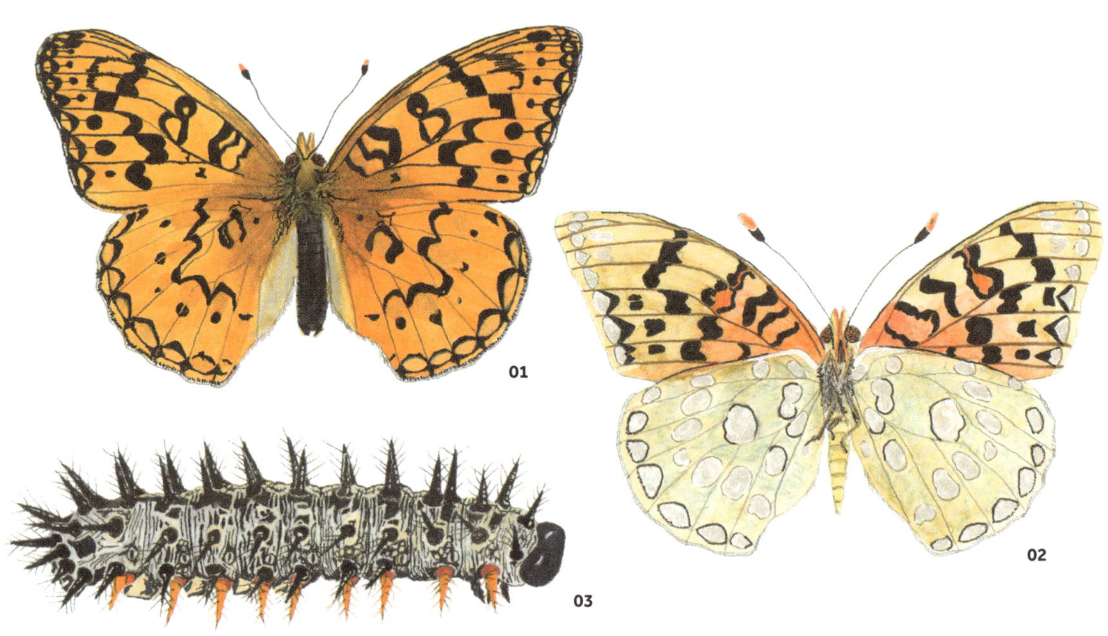
01
02
03

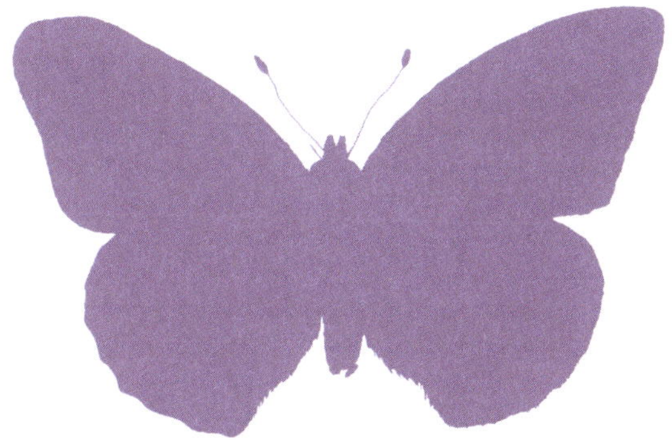

CROWN FRITILLARY

W. H. Edwards 1864

Argynnis coronis

also known as **CORONIS FRITILLARY**

This one is rather easy to key away from the others, as it is larger than other Silverspots and even the silver spots are larger. A definitive shot at a flower is always best. (Butterflies usually hold their wings up while nectaring, giving one a clear shot of the ventral markings.) Crowns are still on the wing even after Callippe is done for the season. They sympatrically fly with Callippe and Beheren's at Sonoma Raceway (Sonoma County).

Habitat: Forest edges, glades of evergreen woodlands

Host Plants: Various violets (*Viola* spp.)

Life Phases: Pupa white with various dark markings. Dorsal forewings of adults are orange or pale orange. Silver spots on hind wings are large and imposing with caps of pale green or greenish brown on either end. Males patrol all day looking for mates. Univoltine: June–July.

Counties They Fly In: All except San Francisco

Great Places to See Them: Reinhardt Redwood Regional Park, Oakland, (Contra Costa County); Crystal Springs Watershed (San Mateo County)

01: dorsal ♀ and ♂
02: ventral ♀ and ♂
03: mature larva

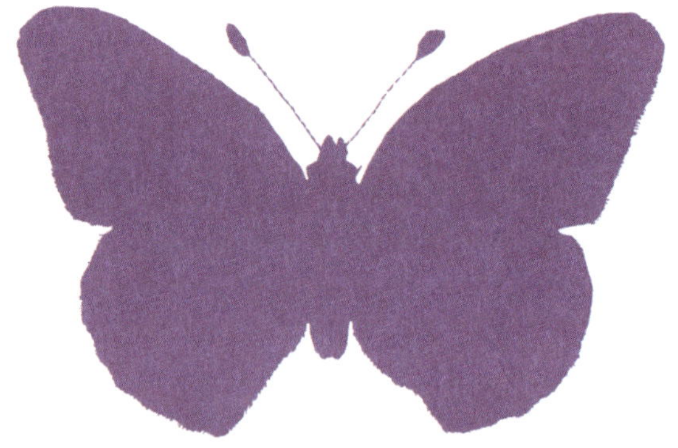

BEHREN'S SILVERSPOT

Emmel, Emmel, & Mattoon 1998

Argynnis zerene behrensii
also known as **BEHREN'S FRITILLARY**
STATUS: ENDANGERED

At the start of 2024, a joint effort between the Mendocino Land Trust, the Bureau of Land Management, the U.S. Fish and Wildlife Service, and the Sonoma-Mendocino division of California State Parks released hundreds of captively bred Behren's Silverspots in undisclosed locations along the coast. They were transferred at the larval stage and were reared by a team at the Sequoia Park Zoo in Eureka. There had been ninety-two sightings of Behren's over the previous fifteen years (Bindman.2024). This will improve things tremendously for the butterfly and give new possibilities for the butterfly enthusiast to actually happen upon it.

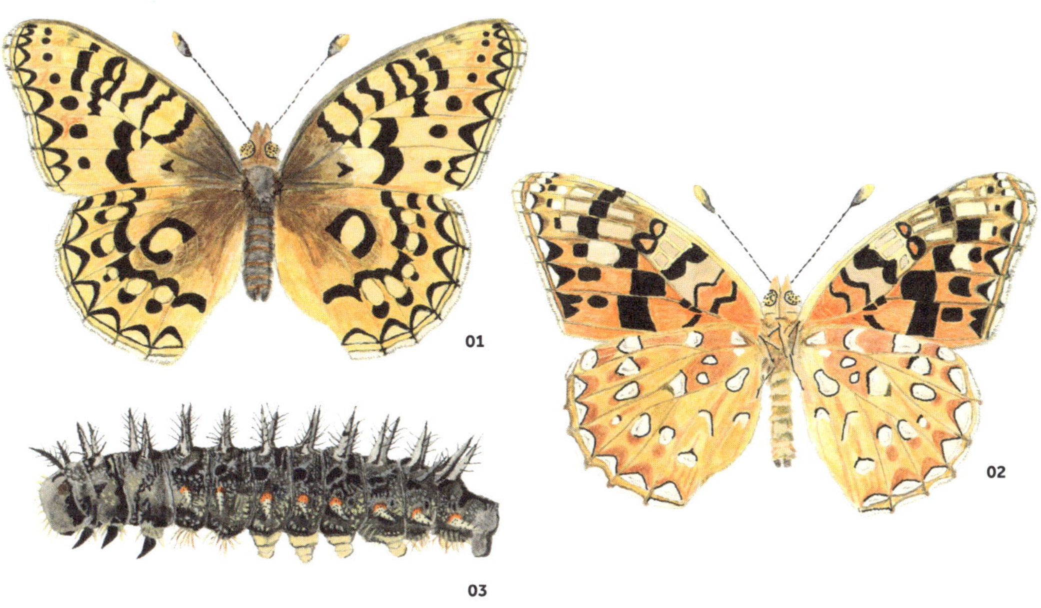

01: dorsal ♀ and ♂
02: ventral ♀ and ♂
03: mature larva

Habitat: Open grasslands

Host Plants: Various violets (*Viola*), western dog violet (*Viola adunca*), and Johnny jump-ups (*Viola pedunculata*)

Life Phases: Larvae brown black with a black horizontal stripe and spines. Pupa orange with black markings. Overwinters as a caterpillar. Lighter and larger than both Callippe and Crown, this butterfly gives the experts headaches even trying to key it from other Zerene Fritillaries. Lucky for us the other Zerene discussed in this book does not fly with Behren's. Univoltine: May–July.

Counties They Fly In: Mendocino, Napa, and Sonoma

Great Places to See Them: Limited to a small area near the Sonoma Raceway (Sonoma County) that is off-limits to the general public. So . . . no, no great places to see them. Reported populations in Cape Mendocino (Humboldt County) (Howe.1975.490).

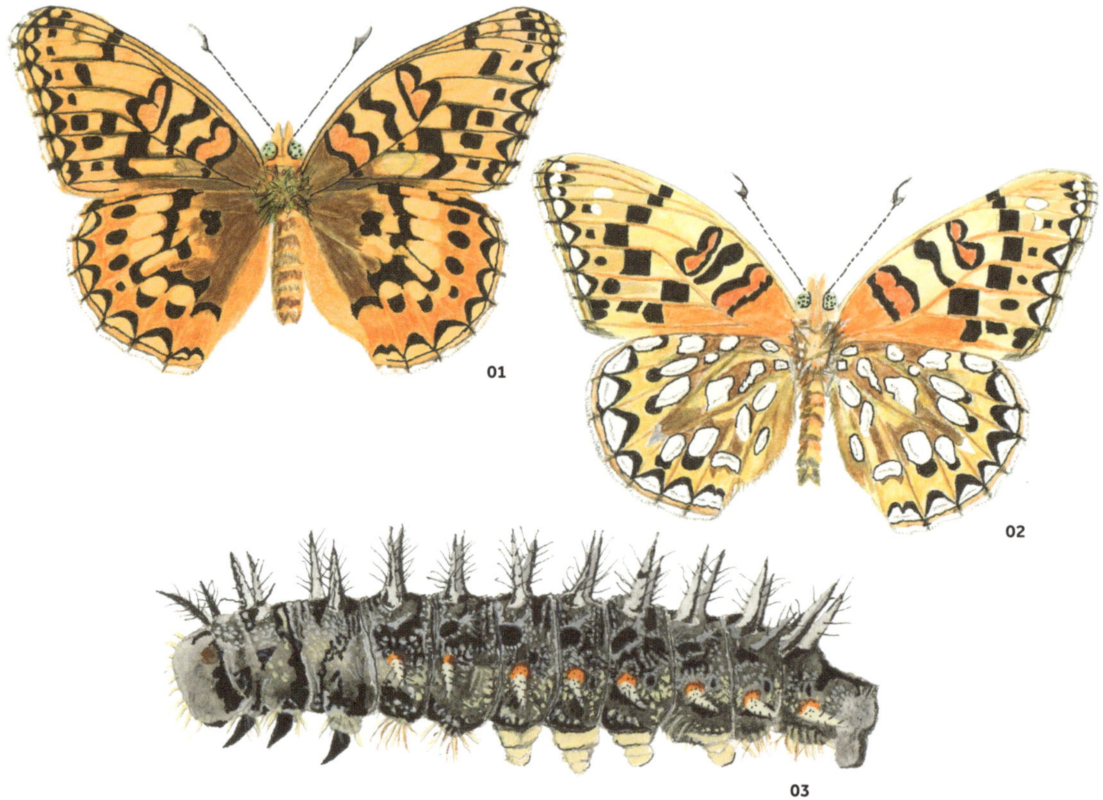
01
02
03

POINT REYES SILVERSPOT

Emmel, Emmel, & Mattoon 1998

Argynnis zerene puntareyes

also known as **MYRTLE'S SILVERSPOT**

STATUS: ENDANGERED

You can easily get skunked out by the marine layer in Point Reyes looking for this bug. But there is usually a brief, sunny window midday even on the foggiest trip, so keep your hopes up. Goes to flowers regularly, especially bull thistle. In a dazzling array of splitting, it was decided by brother lepidopterists Tom and Jon Emmel sometime in the 1980s that its prior name Myrtle's belonged to a now extinct butterfly and therefore needed a change. Hence the Point Reyes Silverspot.

Habitat: Coastal meadows, coastal scrub, dunes, open glades in forests
Host Plant: Western dog violet (*Viola adunca*)
Life Phases: Eggs laid near the host plant (Scott.1986.334). Pupa orange with black markings. Dorsal forewings and hind wings are both orange brown with darkening shade toward the thorax. Ventral side tan to brown, yellow at edges. Univoltine: June–August. Males appear first.
Counties They Fly In: Marin, Mendocino, and Sonoma.
Great Places to See Them: Point Reyes National Seashore (Marin County). Take Sir Francis Drake Boulevard out toward the Point Reyes Lighthouse. Take the South Beach exit to the right. Hike the dunes to the north of the parking lot. This butterfly always seems to be on the wing for me there. Drake's Estero Trail as well. Bull Point Trail walking east leading out toward the water, usually on blooming bull thistle out there.

01: dorsal ♀ and ♂
02: ventral ♀ and ♂
03: mature larva

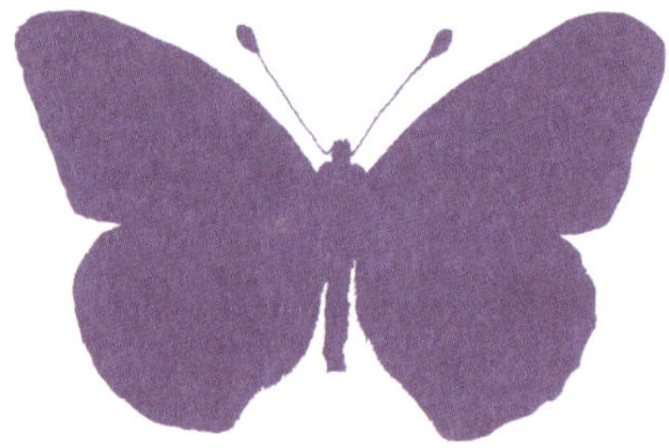

UNSILVERED FRITILLARY

W. H. Edwards 1864

Argynnis adiaste

Including other subspecies: *Argynnis adiaste adiaste* and *Argynnis adiaste clemencei*

I have a dopamine rush every time I see this special butterfly—the Unsilvered is, true to its name, the only *Argynnis* that's has no pale spots below (other *Argynnis* in the country have a somewhat muted silvering). It also breaks down to two subspecies within the span of this book—San Mateo, Santa Clara, and Santa Cruz Counties (*A. a. adiaste*), and Monterey (*A. a. clemencei*). Once one keys this away from the other fritillaries flying about, it's a real breakthrough. One of the rarest in any personal butterfly collections.

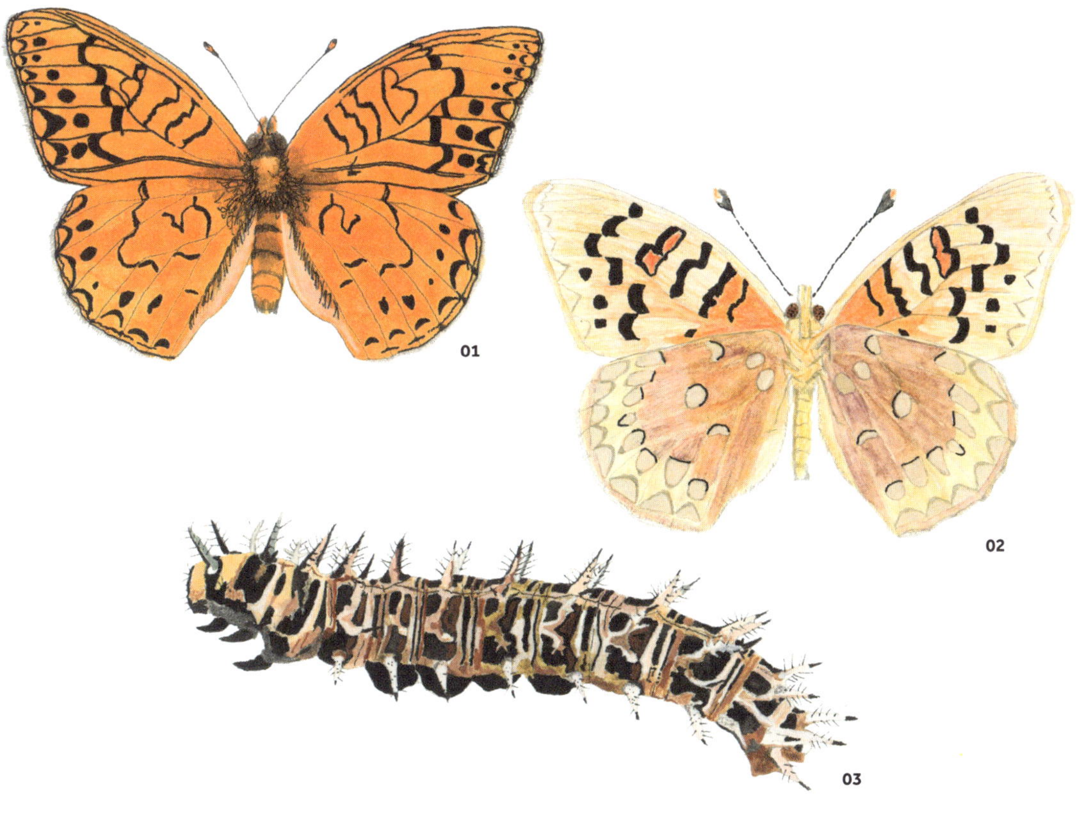

01: dorsal ♀ and ♂
02: ventral
03: mature larva

Habitat: Open chaparral, edges of conifer forests
Host Plants: Various violets (*Viola* spp.)
Life Phases: Early life unknown in detail (Emmel & Emmel.1973.29) Pupa is like Callippe's. Adults are orange reddish above with a violet tinge below. *Their spots are not silver.*
Counties They Fly In: Monterey, San Mateo, Santa Clara, and Santa Cruz
Great Places to See Them: We are fortunate enough to have a couple of solid locales to see this elusive creature. The subspecies *A. adiaste adiaste* can be out Skyline Boulevard/Highway 35 (San Mateo County). Near the Stevens Creek Trailhead, check all the blooming buckeye trees. Peters Creek Trail and Long Ridge Open Space Preserve, both in San Mateo County, have revealed the Unsilvered Fritillary. The subspecies *clemencei* is a Chews Ridge denizen (Monterey County) (see **Best Butterfly Walks** for directions). Park on the right side at the summit and follow the trail out north. Watch all thistles in bloom for this butterfly. It will be flying there with both Comstock's and Crown Silverspots.

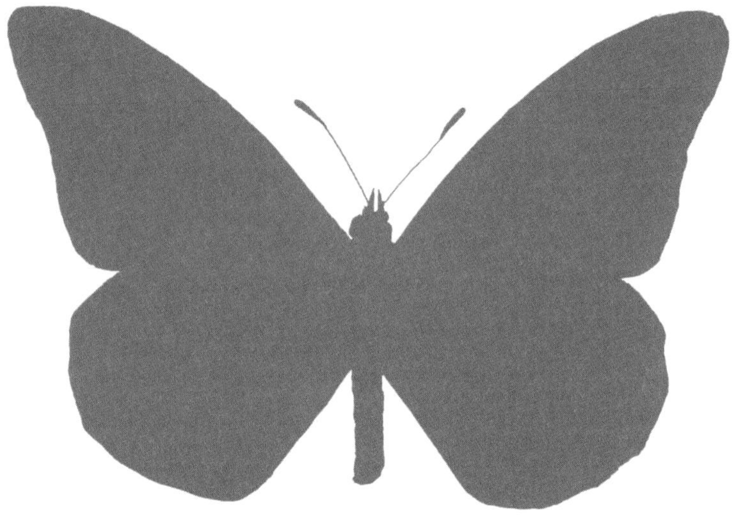

MONARCH

Linnaeus 1758

Danaus plexippus

The most celebrated butterfly in the world, by far. The migration in the West starts around Idaho and Washington and goes as far as San Diego. The migration in the East begins in the Canadian Maritime Provinces and drops south, rounding and crossing the gulf and terminating in Michoacán, Mexico. I've monitored its return to San Francisco County during the fall for the last few decades. In 2021 the butterfly population plummeted along the West Coast during its overwintering to numbers we thought it could never recover from. But then it rallied to a count north of 300,000 in 2022 and 2023. The numbers crashed again in 2024—second lowest. What I believe it is teaching us is that we still have a great deal to learn from this very famous creature.

There was a story a few seasons back that a group of orioles in Mexico had been found to pluck the wings off the Monarchs (the most toxic parts of the butterfly) and feed their chicks the middle section of the head, thorax, and abdomen. Seems like those smart birds have cracked the aposematic coloration code. Amazing.

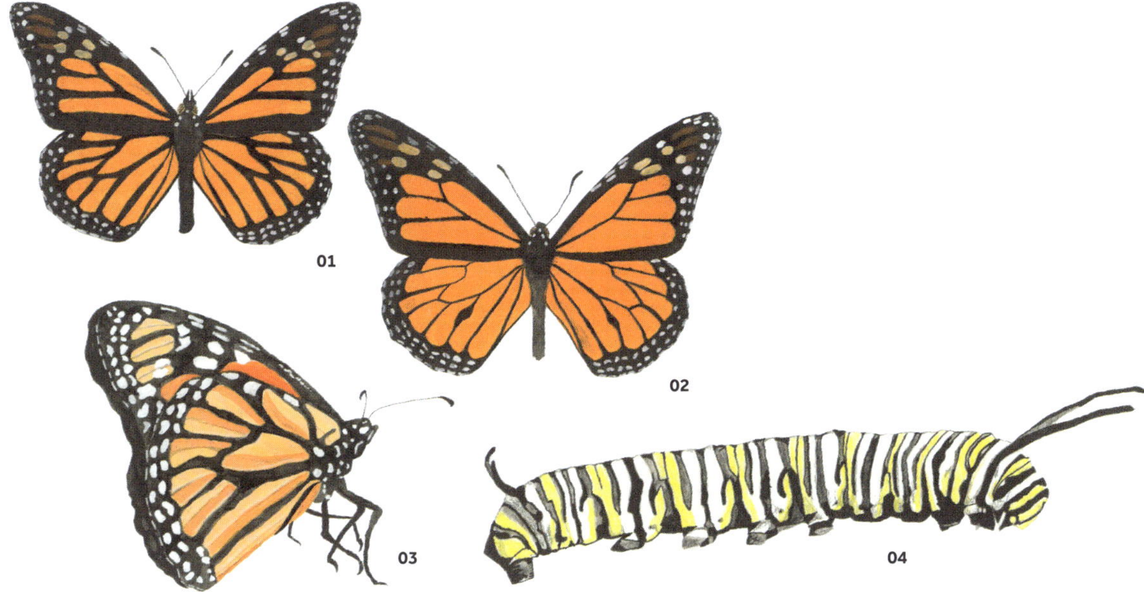

01: dorsal ♀
02: dorsal ♂
03: in situ
04: mature larva

Habitat: Community gardens, meadows. During their migration they can be found or passing through many ecosystems, mostly in sunny open space.
Host Plants: Milkweeds (*Asclepias* spp.)
Life Phases: Eggs are laid on the underside of leaves. Larvae imbibe the poison cardenolides from their host. The pupa is referred to as a mermaid's purse. Adults exhibit aposematic coloration: black and orange, warning consumers its full of a noxious poison. Four generations a year hopscotch the coming-and-going migration. Males can be distinguished on the upper side by the large black scent patch along the central hind-wing vein. The winter reveals large sleeping congregations of Monarchs that roost mainly in eucalyptus trees but also Monterey pine.
Counties They Fly In: All
Great Places to See Them: Natural Bridges State Beach, Santa Cruz (Santa Cruz County) is the go-to place to see them amassing by the thousands in the fall. Pacific Grove (Monterey County) has a protected sanctuary that the butterfly returns to annually.

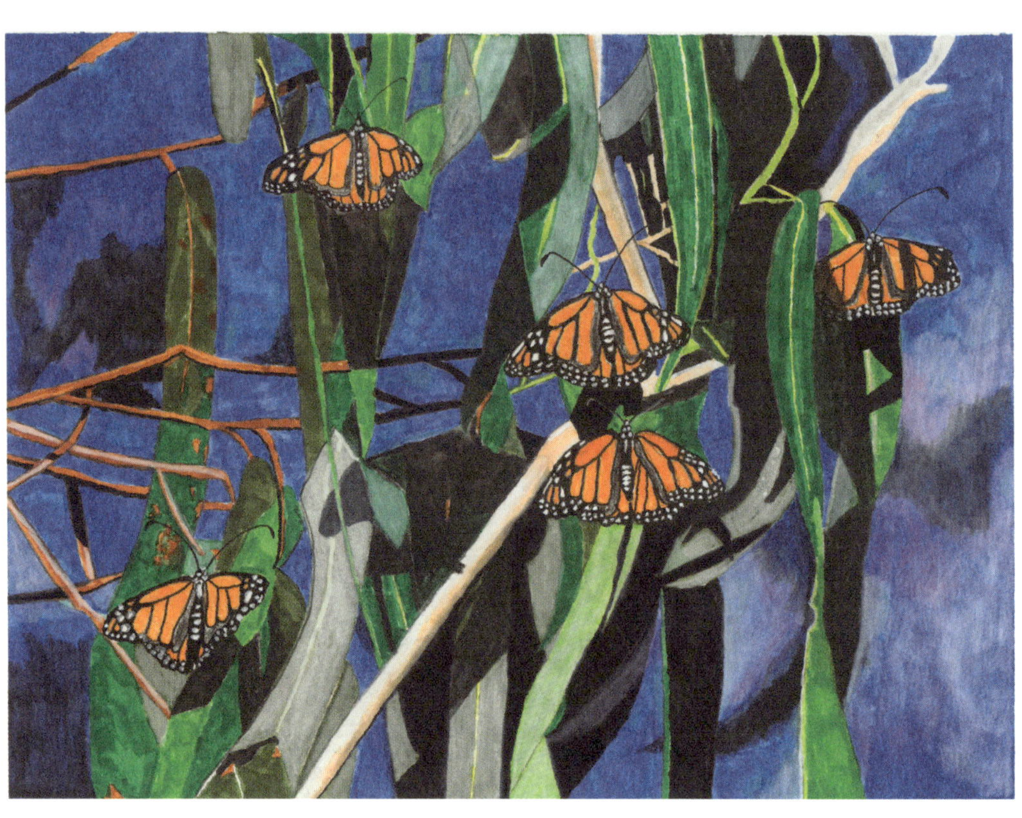

INTERVIEW WITH MIA MONROE

Before I met her, I must admit, I was already slightly obsessed with her. A woman with a passion for butterflies (at least one butterfly in particular) as notorious as mine. I kept putting myself in her general vicinity in order to run into her. Not sure if that was technically stalking, but we eventually met at the Muir Woods National Monument in Marin County, where Mia Monroe had worked for decades as a park ranger for the National Park Service. She is now retired.

We sat down for an interview September 29, 2023, in the resplendent Fort Mason Community Garden in San Francisco—a place rife with butterflies on this day and a known cluster location for Monarchs (*Danaus plexippus*). Mia is the co-founder of the Western Monarch Count and Bay Area liaison to the Xerces Society, a group formed by Robert Michael Pyle dedicated to the conservation and protection of threatened and endangered invertebrates and their habitats.

Liam: We are actually in the county (San Francisco) where the Monarch came to science. Can you tell us a little about that story?

Mia: Well, I learned that it's a story that they really hadn't been able to authenticate at first. What I had heard was the voyage of the Russian ship *Rurik* in 1816 had famously landed off our coast with several prominent naturalists aboard, including Adelbert von Chamisso and J. F. Eschscholtz, and it was then that the first scientific collection of a Monarch occurred. The story was finally authenticated when this collection once upon a time came to the California Academy of Sciences with all the type specimens, and there was the Monarch butterfly on display. The place was recorded as the Presidio, where they collected it.

Rob Hill, gouache, 11" x 17"

Liam: Mia, do you remember your first Monarch?

Mia: I was fortunate my grandmother lived in San Francisco. She would take the train down to San Carlos, and as we would wait for her, we would run up and down the tracks, which were full of milkweed. We'd gather Monarch caterpillars and rear them at home. I grew up loving the abundance and beauty of nature. We'd take occasional trips to Pacific Grove to see the Monarchs overwintering, so I grew up on a first-name basis with them.

Liam: Was there any seminal moment as an adult that made you step up to what you are doing now?

Mia: Sure. So, I moved at an early stage of my National Park Service career to Muir Woods National Monument. Two things happened. My first year there, the park celebrated its seventy-fifth year, and a photo contest was held, and the winning photo was a cluster of Monarchs on a pine branch in Muir Woods, and I'd never seen anything like that at the park. I knew of the famous Terwilliger overwintering site down the road at Muir Beach, but nothing within the woods.

And then the next seminal thing happened. The park received a Xerox (a copy) inquiring about overwintering sites and observations because the Xerces Society had received a World Wildlife Fund grant that brought aboard prominent scientists. Because I'd expressed interest in this photo contest, the chief ranger said, "Okay, Mia. Monarchs. Run with it." The team was composed of Walt Sakai, Chris Nagano, and John Lane. So I was kind of put in charge of Marin. Walt Sakai became my mentor and I traveled with him up and down the coast learning how to locate overwintering clusters and count them, how to understand the sites' conditions and habitat, and it soon became part of the cycle of my life.

I joined the Monarch Program in 1993 because at the time it was based in Encinitas in Southern California, and they were the go-to for Western Monarch information. I went to their little workshops and met people and thought we ought to make this more rigorous. I was talking to enthusiasts who loved going out already to see overwintering sites, so forms went out and that was the creation of the Western Monarch Count. (*Author note: this was right about when I joined up to cover the San Francisco County sites.*) We volunteers started to get more scientific in our approach. And then, as often happens, the Monarch Program rose and fell, and the Xerces Society said, "Yes, we will welcome the Monarchs back into our portfolio." For the longest time it was called the Thanksgiving Count with one visit per each site, but we learned there is so much more to learn with multiple site visits. Then it expanded to the Western Monarch Count with visits back to the sites in the New Year. And thanks to iNaturalist, we started to encourage people every time they saw a Monarch to document the sighting. This helped us understand where they dispersed. And then there are people that garden and have milkweed, which inspired the Western Monarch Milkweed Project that documents what's happening in your milkweed patch. For me, it was all about developing public awareness and connecting everyone to deeper ecological concepts. It's not just about planting milkweed; it's what is going on in your garden. I'll never forget, Liam, when the big issue for us was Monarch releases. And then we kinda graduated to the right plant at the right place and then getting folks to stay away from tropical milkweed (*Asclepias curassavica*). You and I have seen some big problems.

Liam: Yeah, like when they crashed so incredibly and then rebounded the next year. The World Wildlife Fund put out a survey many years ago and they were trying to find out: if money was no object in one's life, how much money would one donate to these particular creatures. I think the Panda Bear was first, the Polar Bear was second, and the Monarch was third.

Mia: Wow.

Liam: And it said in general a person (and again if money were no object) on average would donate $350,000 a year to save Monarchs. Fascinating little factoid. I'm just curious what you think. What do you think it is about this butterfly?

Mia: Well, I feel like Monarchs, first of all, are so beautiful, and they are a part of nature almost everyone recognizes. You hear that other study about how we know a zillion logos—Coca-Cola, Nike, yet most people can only recognize a few things in nature, and Monarchs are one of them.

Liam: Wow.

Mia: And so you feel good. It's not just a butterfly, it's a Monarch! And the other thing I've learned is almost everyone has a connection to Monarchs: they've reared them as a child; they saw them moving through their home in Kansas; they went to an overwintering site. Another study I've read shows that butterflies appeal to how we want the world to be: flying free, with flowers, beautiful . . .

Liam: Beauty.

Mia: . . . and of course Madison Avenue they put it on everything now. If you look, you'll see a Monarch on shampoo.

Liam: My favorite was in the span of an afternoon. I once saw them both on the cover of a health care magazine and later on a carton of cigarettes. The juxtaposition!

Mia: And you know, I feel like we need a symbol. In the last few years when there was so much angst in the world. Is it coincidental or not? While we were all fools distracted by the pandemic, the Monarch was declining too and about to blink out, so we needed something to focus our existential angst, and so people's pandemic projects became pollinator gardens, eating healthier for us and having a healthier world for the Monarchs. An interesting juxtaposition as well. As I always tell people, with Monarchs it's been a mystery for so long. Where did they go in Mexico? The mystery of metamorphosis. There are so many mysteries out there.

Liam: One of the mysteries for me has been my love-hate relationship with this creature. I think I've told you about this. I finally landed on just maybe it is the aposematic coloration (warning colors): I find it gaudy and off-putting somewhere in my primitive monkey brain—orange and black. It's a little out there, but I'm running with it these days. And as an artist, I seem to not be able to find anything new to bring to the Monarch, it's, for me, almost a cliché of beauty, inherently saccharine. But through the years of doing this work with you, I've definitely come around to it . . . and it's not its fault I find it gaudy in my mind. One of the more interesting manifestations of this dilemma was I entered a national art contest and . . . the only way I could paint a monarch was I literally found one on its last day of life. Greased out and wings tattered. [*All butterflies emanate an enzyme soon after emerging that literally is killing them on the wing. This is why they are in a daily race with time to mate and lay eggs. When you see one in this disheveled state with bird strikes and barely any wings left, that is what is going on, and the end is near for this adult.*] I had to deconstruct the Monarch in order to construct it at all. Again, this is my own journey with it. I think it gets a great deal of attention, and sometimes I feel that this is to the detriment of other species. It's . . . complicated.

Mia: Well, frankly, I love it. I love the complications. For me, of course, eucalyptus is enemy number one in my National Park Service world. And yet I've had to come up with a different feeling and a different way of representing how we throw a lifeline to another creature when we are developing a deeper understanding or a better response. And saying to my eucalyptus-loving friends, "Let's study what the Monarch needs here so that the rest can go."

Liam: I say in my talks, Mia, that not only am I interested in the butterflies themselves, but I'm equally enthralled by our relationship to them. Nothing presses more love buttons than a Monarch. Okay . . . maybe a panda. In a recent article you alluded to a "Monarch movement"—could you expand on that?

Mia: Well, what I found is that Monarchs have captured the attention and the urgent need to do something by everyone I know, and everywhere I turn . . . it's actually overwhelming. There's Monarch corridors, there's wayside gardens, there's Miles for Monarchs, there's welcome back Monarch festivals . . .

Liam: . . . don't forget Monarch Mamas.

Mia: It's overwhelming the amount of interest, and what is so encouraging to me, is so profoundly moving to me, is schools are now moving from a vegetable garden to a pollinator garden, and I can't help but think that moving toward something for another creature is a profound thing for a young person. And the fact that the Xerces Society and Monarch Joint Venture are working with tribal nations. Native American peoples are embracing the lands along the migratory corridors, and the symbolism that must have! And also working with PG&E and General Mills and working with agriculture and wildlife preserves on these vast scales is so encouraging to me and they are so deeply into it. "We've got to protect the overwintering sites," "We've got to do everything we can to restore the breeding habitat; it has to be a healthy world where there are no pesticides."

Liam: The scope!

Mia: The scope is amazing! And finding something that everybody at every scale can do, from a corporation to a library to a home garden. And stepping up—you were talking about if money were no object? It almost feels like money is no object right now, because there are the federal dollars, there's the state dollars, there's grants, yet there is still good oversight, there's still good messaging that you really need "right plant, right place."

Liam: Is there anything about all that you are involved with that gives you sleepless nights?

Mia: I haven't been able to effectively communicate anywhere, Liam, that we have to do better for our overwintering sites. People can love Monarchs, they can be proud that they have them on their property. It can even be the National Park Service. But there isn't an example out there of proactive implementation of protection enhancement, better practices.

Liam: Well, it seems like Pacific Grove in Monterey is attempting . . .

Mia: Attempted, right. And thank god for Stu Weiss who has given them a plan, as he has helped them implement it. Not only is the reward that the monarchs have returned, but they returned abundantly. Liam, I visited Pacific Grove so often last winter with one storm after another storm. There were so many places of refuge at Pacific Grove. If the storm comes from this direction, the Monarchs tuck here. If it came from the south, the Monarchs would be able to tuck over there. If the rains continued forever like it seemed last year, then they moved deep into the grove and the lichens. It's so amazing to see a recovering grove do its job. They have volunteers there who work every day going in and documenting. They are implementing Stu's plan despite renegades there who vilify the plan.

Liam: What was so exciting about that Treasure Island site in San Francisco was seeing so many Monarchs in that one area, before the trees they used were removed to make way for new buildings. Mia, looking back now, was there anything we could have done to save that site?

Mia: There is a hope that the Monarch will be listed as an endangered species under the U.S. Fish and Wildlife Service. If that had been in place when we made our observation on Treasure Island, that would have required mitigation there or maybe it would have been like the Gill Tract in Berkeley, where the university changed the apartment building plans so as not to impact the overwintering site there.

Liam: Really?

Mia: They are doing the Stu Weiss plan at Gill Tract. The apartment building is still there, but it's situated in such a way that it provides a buffer from the wind and the creek is day-lighted so there is a surface source of water, and so for those people who said no to development, well, they still got some, but it was done with Monarchs in mind.

Liam: That's wonderful. That's something I'd never heard before. Last question. If you could get people to do one thing, one thing as your legacy request, what would that be?

Mia: I guess I'd like to see some real model overwintering sites. I'd love to see the park services say we are going to do everything we can to bring this Fort Mason site back [*severely damaged by falling trees in a storm and random tree pruning, opening up the "bowl aspect" they need to overwinter into a wind tunnel*], that we are going to do everything we can to bring the Stinson, Muir Beach, and Bolinas sites back. That would be my dream partly because it's the flipside of my greatest source of frustration. I don't want to feel like a failure. I'm still full of energy. I feel these are dreams within our lifetime.

Liam: Thank you, dear friend.

Mia: Thank you, Liam.

A final note: There is a misconception that the butterfly was recently listed as endangered. The International Union for Conservation of Nature (IUCN) in July of 2022 made a change to its classification of the Monarch from declining to endangered. However, this is more symbolic than anything and does not have the legal teeth that a federal listing under the Endangered Species Act would. It's given a great deal of Monarch enthusiasts a false sense of hope. In the winter of 2024 people fighting for some protection for this butterfly received good news. The federal U.S. Fish and Wildlife announced they would propose that this creature should receive threatened status, one notch down from an endangered listing. On a follow-up call to Mia, she seemed satisfied that this listing would put the Monarch on the path for endangered listing in the future if anything becomes more cataclysmic with its numbers. It seemed like perfect timing as the Western Monarch Count for 2024 posted its second-lowest numbers for Monarchs observed, with around 9,000 seen in total during the count season. What will be its tipping point when we will not see Monarchs any longer in the West? And are we already there?

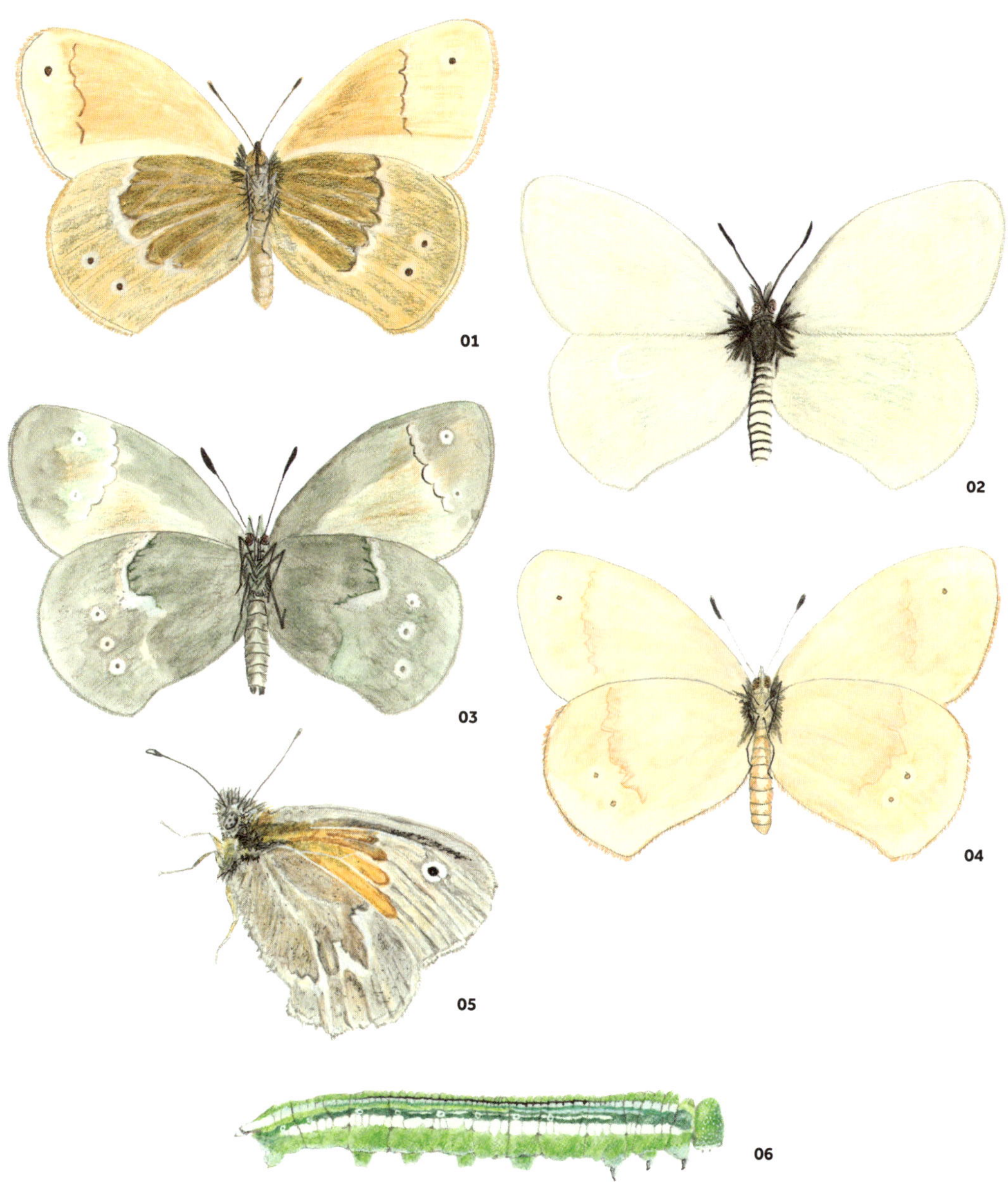
01
02
03
04
05
06

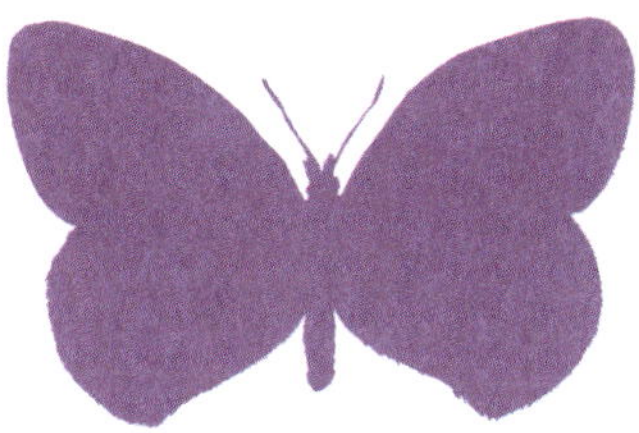

CALIFORNIA RINGLET

Mueller 1764

(Coenonympha california)

The Ringlet is one of our most common California butterflies. Its lazy, loping gait over grass means it is sometimes dismissed by the uninitiated as being a moth. The butterfly used to be divided up into many subspecies, but the lumpers ruled the day and now it's all considered *Coenonympha california*. It's one of my favorite butterflies to paint as its ventral side is a living watercolor unto itself.

Habitat: Open grasslands, edges of coastal scrub
Host Plants: Grasses (Poaceae), both native and non-native
Life Phases: Larvae are pale green streaked with brown. Pupa brown or green with nine stripes. The spring generation adults are mossy green—the color of newly emerged grass. This butterfly enters a reproductive pause during its summer-months phase (estivation), and can delay laying its eggs to early fall. The later adult flight is light brown to beige to ghost white—the color of hollow, dead grass. Overwinters as a larva. Bivoltine: January–November.
Counties They Fly In: All
Great Places to See Them: Ringlets can be found in most of our open spaces. Marin Headlands (Marin County); San Bruno Mountain (San Mateo County); Mount Diablo State Park (Contra Costa County). A team led by Jonathan Young of the Presidio Trust recently restored them to the Presidio. Prior to that, I was the last person to see them there in 2009.

01: ventral, summer
02: dorsal, spring
03: ventral, spring
04: ventral, fall
05: in situ
06: mature larva

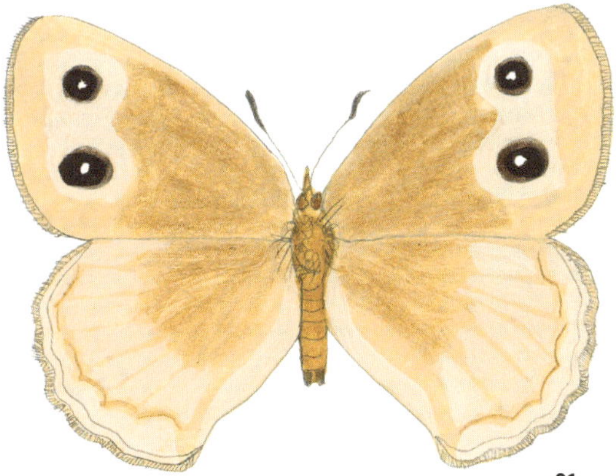

01

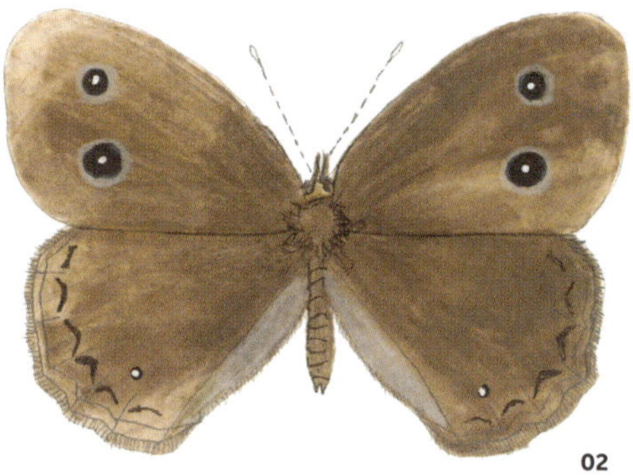

02

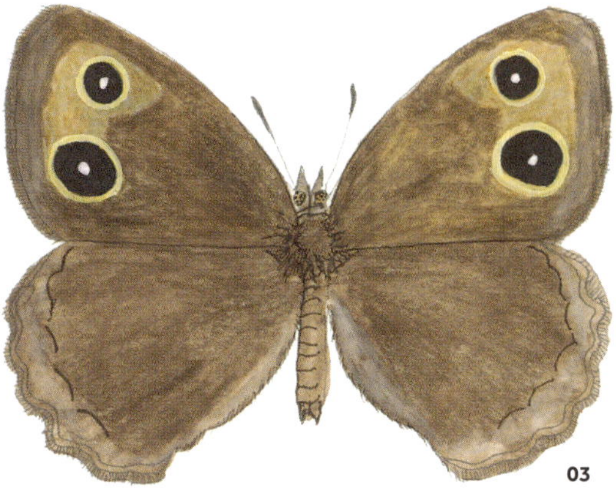

03

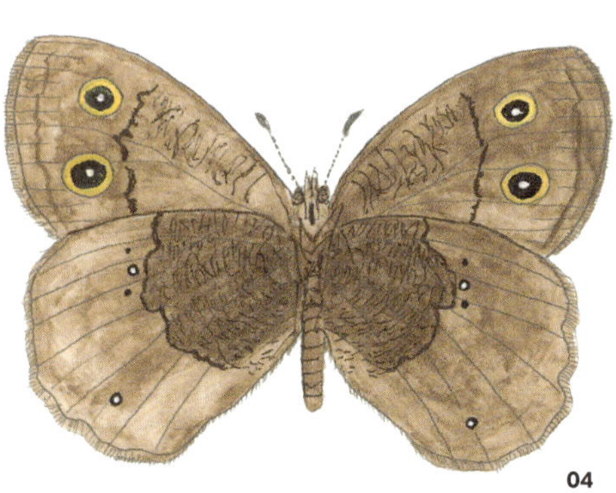

04

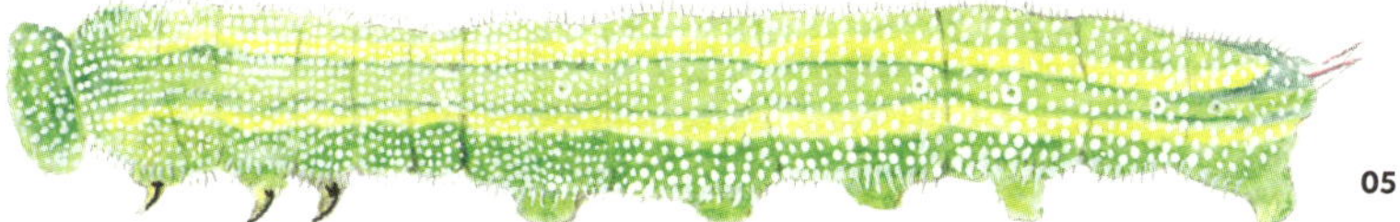

05

COMMON WOOD NYMPH

Behr 1864

Cercyonis pegala boopis
also known as the **LARGE WOOD NYMPH**
also known as the **OX-EYED SATYR**
also known as the **BLUE-EYED GRAYLING**

Like the California Ringlet, the Wood Nymph is also a denizen of the native grass community but is considerably larger than a Ringlet. The Wood Nymph also has a very pale form, "stephensi." Common Wood Nymphs are also bigger than others in the genus *Cercyonis*, the Great Basin Wood Nymph, and come out earlier in the season than Great Basins (there is, however, a slight overlap). When it gets really confusing is when they fly together (see **Appendix B, Tableau 7** for differences).

Habitat: Open grasslands, edges of forests
Host Plants: Grasses (Poaceae)
Life Phases: Larvae are green with four light yellow horizontal stripes and are known to be nocturnal feeders (Steiner.1980.149). Pupa are yellow and green. Univoltine: June–August.
Counties They Fly In: All but San Francisco
Great Places to See Them: Ring Mountain (Marin County); Jack London State Historic Park, Glen Ellen (Sonoma County); Sierra Azul Open Space Preserve (Santa Clara County); Mount Diablo State Park (all trails) (Contra Costa County)

01: dorsal ♀, form boopis
02: dorsal ♂
03: dorsal ♀
04: ventral ♂
05: mature larva

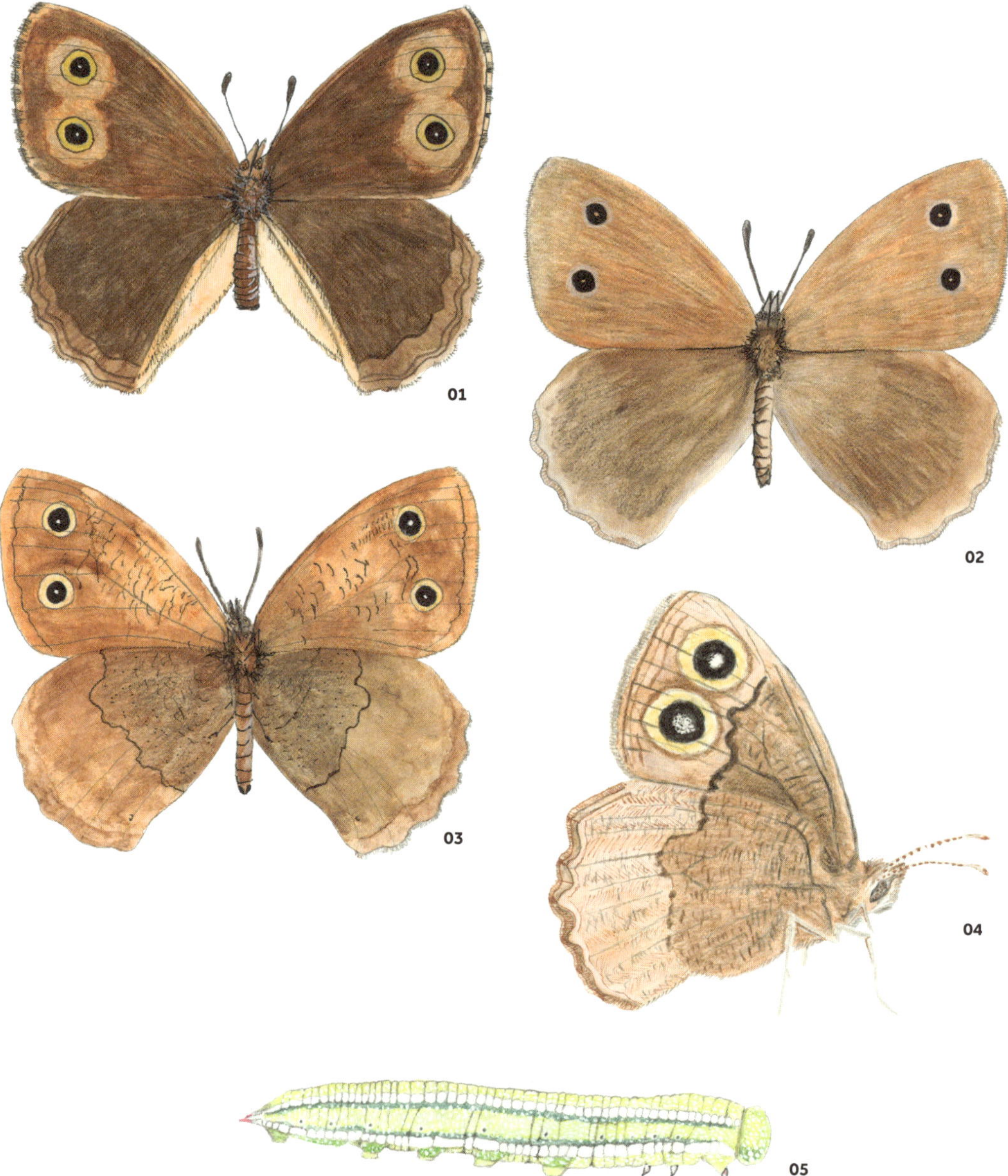

01
02
03
04
05

GREAT BASIN WOOD NYMPH

Edwards 1862

Cercyonis sthenele behri

Much smaller than the Common Wood Nymph, Great Basin's wings are smaller as well. It's not the easiest bug to identify, as they hop quickly away in the tall grass when approached. Plus, their wing patterns are so similar that I find the best way to tell them apart is behavior—the Common Wood Nymph is more abundant, and the Great Basin emerges later in the season. Instead of memorizing, just get out there and observe. The size difference will come to you only through repetitive viewing.

The Sthenele Satyr (*Cercyonis sthenele sthenele*) once flew in the county of San Francisco but is now extinct. Livestock grazing out the majority of native grasses is thought to have been the reason. This butterfly disappeared before the Xerces Blue's extinction.

Habitat: Coastal scrub, grasslands, and oak woodlands
Host Plants: Grasses (Poaceae)
Life Phases: Caterpillar light green with dark green horizontal stripes. Pupa the color of fresh grass. Adult males emerge before females (protandry) in May, then more abundant in July. Univoltine.
Counties They Fly In: All but San Francisco, San Mateo, and Solano
Great Places to See Them: Mount Diablo State Park (Contra Costa County); Pine Mountain Fire Road (Marin County); Sierra Azul Open Space Preserve (Santa Clara County)

01: dorsal ♀
02: dorsal ♂
03: ventral ♂
04: in situ
05: mature larva

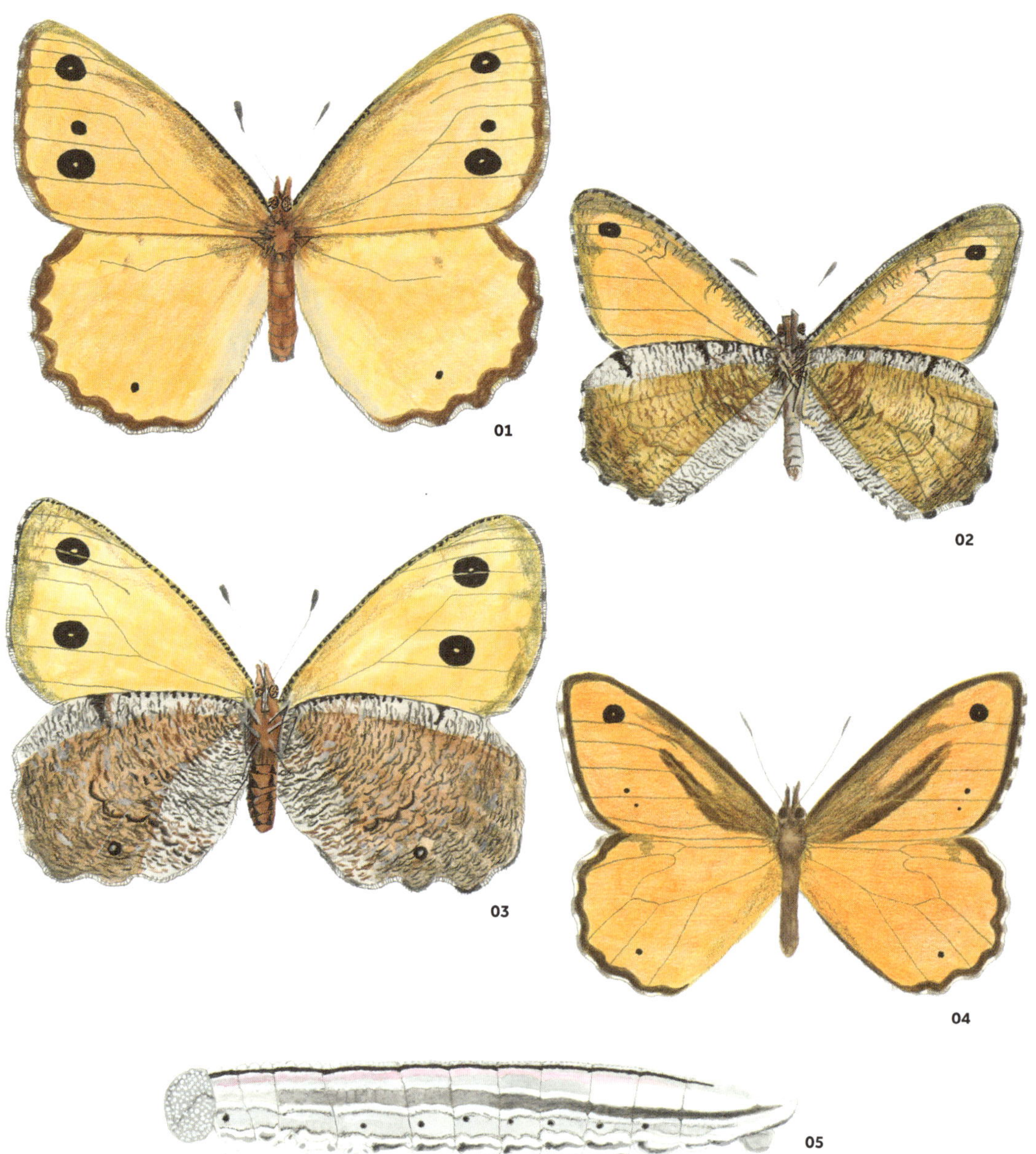
01
02
03
04
05

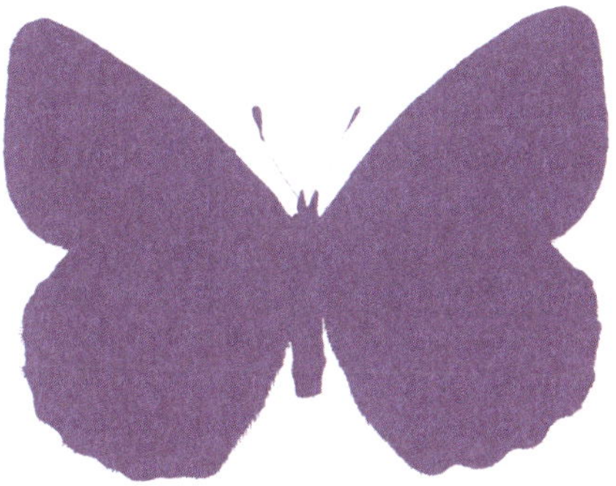

GREAT ARCTIC

W. H. Edwards 1874

Oeneis nevadensis

also known as **PACIFIC ARCTIC**

The largest member of this genus, the Great Arctic does a really cool thing it shares with all other Arctics: when they land on the ground, they fall over to one side. The markings ventrally are not too far off from the colors of dirt and pebbles, making for outstanding camouflage.

Habitat: Forest clearings and glades, mountain summits
Host Plants: The host plant is unknown but is probably grasses. (Scott.1986.247)
Life Phases: The larva comes in an array of colors as it molts and is biennial, with two overwinters before pupation. Adults are known to land on logs. Univoltine: spring and summer.
Counties They Fly In: Mendocino and Sonoma
Great Places to See Them: The town of Plantation (Sonoma County) is one possibility. A quick iNaturalist glance reveals no sightings within the scope of this book. It certainly doesn't mean they aren't there and probably has more to do with their every-other-year emergence. No one has happened upon them. An expedition to Plantation by any of you reading this might put one on the map for all of us.

01: dorsal ♀
02: ventral ♂
03: ventral ♀
04: dorsal ♂
05: mature larva

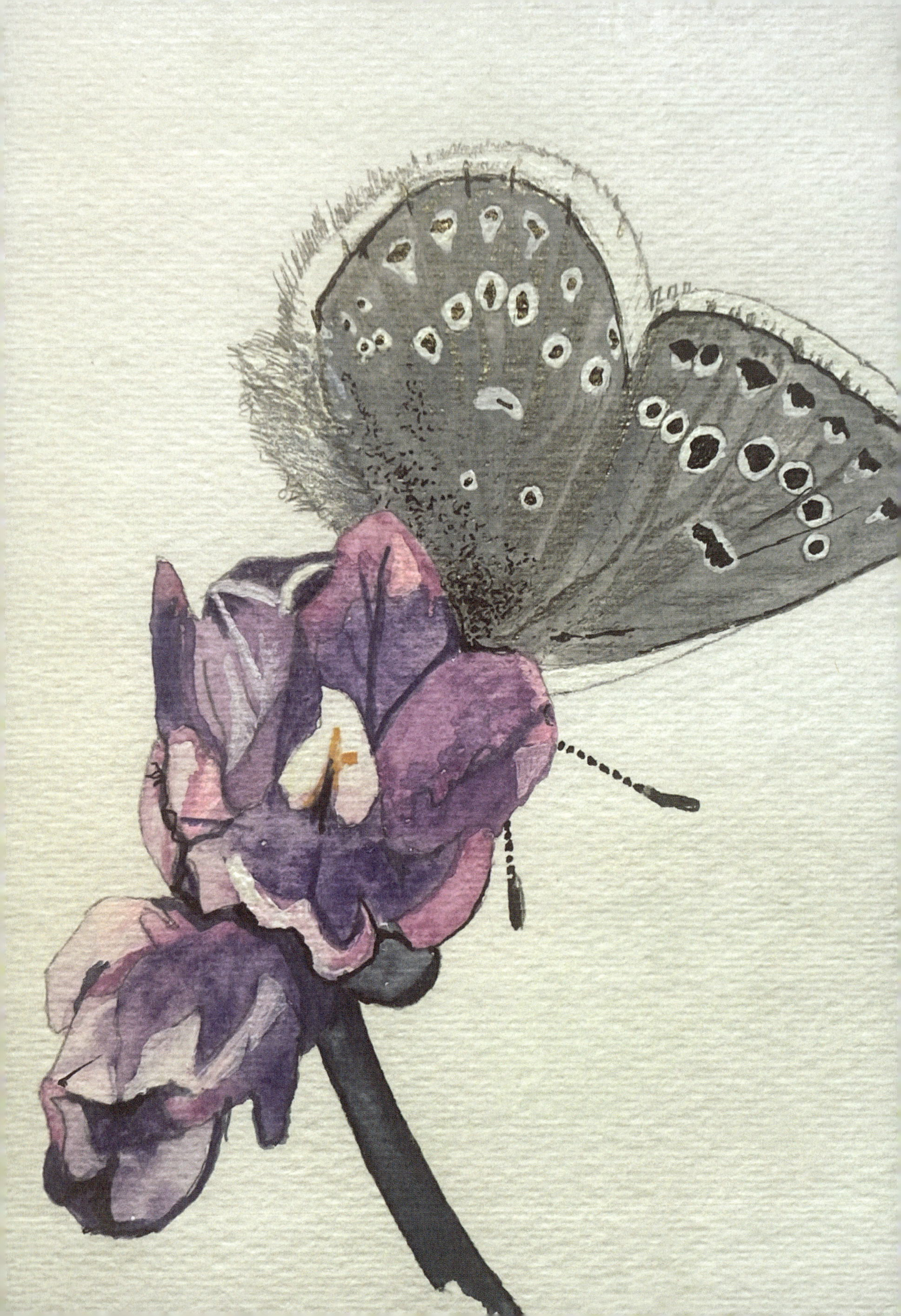

The Gossamer-Winged: Blues, Coppers, and Hairstreaks (*Lycaenidae*)

I have to fight calling this group "little jewels," as it objectifies butterflies, and butterflies are living creatures, not things. Suffice to say, they are among some of the most beautiful butterflies and some of the smallest. There are thousands of species in all regions of the globe, all about the size of a nickel or dime. Both sides of the hind wings usually hold some sort of iridescent coloration, and many undersides are an elaborate pattern of "false heads" (like the swallowtails) evolved to trick the predator just long enough for a quick exit.

The blues are usually found in grasslands, dry alluvial fields, and dunes. Many of them have a relationship with ants, often tended by them, and the ants received a packet of honeydew for keeping predators off the host plant. This type of relationship is called mutualism. Even in vacant lots in a city, blues can be found. Females are uniformly brown, no doubt to blend into the environment for the protection of the eggs. The males generally are the color of the sky on a sunny day. Nineteen species are discussed on the following pages.

We only have five coppers in our Greater Bay Area, making their sightings rare and thrilling. Hands down the best place to see a few of them in a day is Pinnacles National Park (San Benito County) in June, where they seem to thrive in this xeric plant community. One of my favorites to paint is the Tailed Copper (*Tharsalea arota*).

Hairstreaks usually have a long, broken line that transects both the undersides of the fore and hind wings, like a "streak of hair" (get it?). The zenith of their complex patterns is reached in the tropical Americas. We've got some stunners here as well.

Mission Blue on Blue Dicks, watercolor pencil, 5" x 8"

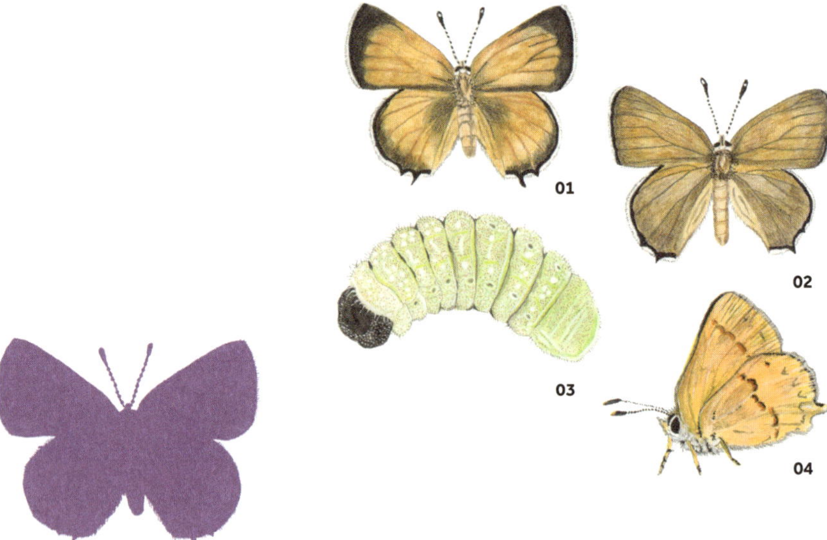

GOLDEN HAIRSTREAK

Field 1938

Habrodais grunus

A very strange butterfly. It emerges, flies for a bit, goes to sleep as an adult, then re-emerges to mate. They seldom visit flowers but have been recorded at mud. They receive their nutrients at the larval stage. They hover around their oak tree hosts and are quite approachable on the lower branches for photography.

Habitat: Oak forests
Host Plants: Canyon oak or canyon live oak (*Quercus chrysolepis*), tan oak (*Notholithocarpus densiflorus*), chinquapin (*Chrysolepis chrysophylla*)
Life Phases: Eggs laid on oak twigs (Garth & Tilden.1986.120). Pupa is pale greenish blue with hairs and brown splotches. Many adults can be found on the host, and they seem to have a crepuscular lifestyle. Adults brown below and golden above. Univoltine: July–September.
Counties They Fly In: All but San Francisco and Solano
Great Places to See Them: Sierra Azul Open Space Preserve, Mount Umunhum (Santa Clara County); Foothills Nature Preserve (Santa Clara County); Martin Road near Bonny Doon (Santa Cruz County); Huckleberry Botanic Regional Preserve (Contra Costa County). At Huckleberry take the trail to the right in the parking lot, go to plant marker 11, then make a left and go to where the trail dead-ends at a live oak. Throw a couple rocks at tree branches to scare up the butterflies. (This has been the historic site to see them on the Berkeley Butterfly Count.)

01: dorsal ♀
02: dorsal ♂
03: mature larva
04: ventral, in situ

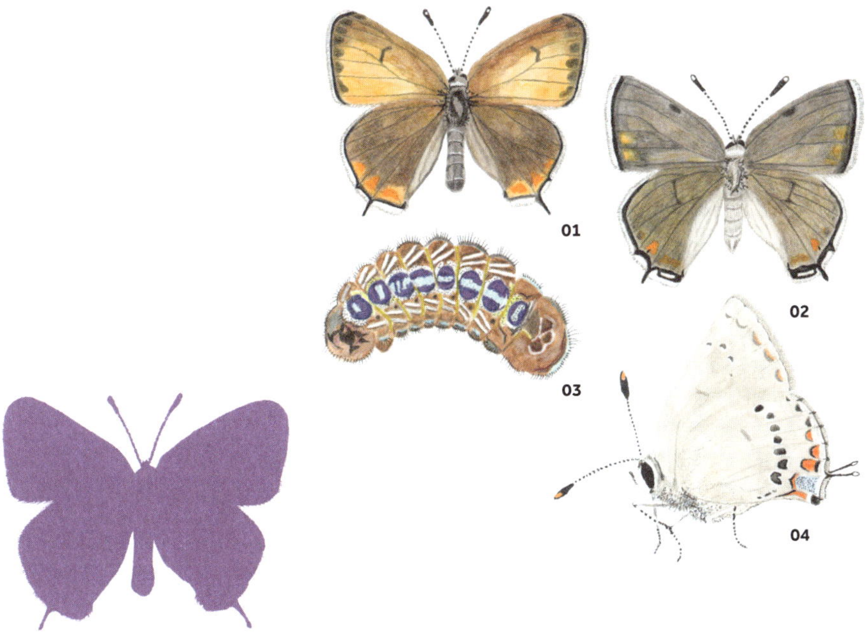

CALIFORNIA HAIRSTREAK

W. H. Edwards 1862

Satyrium californica

01: dorsal ♀
02: dorsal ♂
03: mature larva
04: ventral, in situ

Easily confused with the Sylvan Hairstreak, the California Hairstreak likes to perch on the host tree and can sometimes be seen in significant numbers. Males also patrol oak canopies looking for mates. This butterfly is a moderate hill-topper.

Habitat: Oak woodlands
Host Plants: Oaks (*Quercus* spp.), California lilac (*Ceanothus* spp.), mountain mahogany (*Cercocarpus betuloides*), Western chokecherry (*Prunus virginiana*)
Life Phases: Overwinter as eggs. Larvae eat leaves. Pupa is reddish brown with black spots. Adult butterflies have a darker grayish brown overall field to ventral underside and a single row of orange spots that extend from hind wing to forewing. The similar-looking Sylvan Hairstreak (*Satyrium sylvinus*) has a light gray field and has a single orange spot adjoining a blue spot ventrally near the tail (see **Appendix B, Tableau 8**). Univoltine: April–July.
Counties They Fly In: All except San Francisco
Great Places to See Them: Pinnacles National Park (San Benito County)—check the buckeye tree blossoms around the campground. Mitchell Canyon Trail, Mount Diablo State Park (Contra Costa County); Sunol Regional Wilderness (Alameda County).

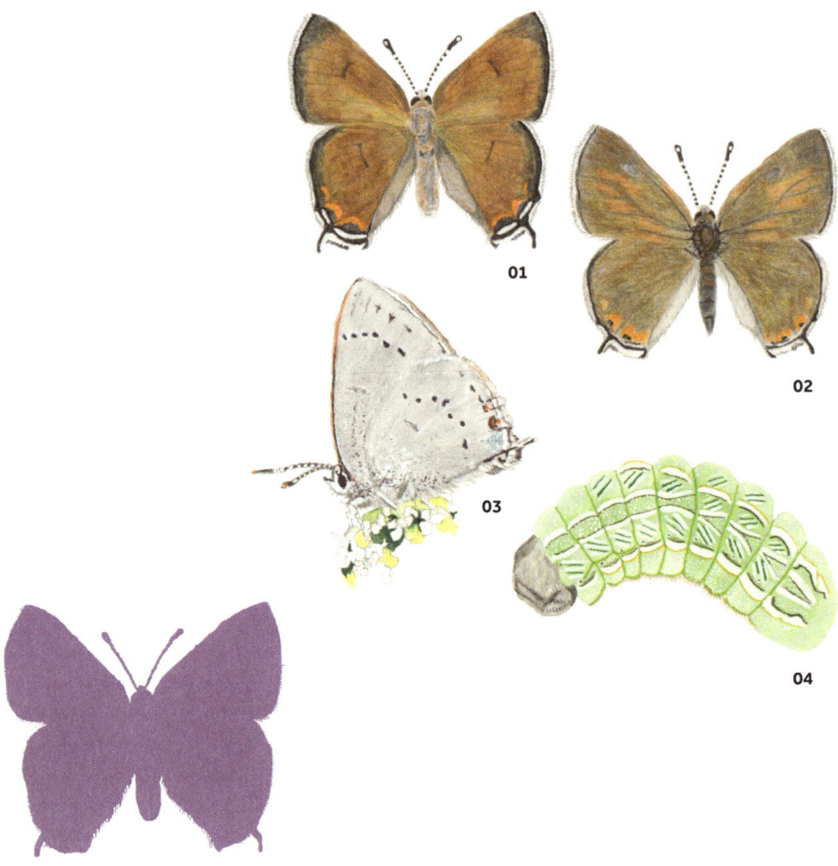

SYLVAN HAIRSTREAK

Satyrium sylvinus

Boisduval 1852

Easily confused with the California Hairstreak (see **Appendix B, Tableau 8**), it's only recently that the Dryope Hairstreak (*Satyrium dryope*) has been split off from the Sylvan Hairstreak and given full species status; one has a tail and one doesn't: Sylvan tailed, Dryope tailless. The butterfly presents various sizes of tails throughout its range.

Habitat: Riparian woodland, streamsides
Host Plants: Willows (*Salix* spp.)
Life Phases: Overwinter as eggs. The larvae are cryptically colored the same color as willow leaves. Pupa has both light and dark green mottling. In adults: both males and females are chestnut tan dorsally. Univoltine: May–June.
Counties They Fly In: Solano, Napa, and Sonoma
Great Places to See Them: Check waterways along Suisun Bay (Solano County) and Mix Canyon (Solano County)

01: dorsal ♀
02: dorsal ♂
03: ventral, in situ
04: mature larva

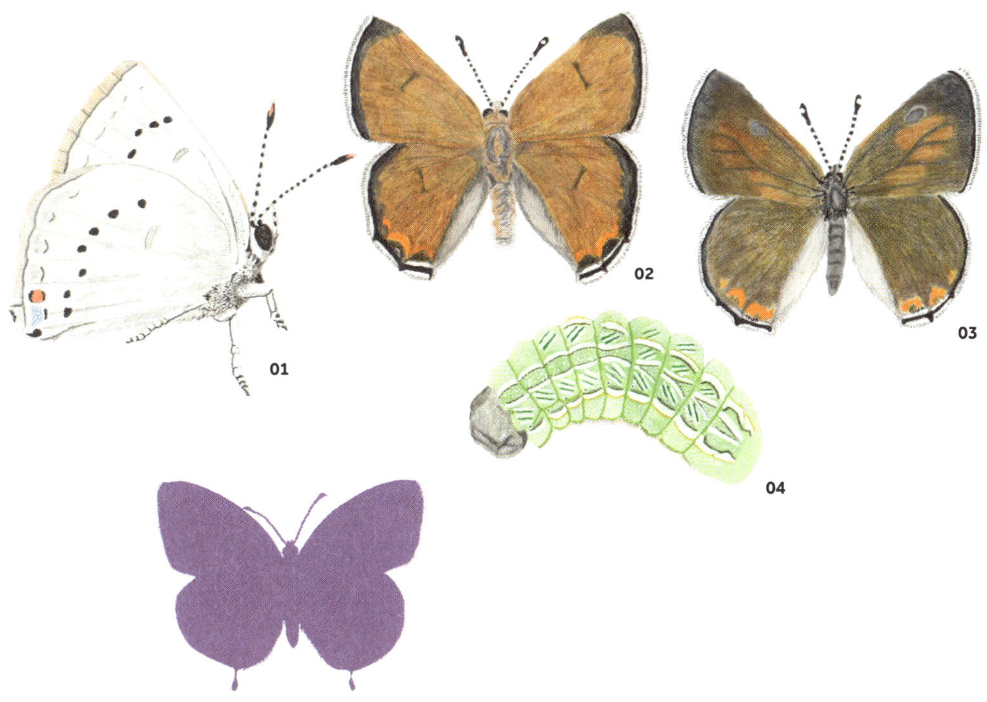

01: ventral, in situ
02: dorsal ♀
03: dorsal ♂
04: mature larva

DRYOPE HAIRSTREAK

W. H. Edwards 1870

Satyrium dryope

Notice this one has no tail. iNaturalist treats Dryope as a subspecies of Sylvan (*Satyrium sylvinus dryope*). It's also been treated as a form of Sylvan in a few books. Usually found within willow thickets along streams, it seems to be highly attracted to milkweed as a nectar source. Paul Johnson recommended a good rule-of-thumb: Dryope rests on bushes at about shoulder height, which distinguishes it from other hairstreaks.

Habitat: Riparian woodland, creeks and streams lined with willow
Host Plants: Willows (*Salix* spp.)
Life Phases: See entry for Sylvan Hairstreak. The adults are tailless hairstreaks with a lighter, more pale underside than in the Sylvan. Its spots ventrally are not as dark and defined as the Sylvan's (Shapiro & Manolis.2007.131). Univoltine: May–July.
Counties They Fly In: All except San Francisco
Great Places to See Them: Vargas Plateau Regional Park (Santa Clara County); Pinnacles National Park (San Benito County); Mitchell Canyon Trailhead (Contra Costa County) parking lot—check the willows that line the overflow lot; Sunol Regional Wilderness (Alameda County)—the creek behind the visitor's center.

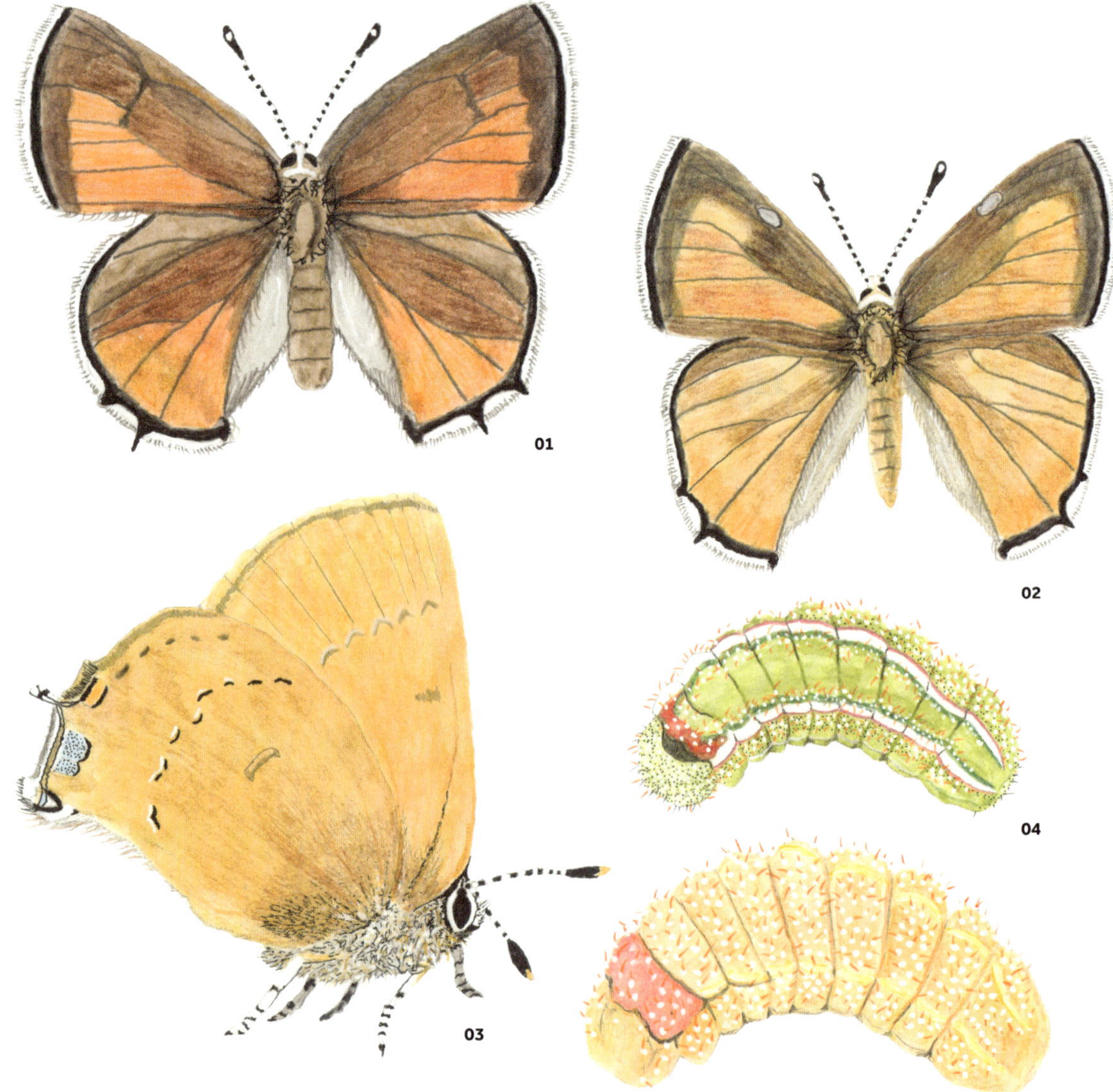
01
02
03
04
05

GOLD-HUNTER'S HAIRSTREAK

Boisduval 1852

Satyrium auretorum

Boisduval named this butterfly after his friend Pierre Lorquin who collected it while "gold hunting" in California during the rush in the early 1850s. Of the 132 species covered in this book, Boisduval named 27 of them first to science—an astonishing record. Lorquin was, it's safe to assume, a busy guy with his net. This one can give you a headache because it's not seen very often and looks a great deal like the blond form of the Hedgerow Hairstreak (*Satyrium saepium*). It took me many seasons just to find it. Surprised Lorquin found it at all.

Habitat: Chaparral, oak woodlands
Host Plants: Interior live oak (*Quercus wislizeni*) and an array of other *Quercus* species
Life Phases: Larvae pupate on oak trees. Pupa tan with pinkish hues. This butterfly is cyclical, having explosive years and years that are hard to find it on the wing (Howe.1975.281). Adults come to buckeye blossoms regularly. Univoltine: May–July.
Counties They Fly In: Contra Costa, Monterey, Napa, San Benito, Santa Clara, and Solano
Great Places to See Them: Near the intersection of Monticello Road and Capell Valley Road in a camping park (Napa County); Laguna Mountain Recreational Area (San Benito County); Coalinga Road, Portola Valley east of Jasper Ridge (San Mateo County); Hastings Natural History Reservation, Carmel Valley (Monterey County). You need an invitation to get into this last one or, better yet, sign up for the annual Hastings Butterfly Count and you get past the gate —look for them on coffeeberry (*Frangula californica*) at the top of "the Arnold."

01: dorsal ♀
02: dorsal ♂
03: ventral, in situ
04: mature larva, type one
05: mature larva, type two

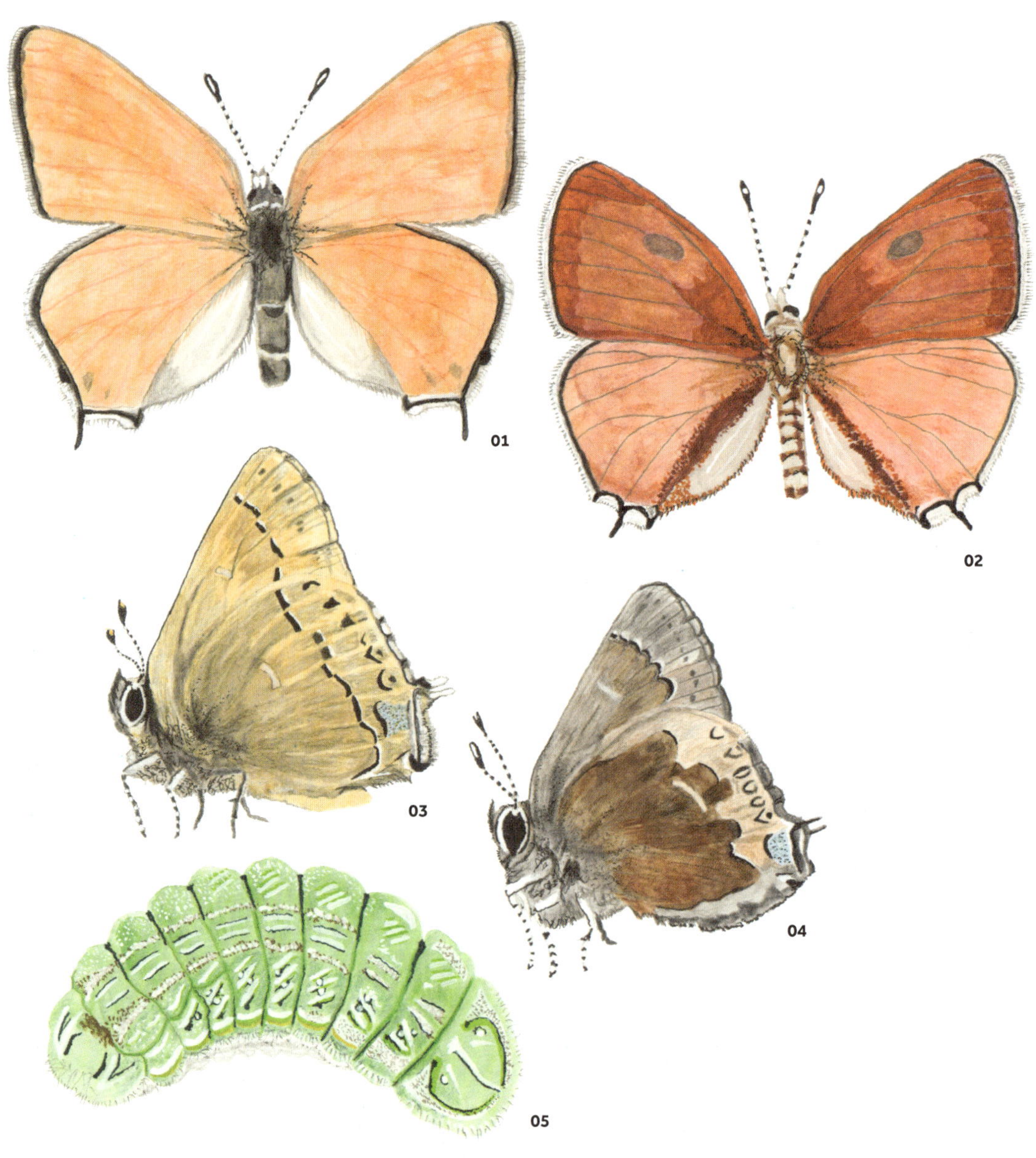
01
02
03
04
05

HEDGEROW HAIRSTREAK

Boisduval 1852

Satyrium saepium

I saw the lighter form of this butterfly at the 2023 Pinnacles Butterfly Count, and it really threw me. I wasn't sure what I'd seen. Paul Johnson, the count leader, later helped me identify it. Though I've never seen them fly together, keep in mind that two other nickel-sized brown butterflies are sympatric on San Bruno Mountain: the San Bruno Elfin should be done with its flight by the time the Hedgerow comes out, but the Western Brown will most assuredly be on the wing. The Hedgerow has a prominent blue spot from below, near its small tail. The Western Brown does not. It is a moderate hill-topper on hills and ridges.

Habitat: Oak woodlands, chaparral, coastal scrub
Host Plants: California lilac (*Ceanothus* spp.), mountain mahogany (*Cercocarpus betuloides*)
Life Phases: Overwinter as eggs. Pupa has black mottling covering a brown field. This is another perching butterfly that darts out to any possible mate. Adults come in a dark brown or blond form. Univoltine: May–September.
Counties They Fly In: All except San Francisco (historic records have it listed there, but not as of recent)
Great Places to See Them: San Bruno Mountain (San Mateo County); Sierra Azul Open Space Preserve, Mount Umunhum (Santa Clara County); Mount Diablo State Park (Contra Costa County); Pinnacles National Park (San Benito County)

01: dorsal ♀
02: dorsal ♂
03: ventral, in situ, variation one
04: ventral, in situ, variation two
05: mature larva

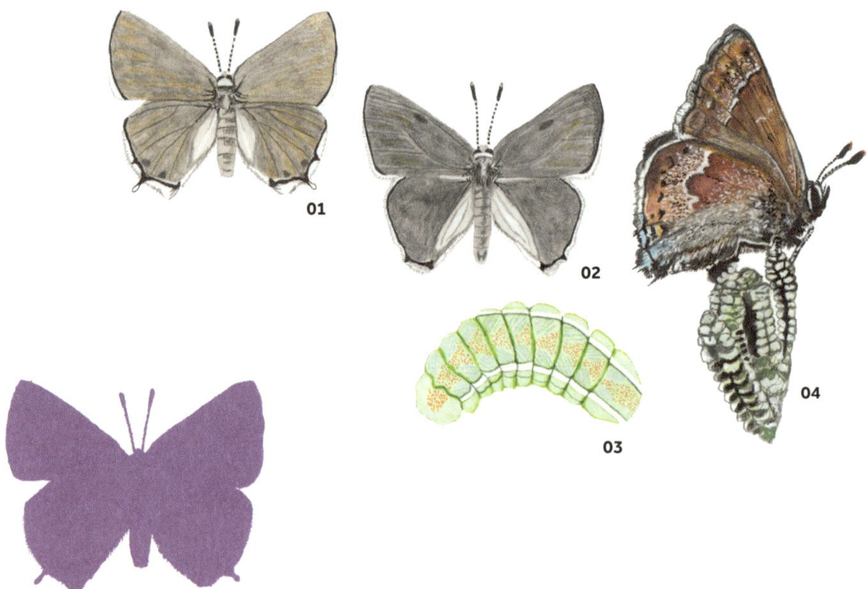

MOUNTAIN MAHOGANY HAIRSTREAK **W. H. Edwards 1870**

Satyrium tetra

Most hairstreaks are tiny triangles the size of a dime. *Satyrium tetra* sets itself apart by being a rather large isosceles triangle the size of a nickel, with the apex of the forewing arched high above its head. Even so, it can be hard to key away from the Hedgerow Hairstreak (*S. saepium*) when they are on the wing together. A moderate hill-topper like the Hedgerow, the Mountain Mahogany can readily be seen at buckeye tree blossoms, and I have observed one nectaring on yerba santa (*Eriodictyon californicum*).

Habitat: Riparian and oak woodland, chaparral
Host Plant: Mountain mahogany (*Cercocarpus betuloides*)
Life Phases: Overwinter as eggs. A mottling of black covers a tan pupa. Male adults perch all day in nearby bushes to await females (Scott.1986.365). Univoltine: June–July.
Counties They Fly In: All except Marin, San Francisco, and Santa Cruz
Great Places to See Them: Mitchell Canyon Trail, Mount Diablo State Park (Contra Costa County); Sierra Azul Open Space Preserve—along the trail leading to the summit of Mount Umunhum (Santa Clara County); Pinnacles National Park (San Benito County); Hastings Natural History Reservation (Monterey County). (I've personally seen it twice there, but it might actually be hard to find there.) If you sign up for the Hastings Butterfly Count, you get beyond the gate; otherwise one needs an invitation to enter Hastings.

01: dorsal ♀
02: dorsal ♂
03: mature larva
04: ventral, in situ

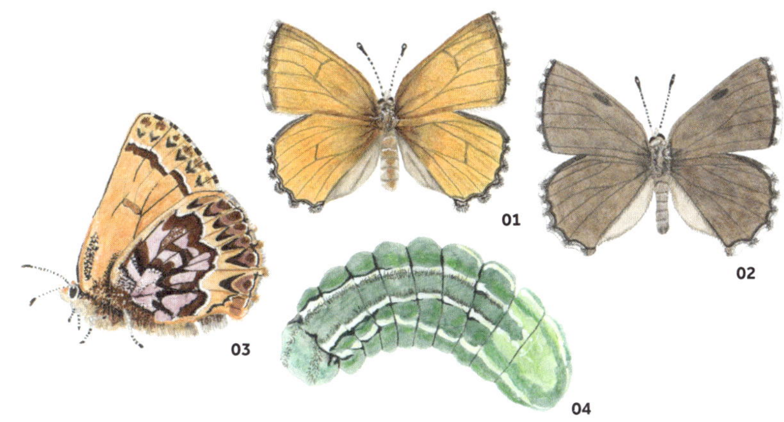

WESTERN PINE ELFIN

Boisduval 1852

Callophrys eryphon

The Western Pine Elfin is always difficult to see in the Greater Bay Area, for it is up against many places now removing their non-native pine trees. Once I observed a stray into San Francisco from Kirby Cove in Marin. It had flown due south across the mouth of the Golden Gate. One of the more difficult butterflies to paint, it reminded me of a Navajo rug. They are quite cooperative for photos.

Habitat: Pine forests, pine within chaparral
Host Plants: Pines (*Pinus* spp.)
Life Phases: Larvae eat the softest parts of the needles (Steiner.1980.93). Pupa black spots on a brown field. This butterfly as an adult cannot be mistaken for any other hairstreak with its spectacular underside of mauve and brown chevrons and black arrowheads. Overwinter as pupae. Univoltine: March–June.
Counties They Fly In: Contra Costa, Marin, Monterey, and Santa Clara
Great Places to See Them: Kirby Cove, Marin Headlands (Marin County); Point Reyes National Seashore (Marin County); Valenzuela Road, Carmel (Monterey County); Miller/Knox Regional Shoreline (Contra Costa County)

01: dorsal ♀
02: dorsal ♂
03: ventral, in situ
04: mature larva

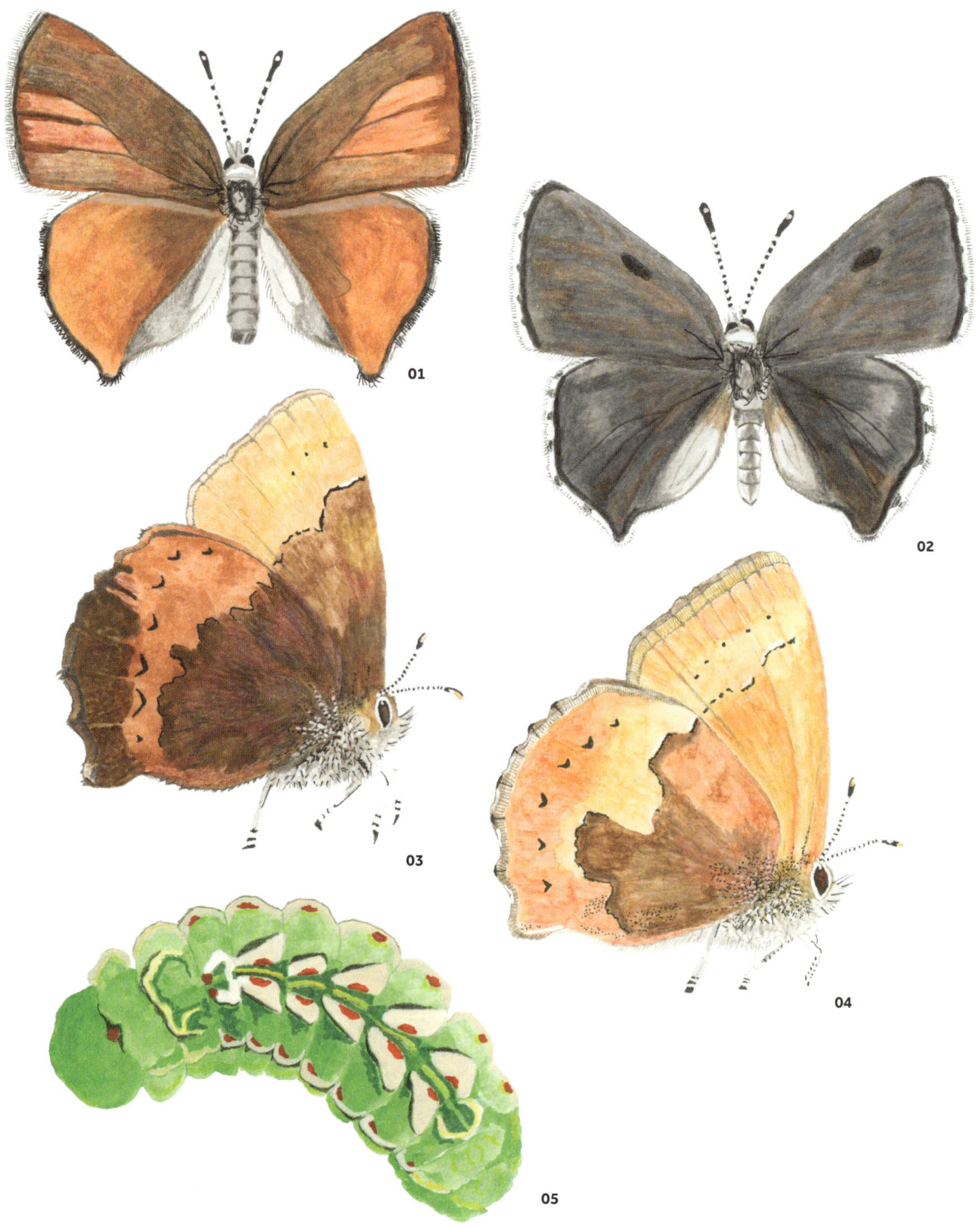
01
02
03
04
05

WESTERN BROWN ELFIN

Boisduval 1852

Callophrys augustinus iriodes

Easy to identify, the Western Brown Elfin comes in a light and dark form. Darker on the basal and outer edges. Males perch shoulder height along a trail. Deep brown shade (and just a hint of violet at certain angles in the sun) give it a smoky appearance. Think of elfins as tailless hairstreaks.

Habitat: Coastal scrub, oak woodlands
Host Plants: This species is notable in the wide variety of unrelated host plants it feeds on. California lilac (*Ceanothus* spp.), dodder (*Cuscuta* spp.), madrone (*Arbutus menziesii*), buckbrush (*Ceanothus cuneatus*).
Life Phases: Five molts produce five instars before pupation. Unique within the genus, the larvae are not limited to a single food plant (Dornfeld.1980.91). Pupa is light brown with black spots. Bivoltine: April–June and again July–September.
Counties They Fly In: All except San Francisco
Great Places to See Them: Any good coastal scrub trail. San Bruno Mountain (San Mateo County). Careful there—when the San Bruno Elfin is in flight there; the two can easily be confused. Marin Headlands (Marin County). Can appear in many open spaces. Widespread distribution. Almost never abundant.

01: dorsal ♀
02: dorsal ♂
03: ventral, in situ, variation one
04: ventral, in situ, variation two
05: mature larva

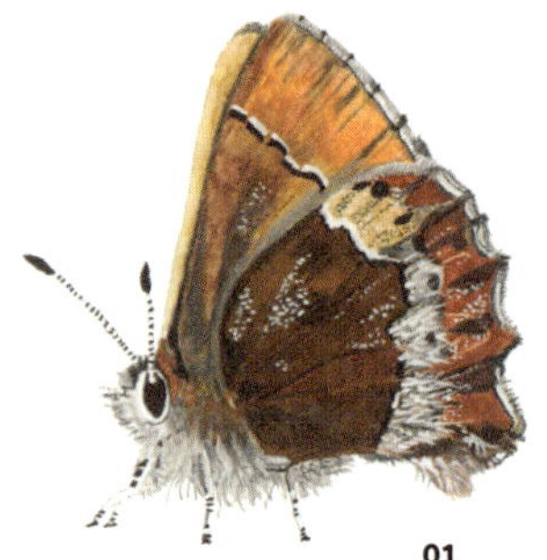

SAN BRUNO ELFIN

R. M. Brown 1969

Callophrys mossii bayensis

STATUS: ENDANGERED

Remember folks: When you are looking at elfins and hairstreaks, they are lateral baskers with the underside primarily exposed. One rarely gets to see their dorsal side—glimpses of it can be seen when the creature flies.

I've learned through personal conversations with the folks at Creekside Science who monitored this species till recently, this endangered butterfly (federally listed in June of 1976) has maintained a level population through the years of surveying (Kobernus.2000.pers comm.).

Habitat: Rocky outcrops, fog-shrouded summits
Host Plant: Stonecrop (*Sedum spathulifolium*)
Life Phases: See entry for **Moss Elfin**. The frosting on the underside hind wing, heavier in this species more than others, disappears after a few days on the wing. Univoltine: February–early April.
Counties They Fly In: San Mateo
Great Places to See Them: San Bruno Mountain used to be the only place in the world to see this special butterfly, but recently discovered Moss Elfins (*C. mossii*) on Montara Mountain in the Golden Gate National Recreation Area (San Mateo County as well) have now been given this moniker too. (Shouldn't they be Montara Mountain Elfins?) On San Bruno, look for carpets of stonecrops on the ground up there among the radio towers. Males perch high on surrounding bushes, viewing the females on the host below.

01: ventral, in situ

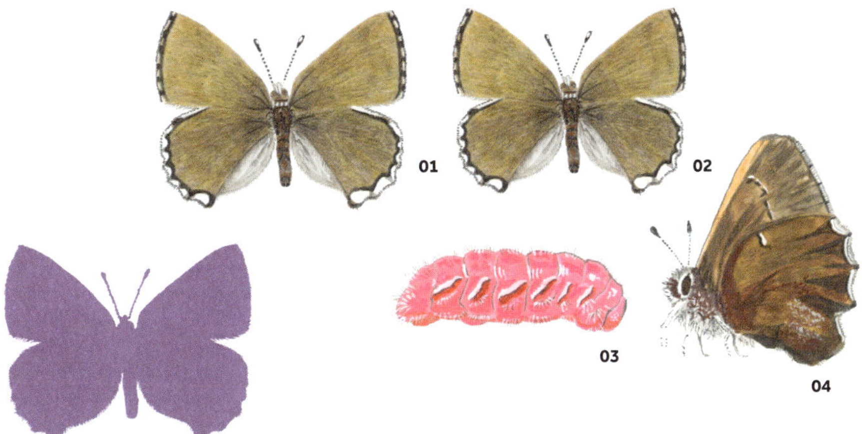

MOSS ELFIN

Hy. Edwards 1881

Callophrys mossii
including other subspecies *Callophrys mossii marinensis*, *Callophrys mossii doudoroffi*, and an unnamed Moss Elfin ssp. from Mount Diablo State Park (Contra Costa County)

You get all of that splitting? How about we just back this car up and call them all Moss Elfins (*Callophrys mossii*) and call it a day. This is one of our earliest-flying butterflies each season.

Habitat: Rocky outcrops (with the Marin Elfin being in redwoods)
Host Plant: Stonecrop (*Sedum spathulifolium*)
Life Phases: Being Pepto-Bismol pink on the yellow flowers, the larvae are quite conspicuous. (Why would they want to stick out like that?) Pupa chocolate brown with reddish lines. The white line on the underside of the adults is quite noticeably dividing the red brown on the outer border from the frosting toward the thorax. Adult males perch high over the low-growing stonecrop to await mates. Univoltine: March–April.
Counties They Fly In: *C. m. marinensis*: Marin; unnamed subspecies: Contra Costa; *C. m. doudoroffi*: Monterey County.
Great Places to See Them: *for C. m. marinensis*—Leo T. Cronin Fish Viewing Area off Sir Francis Drake Boulevard just beyond the town of Lagunitas, (Marin County). Park in the parking lot off Shaffer Fire Road, walk back out onto the boulevard and cross the bridge (not the road) to the other side of the creek, bringing you to Peters Dam Road. One should start to see the Elfin's host plant, stonecrop, on the cliff to the left. *For C. m. doudoroffi*—Landels-Hill Big Creek Reserve along Highway 1, Big Sur (Monterey County), where entrance is by invitation only.

01: dorsal ♀
02: dorsal ♂
03: mature larva
04: ventral ♀ and ♂, Marin Moss Elfin, in situ

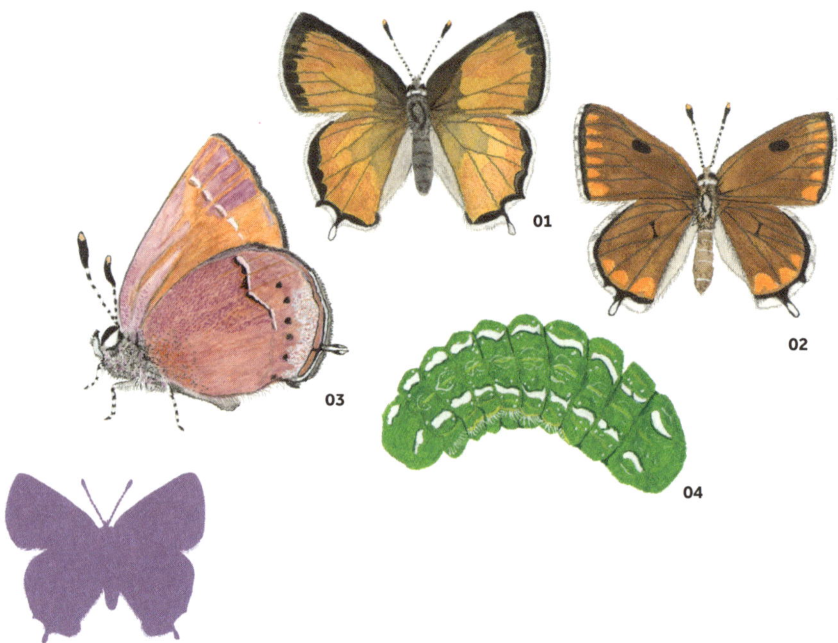

NELSON'S HAIRSTREAK

Callophrys nelsoni

Huber 1819

This and the next butterfly, Muir's Hairstreak (*C. muiri*), were both split off of the Juniper Hairstreak Complex and given full species status. But some splitters aren't convinced we've gone far enough. Robert Michael Pyle writes, "The butterfly known as Nelson's probably includes other separate but poorly distinguished species—all with host plants other than incense cedar" (Pyle.1981.436). What we have in the Greater Bay Area is Muir's Hairstreak. There are some differences. Nelson's ventrally is more of a plum, lighter purple whereas Muir's is a deep amethyst. This confusion is limited to Mount Diablo, as far as I know, where they look similar yet there is no incense cedar. This taxonomy is in need of much work.

Habitat: Forests where its host tree grows
Host Plant: Incense cedar (*Calocedrus decurrens*)
Life Phases: Eggs green. Pupa dark brown. Adults have a purplish sheen to them from below with a weak post median white line on an overall brown field. Males perch on host plant boughs in search of females. Univoltine: April–May.
Counties They Fly In: Contra Costa, Monterey, Napa, and Sonoma
Great Places to See Them: Difficult to answer. Some form of this butterfly flies on Mount Diablo (Contra Costa County).

01: dorsal ♀
02: dorsal ♂
03: ventral, in situ
04: mature larva

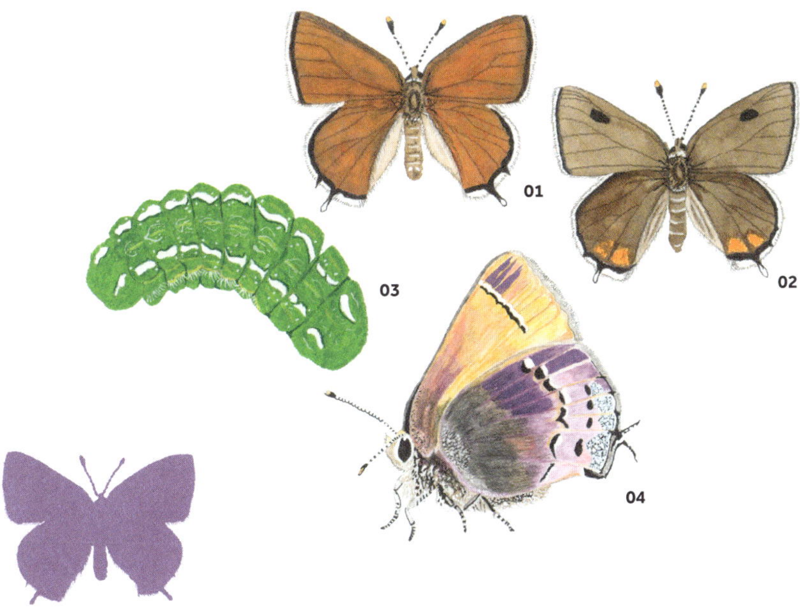

MUIR'S HAIRSTREAK

Edwards 1881

Callophrys muiri

The taxonomy around Muir's and Nelson's remains confusing, despite recent efforts to split the species according to affiliations with different host plants. Stay open to the probability that someone far smarter that myself will figure all of this out somewhere down the road. Though both purple, Muir's exhibits a deep eggplant shade of this color, whereas Nelson's is more of a lighter chestnut brown or grape-soda violet.

Habitat: Serpentine, where the host tree grows
Host Plant: Sargent cypress (*Hesperocyparis sargentii*)
Life Phases: Larvae look like cypress blades. Pupa dark brown. Adults are deep purple from below. Univoltine: March–June.
Counties They Fly In: Alameda, Contra Costa, Marin, Monterey, and Napa
Great Places to See Them: San Geronimo Ridge Road, San Geronimo, (Marin County). Pine Crestline Trail—across from Azalea Hill is the trailhead (Marin County). Walk out this serpentine trail (Pine Crestline) till you find the white *Ceanothus* bush in bloom (early May). You may actually have it on any of the last of the serpentine flowers along the trail. When this butterfly is at the height of its emergence, there can be many on a single bush. Look for the Sargent cypress farther out on the trail. I've also had this butterfly on Mount Saint Helena (Napa County).

01: dorsal ♀
02: dorsal ♂
03: mature larva
04: ventral, in situ

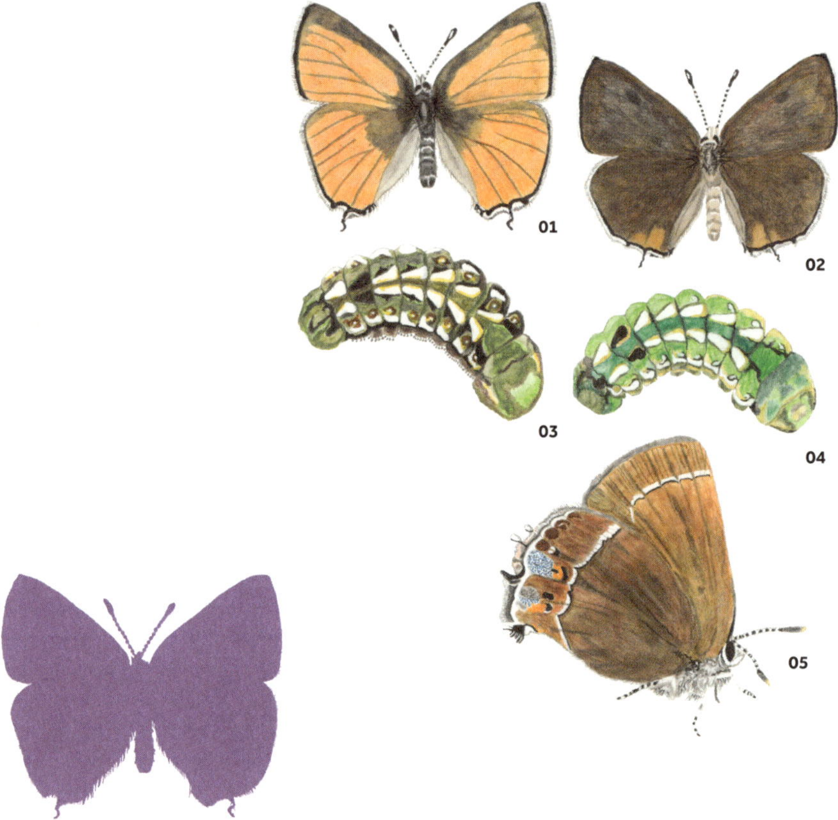

JOHNSON'S HAIRSTREAK

Skinner 1904

Callophrys johnsoni

When I've seen Johnson's Hairstreak, it's been mud-puddling at points way north of this book. The only butterfly it can be confused with is the Thicket Hairstreak (*Callophrys spinetorum*) (see **Appendix B, Tableau 9**). Both are tough to see, as they spend most of their time up in the treetops on their hosts.

01: dorsal ♀
02: dorsal ♂
03: mature larva, brown form
04: mature larva, green form
05: ventral, in situ

Habitat: Douglas fir forests
Host Plant: Pine dwarf mistletoe (*Arceuthobium campylopodum*)
Life Phases: The larvae are highly camouflaged to look like their host in shades of pine green, black, and white. Pupa chocolate brown. Adults drawn to Ceanothus blossoms. Univoltine: May–July.
Counties They Fly In: Napa
Great Places to See Them: The historic record puts this lone sighting into the northern reaches of Napa County. There are no other recorded sightings up to the present.

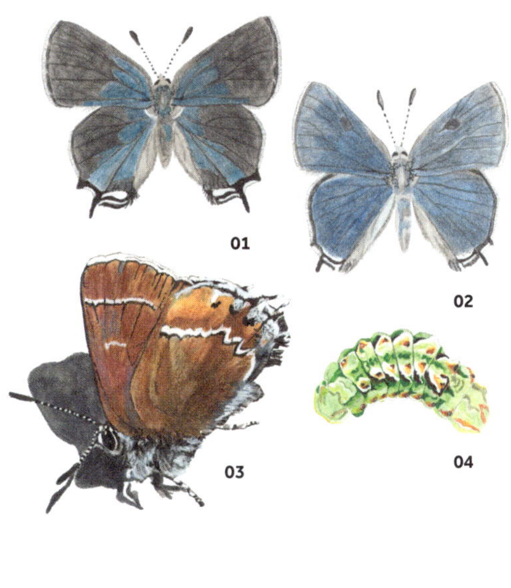

THICKET HAIRSTREAK

Hewitson 1867

Callophrys spinetorum

This one took me decades to see. I finally saw one first at North Peak, Mount Diablo State Park (Contra Costa County) walking with Ken-ichi Ueda, co-founder of iNaturalist. It was sitting on the tippy top of a small oak tree. The only other hairstreak it could be confused with is Johnson's (*C. johnsoni*), (see **Appendix B, Tableau 9**), but there is no habitat for that butterfly here.

Habitat: Pine and oak woodland, ponderosa pine (*Pinus ponderosa*)
Host Plant: Pine dwarf mistletoe (*Arceuthobium campylopodum*)
Life Phases: Larvae are primarily green or reddish depending on the color of the tree. Pupa light brown with dark mottling. Male adults sit in the branches of hilltop trees to await females. Adults are black and gunmetal blue above. White line ventrally forms a W near the tail of the hind wing (Johnson's does not have this).
Counties They Fly In: Alameda, Contra Costa, Monterey, Napa, San Benito, Santa Clara, and Solano
Great Places to See Them: The most consistent place I've seen this species is the summit of Eagle's Peak up Mitchell Canyon Trail, Mount Diablo State Park (Contra Costa County). Check all the needles near the branch tips. They have also been spotted at Lake Hennessey Recreational Area (Napa County).

01: dorsal ♀
02: dorsal ♂
03: ventral, in situ
04: mature larva

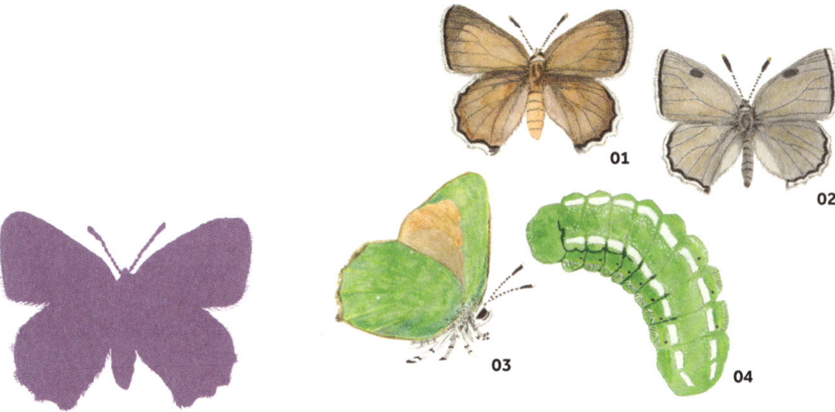

LOTUS HAIRSTREAK

Barnes & Benjamin 1923

Callophrys dumetorum

also known as **BRAMBLE HAIRSTREAK**

Lotus has a prominent brown, triangular notch on the under forewing; Coastal Green does not. I've noticed this butterfly likes to sit on deerweed. When I was participating in the Big Creek Butterfly Count down in Monterey one year, Jerry Powell showed me how even in the rain they can be found on it. To compare with the Coastal Green Hairstreak, see **Appendix B, Tableau 10**.

Habitat: Coastal scrub, chaparral, edges of oak woodlands

Host Plants: Deerweed (*Acmispon glaber*), California buckwheat (*Eriogonum fasciculatum*), coast buckwheat (*Eriogonum nudum*). If both deerweed and buckwheat are present, butterflies seem to show a preference for deerweed for egg-laying (Steiner.1980.84).

Life Phases: Eggs are laid on the flowers. Pupa brown with faint dark markings. Overwinter as pupae. Adults have white lines on ventral forewings and hind wings—lines are broken with few spots. Adults remain on or near the host their entire lives. Univoltine: March–May.

Counties They Fly In: All except San Francisco

Great Places to See Them: Edgewood County Park (San Mateo County); Briones Regional Park (Contra Costa County)—take the trail from Overlook Staging Area north toward Little John Cove, and deerweed should appear along the trail. Pine Flat Road near Bonny Doon (Santa Cruz County); Garrapata State Park (Monterey County)—take a trail going out toward the western valley, where the trail splits. Hike to the end where a treacherous trail takes you up and out of the redwoods. Fields of deerweed appear with trails proceeding toward the summit. It's a killer hike, but Lotus Hairstreaks should be there in June.

01: dorsal ♀
02: dorsal ♂
03: ventral, in situ
04: mature larva

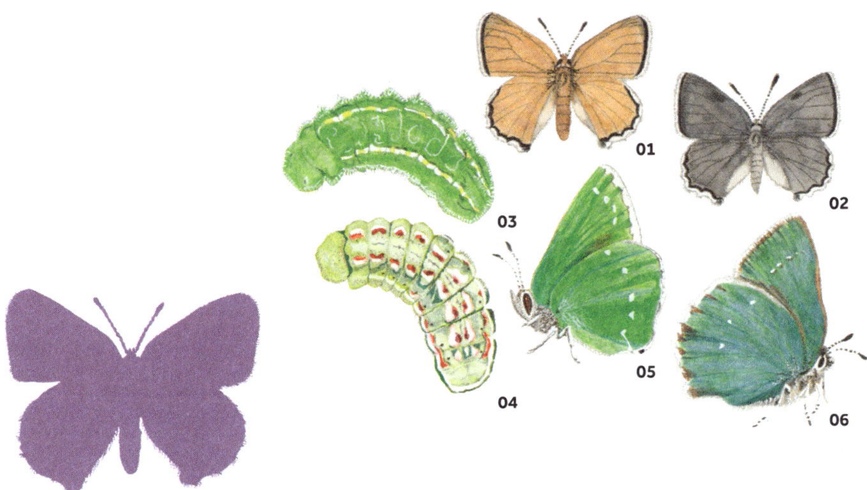

COASTAL GREEN HAIRSTREAK

Boisduval 1852

Callophrys viridis

Freshly emerged "greenies" (as Stu Weiss, a witty lepidopterist who's driven many butterfly projects in the Bay Area, likes to call them) have a blue teal iridescence to their wings, which they lose in subsequent days of life from eclosure. One rarely sees the brown and tan dorsal side—only in flight. The green is the lower side.

I have no objectivity on this butterfly. It flew with Xerces and is still holding on in San Francisco. A conservation effort I created, the Green Hairstreak Project, with the partnership of Nature in the City, continues to this day to help support the well-being of this creature into the future in San Francisco.

Habitat: Coastal scrub, dunes
Host Plants: Coast buckwheat (*Eriogonum latifolium*), deerweed (*Acmispon glaber*)
Life Phases: Females lay a single egg on the inner side of the buckwheat leaves. Larvae come in two forms, one grass green with white, middorsal lines dotted in yellow, one light green with red dots like the pattern of an aging buckwheat blossom. Fourth instar overwinters at the base of the plant among the leaf litter. Pupa orange brown. Univoltine: March–June.
Counties They Fly In: Marin, San Francisco, San Mateo, and Sonoma
Great Places to See Them: Rocky Outcrop, 14th Avenue, next to Grandview Park (San Francisco County); Baker Beach, the Presidio (San Francisco County). There is a later flight (June–July) on the dunes at South Beach, Point Reyes National Seashore (Marin County).

01: dorsal ♀
02: dorsal ♂
03: mature larva, form A
04: mature larva, form B
05: ventral, in situ, regular form
06: ventral, in situ, bluish-green form

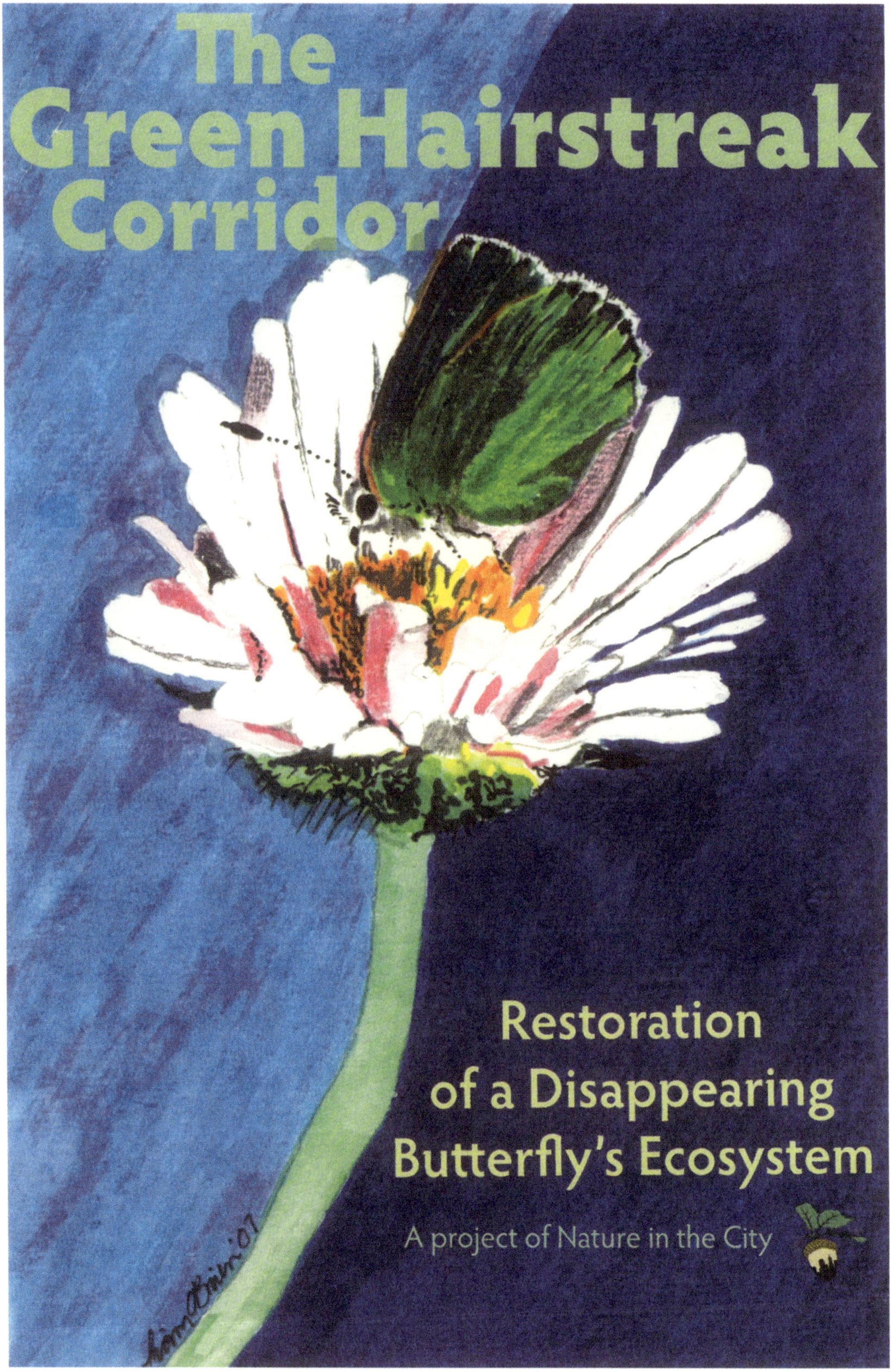

The Green Hairstreak Corridor

Restoration of a Disappearing Butterfly's Ecosystem

A project of Nature in the City

THE GREEN HAIRSTREAK PROJECT

More than a decade into my passion for butterflies, San Francisco, this paved-over, compromised Eden-by-the-Sea, held little interest for me compared with the beauty I encountered on High Sierra visits to pristine ecosystems and the plethora of butterflies flying there. Sure, we had an endangered one, the Mission Blue, but not much else. After hearing myself whine about it for so long, I finally stepped up to Jerry Powell's call to arms: "Learn where you live." It was time to scour the place and find out exactly what I was dealing with.

I wanted to comb the city, traverse every open space, every park, trespass every vacant lot to find what was common and abundant and what was rare and ready to blink out. In 2007 when I mentioned to Barbara Deutsch, a seasoned butterfly enthusiast from San Francisco, that I intended to survey all the butterfly species left in San Francisco County, she said one sentence that truly launched me: "Green Hairstreaks still fly in the Upper Sunset." It seemed as good as any place to start this project

There are four "sky islands" in the Upper Sunset District of San Francisco: Grandview Park, Rocky Outcrop, Golden Gate Heights Park, and Hawk Hill, above Herbert Hoover Middle School. These are part of the crescent-shaped San Miguel Hills, which include Billy Goat Hill, Mount Davidson, O'Shaughnessy Hollow, Twin Peaks, and Mount Sutro. If one actually extended this just a tad farther north, Strawberry Hill in Golden Gate Park would be the final peak. These hills were once the highest dunes from summit to sea level on the continental West Coast of the United States. Much of the sand is gone now, and their rocky, Franciscan chert foundations are left exposed.

Postcard, watercolor pencil, 5" x 7"

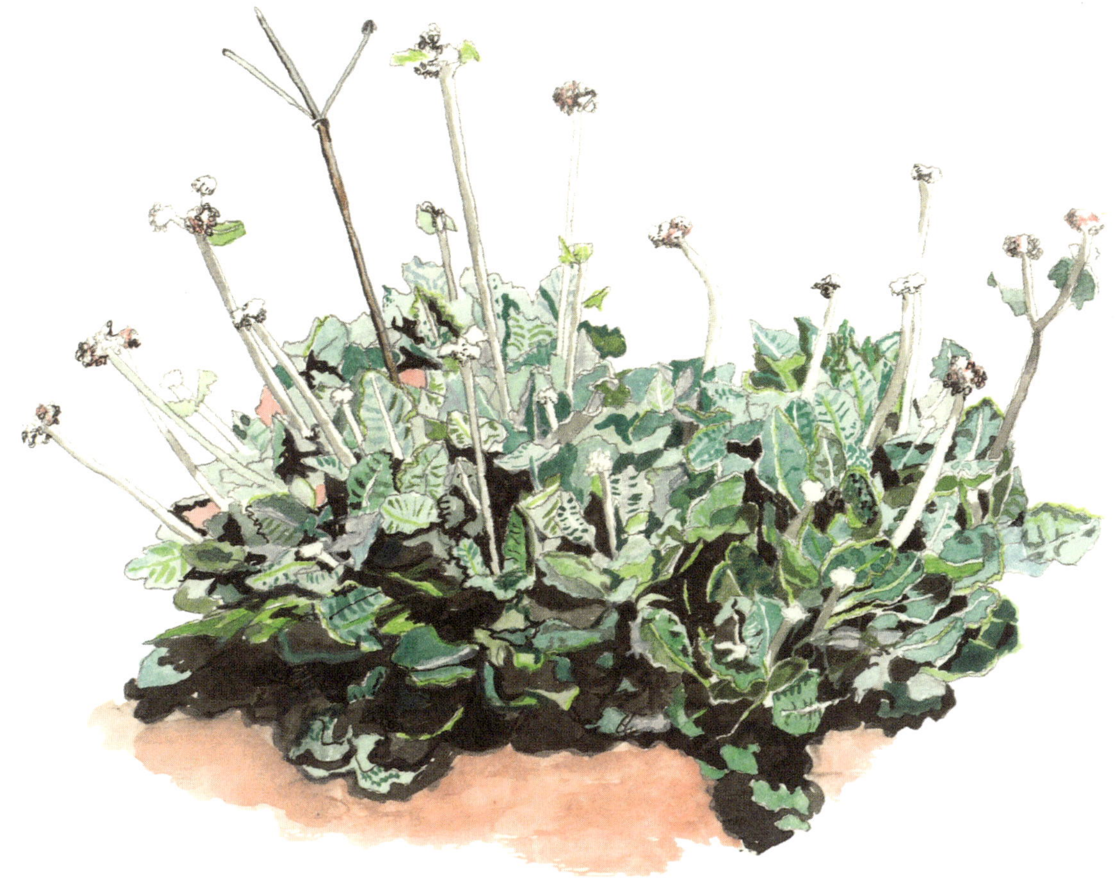

I started with the closest one my bus could get me to. Once called Turtle Hill (I guess from its shape), Grandview Park sticks up and out of the many small houses surrounding it like an anomaly within the neighborhood. The hill had much open space below it until the GIs returning after World War II built homes in all of the remaining open lots in the Sunset. Many native plants still draped its slopes the day I visited. The one I was looking for was coast buckwheat (*Eriogonum latifolium*)—the host plant for the Coastal Green Hairstreak (*Callophrys viridis*). It seemed present primarily on the northern flank of the hill. Wildflowers abounded—San Franciscan wallflower, paintbrushes, and sticky monkeyflower. Taking in the 360-degree view of the city from the summit, I wondered aloud, "I wonder where you are. Are you . . . gone?" Maybe Barbara had seen the last of them a while back.

Coast buckwheat
(*Eriogonum latifolium*)

It was then that I saw the hill's neighbor to the south—Rocky Outcrop, a dramatic cathedral of chert cliffs. I learned later that this piece of land was saved from destruction (for condos with views) by then-senator Dianne Feinstein, who was pressured by the local Yerba Buena Chapter of the California Native Plant Society because of its lichen diversity. It looked more promising to me for the butterfly than where I was standing. Its summit had no apparent trail to the top from its steep slopes, which is probably one of the reasons it has remained so incredibly natural through the decades.

Hand over hand it was getting up to the Rocky Outcrop top. Just before I reached it, a flash of Kelly green crossed my path. One is left with a breathtaking emerald that actually sticks out when it's sitting on its host—no camouflage going on here. It's about the size of a nickel. This is a lateral-basking creature with its wings folded up revealing the underside, the green side. The vast majority of hairstreaks are lateral baskers (as well as sulphurs, Ringlets, wood nymphs), meaning they get their energy from the sun on the ventral side—all other butterflies are dorsal baskers with their wings folded open to reveal their topsides. The topside of the Green Hairstreak is rarely seen. It's an assortment of tan, brown, and chocolate colors, and the sexes are dimorphic. Of course I didn't know any of this at the time. The dopamine rush seeing that green was instant. I'd found my target. A creature that once flew with the Xerces Blue was hanging on still.

I discovered through that year that the Green Hairstreak flew at all four sky islands and there was a population at the Lobos Dunes and at Baker Beach in the Presidio. Going on to discover thirty-three more species that year in San Francisco, I was haunted all the while by an idea that had been growing. What if we . . . flood the neighborhood along 14th Avenue (pretty much where all the sites are) with the butterflies' host plant, the coast buckwheat? This was once a complete coastal dune ecosystem. I'd learned during that year that inbreeding is a real possibility in isolated butterfly populations and that if the female shoots away from this area, the odds of her finding buckwheat in neighborhood gardens is pretty slim. Her belly full of eggs will be for naught. I mapped out my idea by the end of that year and brought it to a non-profit called Nature in the City.

Peter Brastow had left the National Park Service to pursue his dream of building an organization to celebrate and protect plants, ecosystems, and creatures while using his political voice in the halls of city government

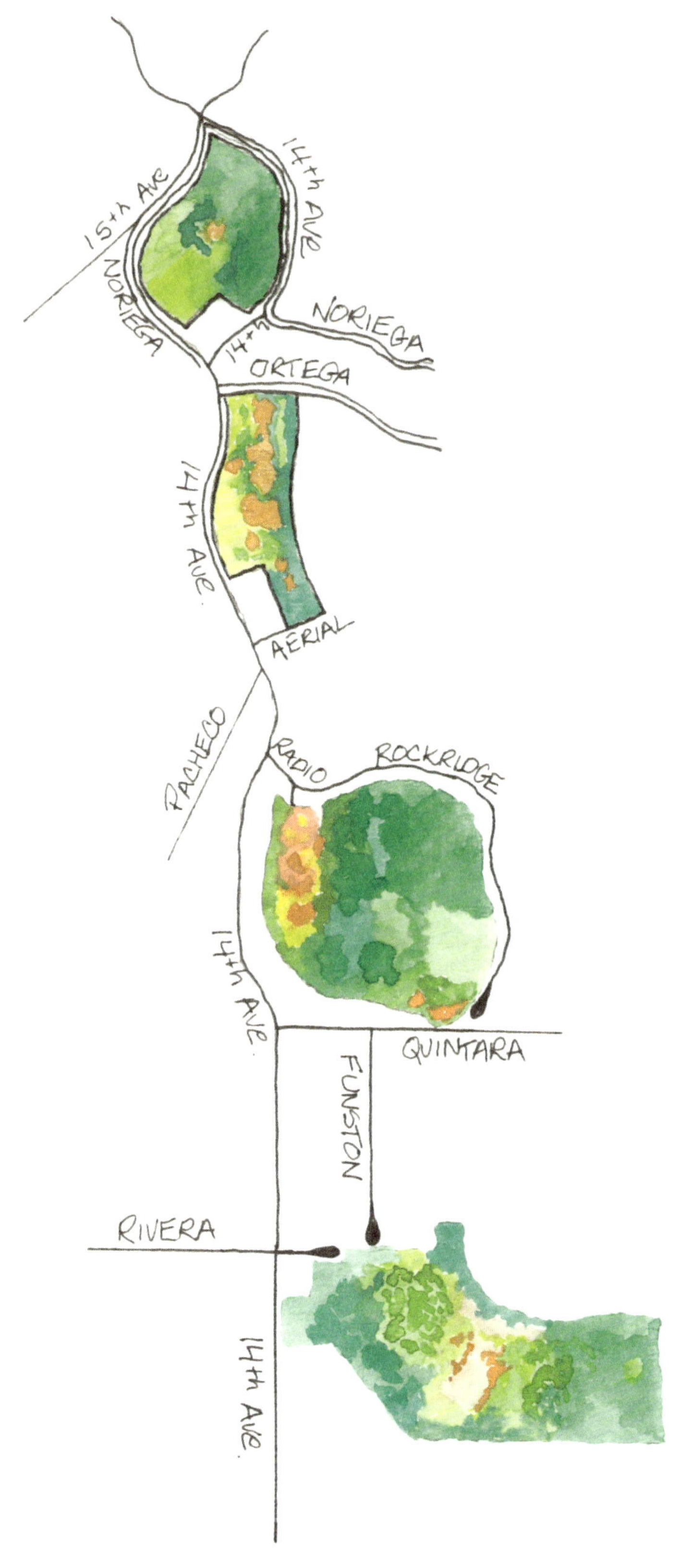
15th Ave
14th Ave
Noriega
Noriega
14th
Ortega
14th Ave.
Aerial
Pacheco
Radio
Rockridge
14th Ave.
Quintara
Funston
Rivera
14th Ave.

on their behalf. He called it Nature in the City, and I first presented my idea for the Green Hairstreak Project in Peter's backyard in Cole Valley in March 2008. He loved the idea. We had our first meeting at Laura Castellini's home not far from Rocky Outcrop. Jake Sigg, Ruth Gravanis, and Amber Hasselbring were all in attendance, all folks who either lived in the neighborhood or were directly involved with Nature in the City.

About the same time, I learned of a program being run by the Golden Gate National Parks Conservancy called Street Parks. It allowed for people to adopt small pieces of unused land in their neighborhoods: corner spots and strips of land along staircases and such.

Fantastic. We already had our eye on a piece of land at 14th and Pacheco and the Aerial Steps across from its location. One of the stipulations in the program is that each site had to have a guardian that lived in the neighborhood. It's one thing to have an idea for a place, and it's a whole other thing to implement it, especially when you don't live there. I might see a vacant piece of ice-plant-covered earth; a neighbor may see this as a place where they walk the dog to pee every night.

A great deal of outreach occurred in those early years. The neighborhood was one of the last places that one could experience seeing a Green Hairstreak butterfly, an experience afforded to all San Franciscans once, when the buckwheat was growing from ocean to bay.

The project took off. We had a slew of great stewards: Mike Belcher, Sarah McConnico, and the late Barbara Kobayashi. I took advice from anyone who had a smattering of a conservation background. But truth be told, having such a charismatic beauty as the poster child did a great deal of the work for us. Laurette Rogers at the time was working for the Bay Institute. She said to me once in exasperation, "A green butterfly? A goddamn green butterfly? Liam, I'm trying to get people excited about mud snails and your project has a green butterfly! That's as good as having a panda bear wandering around a neighborhood." I knew what she was saying—we have a tendency to save pretty things, we humans. I even give the Edwardians a pass on their collection of so many flying jewels; that was before the conservation movement really got going. But no one gets a pass now with the Green Hairstreak. We had to step up because we knew better.

A seminal moment came for me on the project one day while I was watching a bunch of "greenies" zipping about a summit. I sat there like Jane Goodall with the clipboard, noting behavior and total numbers. Rocky

Map of Green Hairstreak corridor

Outcrop, along with Grandview Park, a piece of Golden Gate Heights Park, and Hawk Hill are all managed by the Natural Resources Division of San Francisco Recreation and Parks. It's open land to the public but highly sensitive with unique plant species. Private property butts up against all of these places. A door opened behind me and an older woman, resplendent in her muumuu and sandals, was putting trash into a can. There was an awkward moment of acknowledging each other, as I'm sure she wasn't used to encountering anyone behind her home on the steep slope.

"Excuse me, ma'am. Do you realize you have a green butterfly in your neighborhood?" She came forward as I pointed out the nickel-sized creature nectaring on sea thrift. She used my shoulder to lean on while she was bending down. "Oh, My Lord," she said slowly, "I've lived here for twenty-nine years and have never seen that." Then she got up quickly and sauntered back over to the door. "Larry!" she cried, "bring the kids. They just gotta see this."

As I mention earlier in this book, Peter Brastow said once that the first moment of conservation is learning the name of the thing in front of you (Brastow.pers comm.2007). That woman's excited utterance was the project in a nutshell for me. I can only dream that one of the kids she brought out that day will be the next E. O. Wilson or Dian Fossey and will champion and know how to keep this butterfly flying in the Upper Sunset of San Francisco.

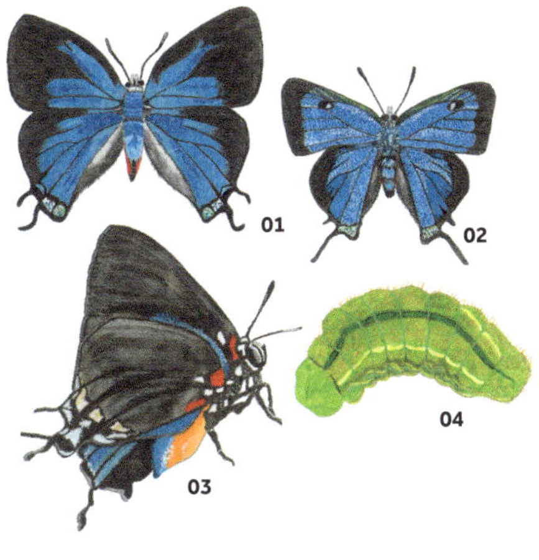

GREAT BLUE HAIRSTREAK

Cramer 1777

Atlides halesus

also known as **GREAT PURPLE HAIRSTREAK**

This is one that just shows up and not a butterfly that one goes looking for on flowers. Except for hilltops—go there to begin looking for them. I identified one once that was sitting on the top of the highest bush on Eagle Peak. *Atlides halesus* is a large, showy hairstreak a little smaller than a half-dollar coin. It'll appear on flowering plants near oak forests, especially on buckeye tree blossoms. Both sexes of this species are a kaleidoscope of colors: blue, green, red, orange, and white set against an ink-black field but, strangely enough, no purple (hence the recent name change from Great Purple Hairstreak to Great Blue Hairstreak—until the recent years of the name change it was a fantastically misnamed butterfly. I'm slightly leery of the new name because its actual primary color is black. I for one will always be utterly entrenched in the misnomer it has carried for 247 years.

Habitat: Riparian and oak woodlands
Host Plant: Common mistletoe (*Phoradendron leucarpum tomentosum*)
Life Phases: Larvae are the same vivid green as the host. Pupa brown with dark markings. Adults are active in the morning and at dusk. Both sexes are teal dorsally with the male being a darker, metallic blue. This side is rarely seen when at rest. Multivoltine: March–October.
Counties They Fly In: All but Marin, San Francisco, and Santa Cruz
Great Places to See Them: Oak forests with flowering plants. Eagle's Peak, Mount Diablo State Park (Contra Costa County); Pinnacles National Park (San Benito County); Santa Rosa Creek (Sonoma County).

01: dorsal ♀
02: dorsal ♂
03: ventral, in situ
04: mature larva

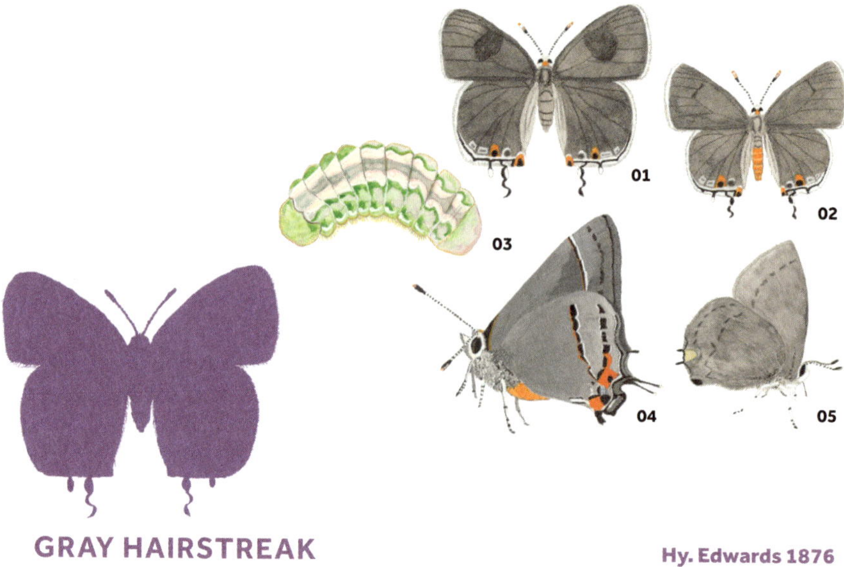

GRAY HAIRSTREAK

Strymon melinus

Hy. Edwards 1876

Millions of dollars are spent every year in the Midwest to kill this butterfly because it's considered a pest on corn fields. (Many of these species we adore are managed as pests by the U.S. Department of Agriculture.) Known in Texas as the Cotton Square Borer, the larvae have the unmitigated audacity to pierce tiny holes in ears of corn and bore in. We humans don't like holes in the ears of corn we buy. So really, it's a cosmetic issue. This butterfly, like the Acmon Blue, exhibits the behavior of "scissoring," twitching its two hind wings together back and forth to draw the predator's eye to the back false head—because it can survive a bite there (like a bird strike), but it cannot survive a bite to the real head.

Habitat: Found in a vast array of communities, from pristine ecosystems to vacant lots

Host Plants: Most butterflies stay within a species or family of plants for their egg-laying. Behold the Gray Hairstreak, who jumps genera in her choices with her larvae, known to feed on over forty-six species of plants (Steiner.1980.94).

Life Phases: Pupa brown with abundant black markings. Adult males are strong hill-toppers and have an orange abdomen, while the female's is gray and white. Multivoltine: (many flights) all year long.

Counties They Fly In: All

Great Places to See Them: Start with your own backyard and branch out from there. They are everywhere. No other hairstreak has such a wide distribution.

01: dorsal ♀
02: dorsal ♂
03: mature larva
04: ventral, in situ
05: ventral, in situ, early spring form

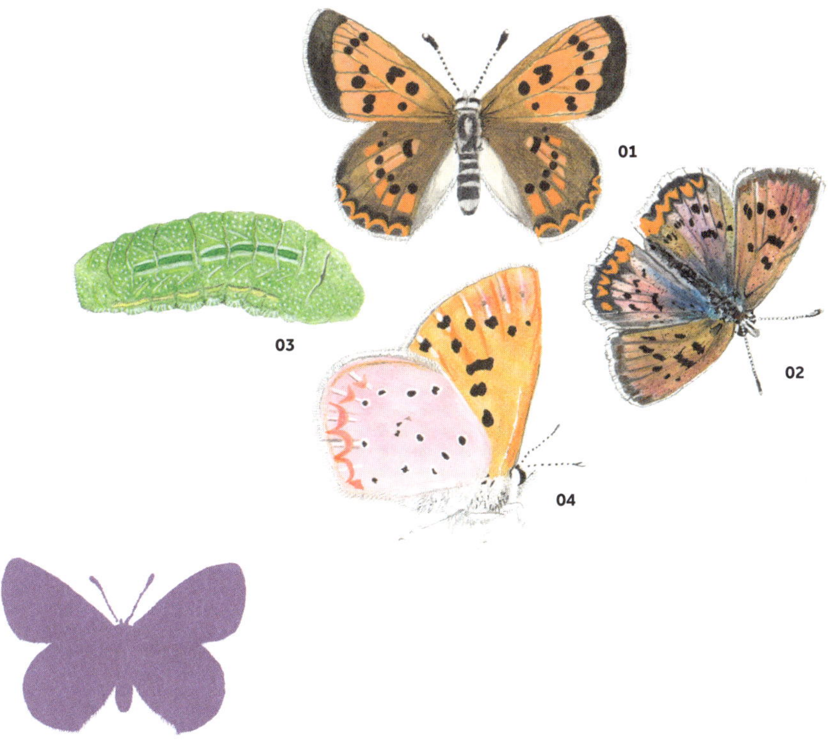

01: dorsal ♀
02: dorsal ♂
03: mature larva
04: ventral, wing up

PURPLISH COPPER

Tharsalea helloides

Boisduval 1852

JoAnn Zlatunich photographed one at the AIDS Memorial Grove in Golden Gate Park in 2018. This sighting of JoAnn's was made all the more amazing because San Francisco is the type locality for this species, but it was not to be known to fly there any longer. It was no doubt one of the multitude of species collected by Pierre Lorquin and carried back by ship to Paris to await Boisduval's naming.

Habitat: Marshes, edges of lakes and ponds, weedy vacant lots
Host Plants: Knotweed (*Polygonum* spp.), dock (*Rumex* spp.)
Life Phases: Eggs are laid haphazardly at the base of the plant (Scott.1986.389). Pupa is vivid green at the thorax with more yellow at the abdomen. Males sometimes sit in large depressions in the ground awaiting females. Multivoltine: March–November.
Counties They Fly In: All
Great Places to See Them: Any of the vineyards in Sonoma County; Big Break Regional Park (Contra Costa County). They can show up just about anywhere, but marshes are a good place to start.

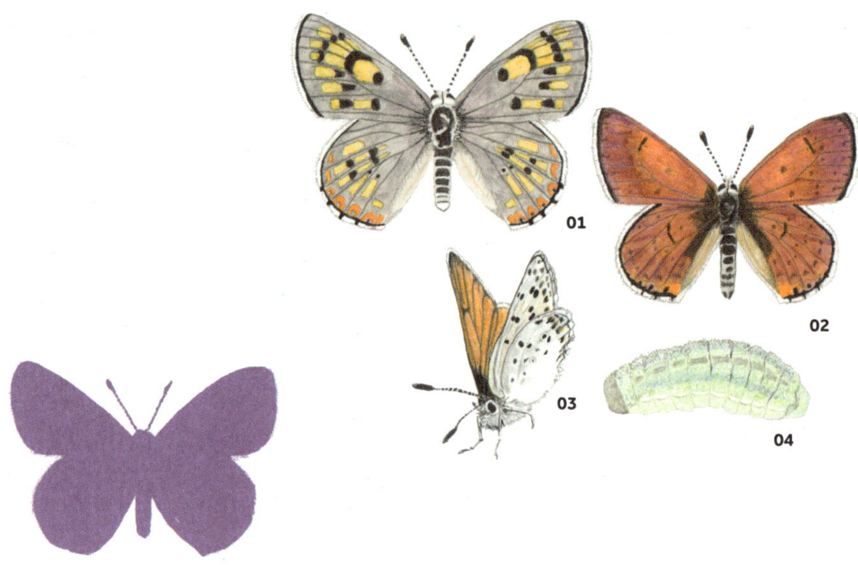

GORGON COPPER

Boisduval 1852

Tharsalea gorgon

An inhabitant of the dry oak and gray pine woodlands of the inner ranges of the California coast. It is rare or absent from the outer coastal range. I usually see them on very hot days in July. Colonies are closely tied to host plants. The bronze, purplish copper luster of the male catches one's eye like a large penny in the grass.

Habitat: Oak woodlands, grasslands, canyons, chaparral
Host Plants: Naked-stemmed buckwheat (*Eriogonum nudum*), coast buckwheat (*Eriogonum latifolium*)
Life Phases: Overwinters as an egg. The leaves are eaten by the larvae from below. Pupa greenish blue with rows of lines and dots. Adult females and males of the Gorgon Copper are boldly different in color and pattern. Female Gorgons can look a great deal like female Great Coppers. Univoltine: May–July.
Counties They Fly In: All except San Francisco
Great Places to See Them: Mitchell Canyon Trail, Mount Diablo State Park (Contra Costa County), toward Eagle's Peak Trail. There is a large, open sloped meadow before you hit the peak that always seems to be teeming with them. Pinnacles National Park (San Benito County); Hastings Natural History Reservation, Carmel Valley (Monterey County); Chews Ridge (Monterey County). Another good place is at Mount Diablo State Park (Contra Costa County) at the Juniper Campground—walk a short distance out the Deer Flats Trail.

01: dorsal ♀
02: dorsal ♂
03: ventral, wing up
04: mature larva

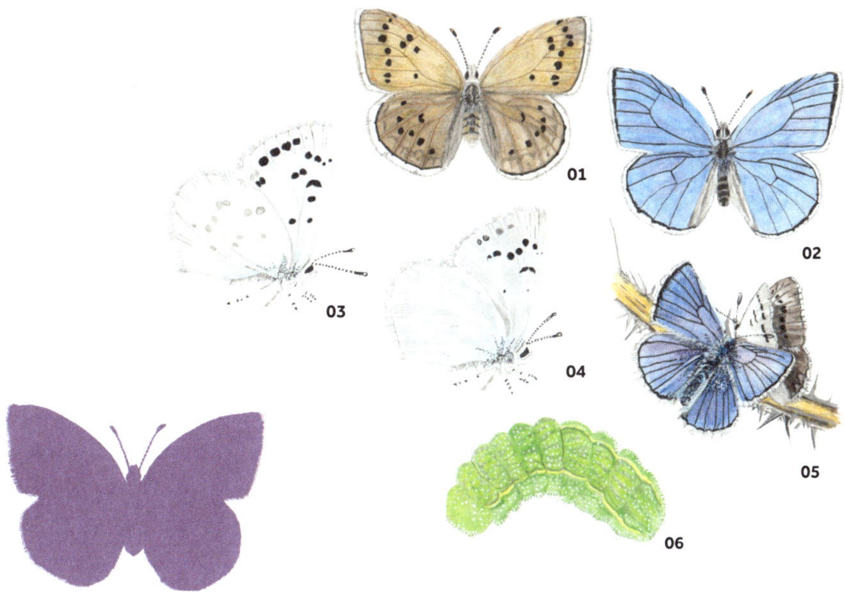

01: dorsal ♀
02: dorsal ♂
03: ventral ♀
04: ventral ♂
05: in situ
06: mature larva

BLUE COPPER

Tharsalea heteronea

Boisduval 1852

Every year I get a picture of a Blue Copper from someone asking, "Is this a Mission Blue?" It is a butterfly that can wreak havoc on the novice's mind. When is a blue not always a blue? It is not part of the Blues Tribe (Polyommatini). "But Liam," you say, "I am looking at a blue butterfly." Indeed you are, but you are in a different tribe: Lycaenini, the coppers. And that, my friends, is how a blue butterfly is not always a blue.

Habitat: Coastal scrub
Host Plants: Naked-stemmed buckwheat (*Eriogonum nudum*), coast buckwheat (*Eriogonum latifolium*)
Life Phases: Overwinters as an egg. "Greenish gray" describes the pupa best. Adult females are grayish brown, and the males are electric sky blue with obvious veining through their upper side. Univoltine: July–August.
Counties They Fly In: Marin and Sonoma
Great Places to See Them: Tennessee Valley, Golden Gate National Recreation Area (Marin County). Look for small cliffs of Franciscan chert rock—the host plant naked-stemmed buckwheat is all over the cliff face. The veining in the fore and hind wings is a giveaway to it being the Blue Copper. You might also find it around River Road to Goat Rock Road to Shell Beach (Sonoma County) and on paths along the Kortum and Coastal Trails (Sonoma County).

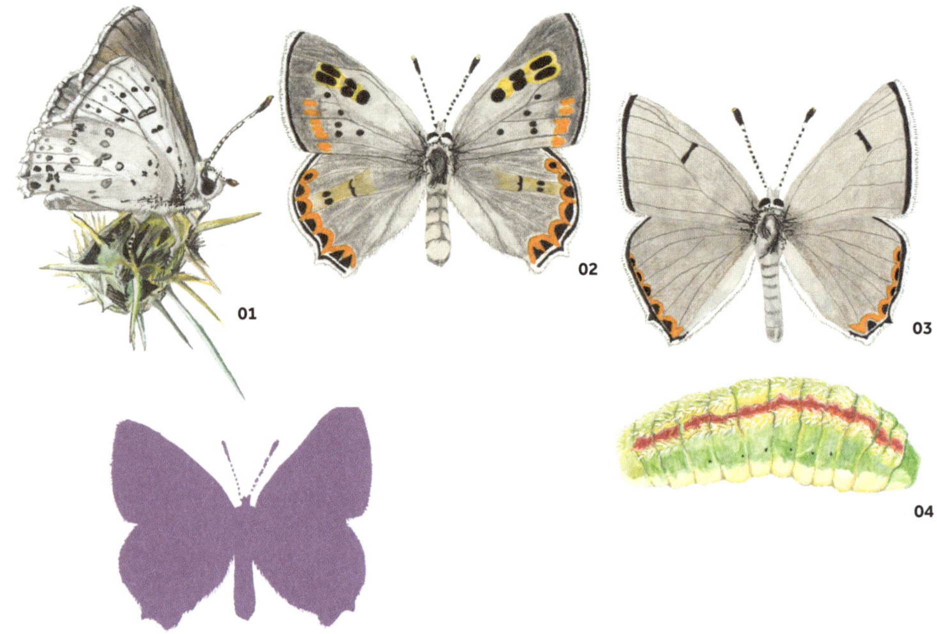

GREAT COPPER

Tharsalea xanthoides

Boisduval 1852

I've always wondered why the San Bruno Mountain population hasn't spread north, considering that dock is in every vacant lot and they're only just on the other side of the San Francisco County line. Turns out they are highly localized and do not move much. True of most coppers.

Habitat: Coastal scrub, grasslands, and oak woodlands
Host Plant: Dock (*Rumex* spp.)
Life Phases: Eggs start out as green but become white. Larvae come in sundried colors. Pupa brown with pink. Adults sit on or near the host plant then dart out to mates passing by (Kaufman & Brock.2003.82). The largest of our coppers, the Great Copper female can be easily confused with the Gorgon Copper female (see page 264). Look carefully at wing shapes and patterns to negotiate this head-scratcher. Univoltine: June–August.
Counties They Fly In: All except San Francisco and Santa Cruz
Great Places to See Them: The Northeast Ridge of San Bruno Mountain (San Mateo County)—go up the small valley where Mission Blue Drive intersects with Guadalupe Canyon Parkway in the city of Brisbane; Pinnacles National Park (San Benito County); Hastings Natural History Reservation (Monterey County); Baylands Nature Preserve, Palo Alto (Santa Clara County); Alum Rock Park, San Jose (Santa Clara County)

01: ventral ♂, in situ
02: dorsal ♀
03: dorsal ♂
04: mature larva

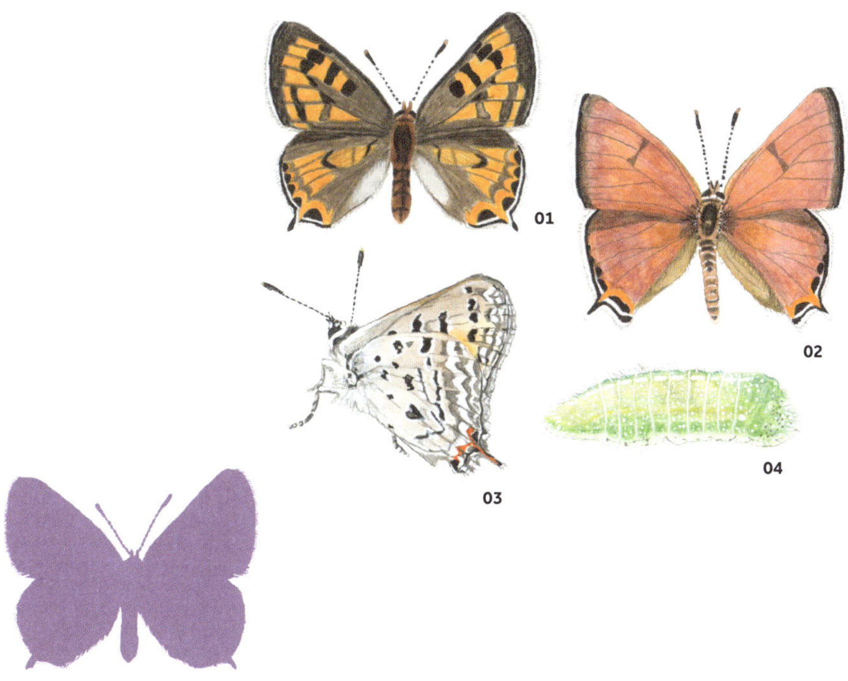

TAILED COPPER

Tharsalea arota

Boisduval 1852

If you do not gasp when this beautiful butterfly lands in front of you, you are not human. Due to the unique underside and tail for a copper, it is easily mistaken for a hairstreak from below. Steiner's survey of Bay Area butterflies suggests a broad distribution of this butterfly that I think is somewhat of an overreach. It's definitely not in San Francisco County, or I would have seen it there. The only place I've seen it with any regularity is at Pinnacles.

Habitat: Riparian and oak woodlands, coastal scrub, forest clearings
Host Plants: Gooseberry (*Ribes divaricatum*), red-flowering currant (*Ribes sanguineum*)
Life Phases: Overwinter as eggs. Pupa heavily shadowed brown to tan. Adult males perch waiting usually 6–10 feet above ground for the opposite sex with wings three-quarters open. Univoltine: May–September.
Counties They Fly In: All except San Francisco and Solano (Steiner.1980.73)
Great Places to See Them: Pinnacles National Park (San Benito County)—any of the trails leading away from the campgrounds. It can appear anywhere in the park but sticks close to its host. Peters Creek Trail (Santa Cruz County); Jasper Ridge Biological Preserve (San Mateo County). Along the road near Hastings Natural History Reservation, Carmel Valley (Monterey County).

01: dorsal ♀
02: dorsal ♂
03: ventral, in situ
04: mature larva

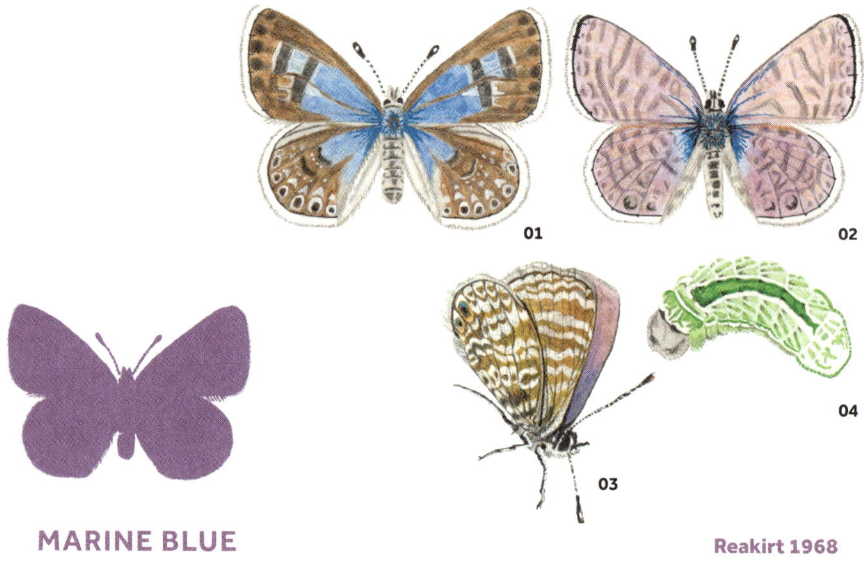

MARINE BLUE

Leptotes marina

Reakirt 1968

Like the Eastern Tailed Blue (*Everes comyntas*), the Marine Blue is a colonial butterfly, pushing north every few years from Southern California where they have a year-long stronghold. With the variety of host plants she'll use, this journey is made all the easier. They haven't, however, taken up permanent residence in Northern California.

Habitat: Open areas, farmland. Shows up in many ecosystems.
Host Plants: A generalist in the pea family (Fabaceae), deerweed (*Acmispon glaber*), sweet pea (*Lathyrus odoratus*), milk vetch (*Astragalus spp.*), false indigo (*Amorpha californica*)
Life Phases: Caterpillars are brown. Pupa is ochre or brown. The adults exhibit a zebra striping from below that is hard to describe: brown and white stripes in a whirling, repetitive pattern. Only Reakirt's Blue (*Hemiargus isola*), another colonial butterfly from Southern California, has the zebra-striped patterning as well. It doesn't fly here. Multivoltine: early spring, early summer, and fall.
Counties They Fly In: Alameda, Contra Costa, Monterey, San Mateo, Santa Clara, Santa Cruz, and Solano
Great Places to See Them: I once netted one in Garrapata State Park (Monterey County) during the Monterey Butterfly Count, but it could literally show up anywhere. Dave Bartholomew, a butterfly enthusiast, had a few in his garden in San Mateo County in 2024, and in the same year Erica Harris reported one in the city of El Granada near Half Moon Bay. They were also all over the Eastern Sierra that same year; many were seen on the Glass Mountain Butterfly Count.

01: dorsal ♀
02: dorsal ♂
03: ventral, in situ
04: mature larva

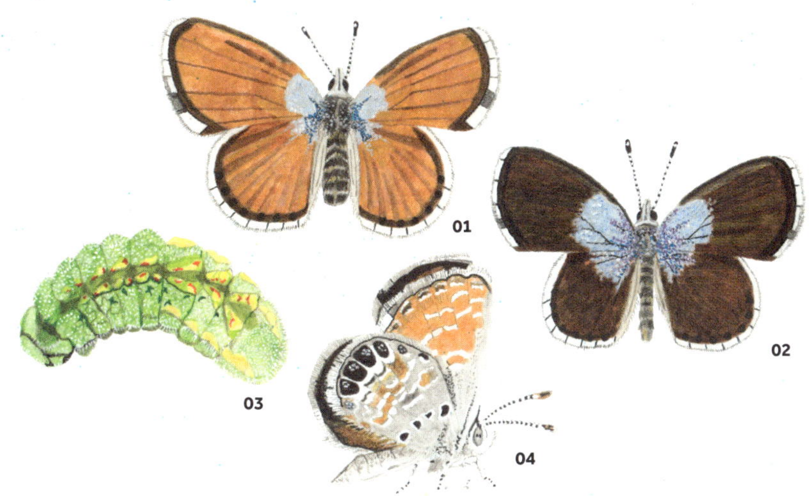

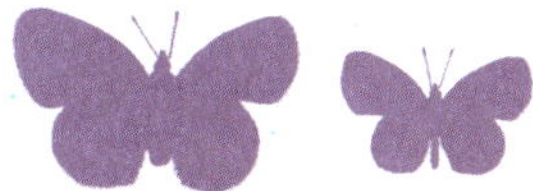

WESTERN PYGMY BLUE

Boisduval 1852

Brephidium exile

In a close race with the Small Blue (*Phiotiella speciosa*) for the title of Smallest Butterfly in the World, I believe the Pygmy will win. The males are half the size of a dime, and both sexes, being so tiny, are easily overlooked. When I point one out to folks they usually respond, "That's a butterfly?"

Habitat: Salt marshes, disturbed roadsides, vacant lots, edges of ecosystems
Host Plants: Pickleweed (*Salicornia pacifica*), saltbush (*Atriplex* spp.), Russian thistle (*Salsola* spp.), pigweed (*Chenopodium* spp.), California seablite (*Suaeda californica*)
Life Phases: Eggs are turban-shaped (Emmel & Emmel.1973.65). Larvae pale, pale green to almost white. Pupae come in a variety of colors: yellow, brown, light green, or a yellow white with a red head. Overwinters as an adult (unique among Lycaenidae). Multivoltine, with the most obvious one occurring in the fall.
Counties They Fly In: All
Great Places to See Them: Pier 94 (San Francisco County); Benicia State Recreation Area (Solano County); Don Edwards National Wildlife Refuge (Alameda County)

01: dorsal ♀
02: dorsal ♂
03: mature larva
04: ventral, in situ

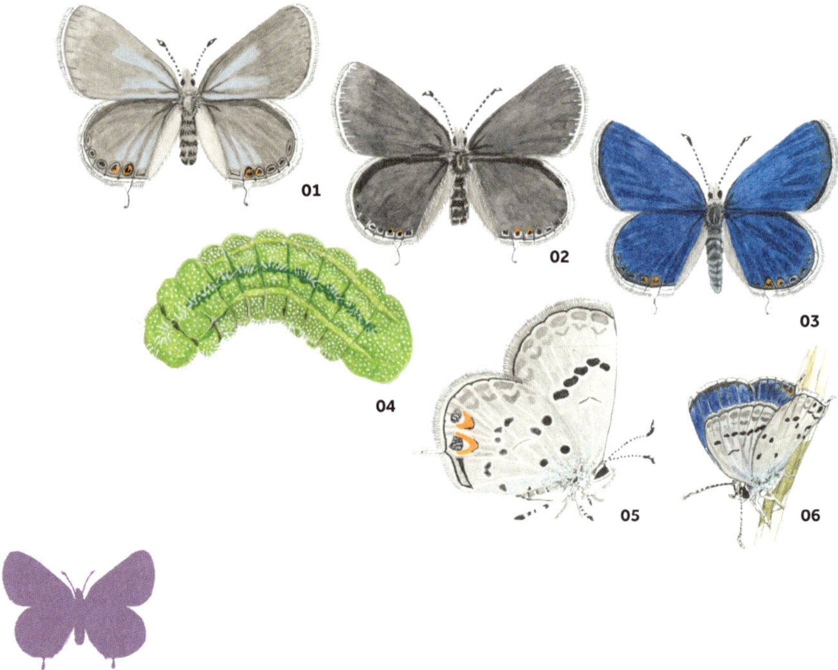

01: dorsal ♀, spring
02: dorsal ♀, summer
03: dorsal ♂
04: mature larva
05: ventral
06: in situ

EASTERN TAILED BLUE

Godart 1824

Everes comyntas

Because this butterfly is referred to as a colonial species (surging north in good years using the non-native bird's-foot trefoil as a host or dying back when it cannot get a stronghold in an area), it could show up just about anywhere. I've seen this butterfly only once in San Francisco County, down along the county line at Fort Funston. This is a blue butterfly that's also a Blue, but with a tail like a hairstreak.

Habitat: Weedy, disturbed places, open spaces
Host Plants: Many in the pea family (Fabaeceae) including clovers (*Trifolium* spp.), bird's-foot trefoil (*Lotus* spp.)
Life Phases: Some of the larval characteristics may vary from molt to molt. (Howe.1975.351) The pupae can go from light green to a dark green. Adults are smaller than Western Tailed Blues and have more orange dorsally. The female comes in a dark, summer form. One might confuse this as well with a Gray Hairstreak (see **Appendix B, Tableau 11**). Bivoltine: (two flights) May–November.
Counties They Fly In: All
Great Places to See Them: Try Marina (Solano County) and Big Break Regional Shoreline (Contra Costa County).

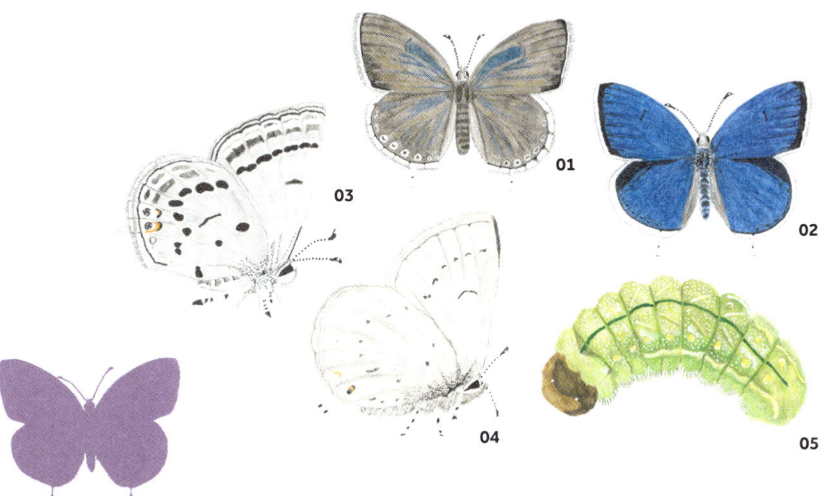

WESTERN TAILED BLUE

Boisduval 1852

Everes amyntula

The Western is notably larger and more of a loner, as opposed to the sizable surges of Eastern Tailed Blues that can happen annually. This butterfly is found more in native habitat than the Eastern (Opler.1999.232). It's never truly at rest once landed, as, like many other blues, it exhibits a scissoring motion with the back of its two hind wings. This action is thought to misdirect the predator's eye to the back in case of a strike.

Habitat: Forest glades, undisturbed sunny places

Host Plants: Chaparral pea (*Pickeringia montana*), vetches (*Vicia* spp.), milk vetch (*Astragalus* spp.)

Life Phases: This butterfly is biennial, taking two full years to complete its cycle. First-season larvae get to the fourth molt, overwintering on or within the dead seed pods of the host plant. Pupa is light tan or olive gray. Pupation occurs the next spring, the structure being pale tan. Adults have only one orange spot (with sometimes a faint secondary one) ventrally while the Eastern Tailed Blue has two strongly delineated ones. Westerns have a chalkier underside with normally very faint markings (see **Appendix B, Tableau 11**).

Counties They Fly In: All

Great Places to See Them: Mount Hamilton (Santa Clara County); Madonna Drive, Suisun City (Solano County). Redwood forest clearings (Sonoma County); Big Break Regional Shoreline (Contra Costa County). There was a population in Thompson Reach in the Presidio (San Francisco County), but I'm afraid not one butterfly has been seen there in years.

01: dorsal ♀
02: dorsal ♂
03: ventral, variation one
04: ventral, variation two
05: mature larva

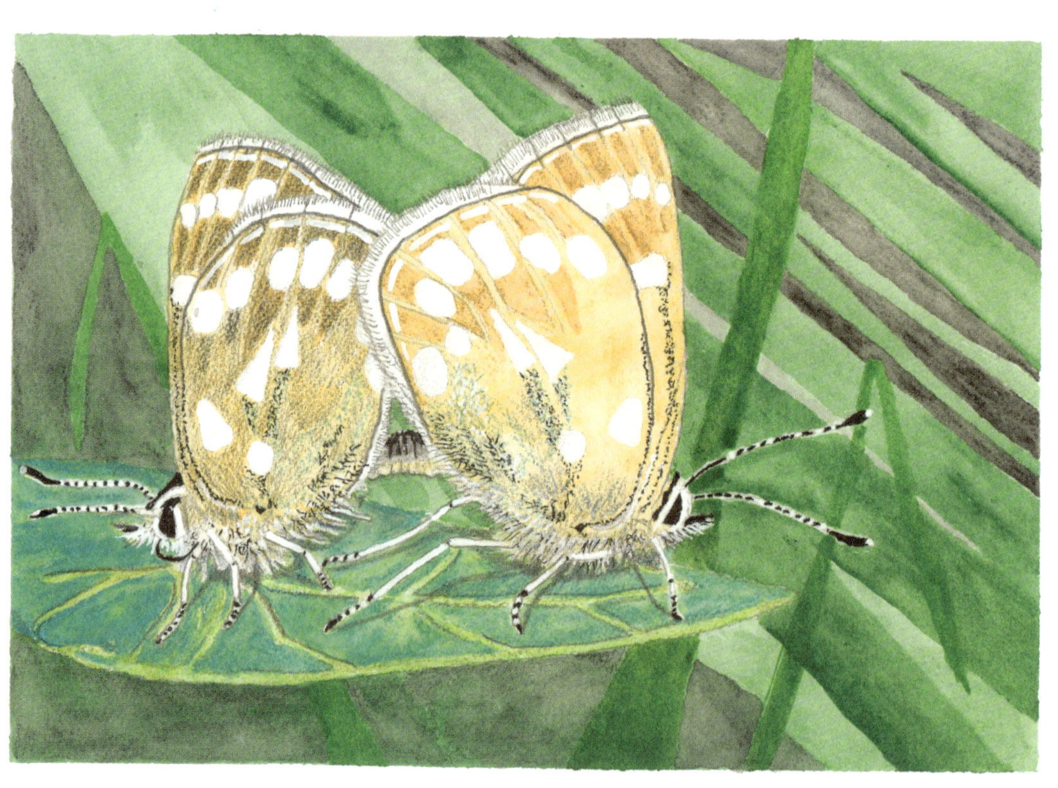

IN THE SHADOW OF XERCES

I went to my first Lepidopterists' Society meeting in 2001. It was held at Oregon State University in Corvallis. Got my check-in packet, peeled off my name tag "Liam O'Brien—San Francisco County," and set off to mingle with my plastic cup full of white wine from a box. I was very intimidated. Lots of older white men who all seemed to know one another. I knew absolutely no one. Except Jerry Powell, who was treated like Elvis there.

"Oh . . . San Francisco?" a guy said.

"Yes," I replied, "Have you been?"

"No, but my wife has always wanted to go." I was astonished how quickly "the wife card" is played, virtually in seconds, whenever any guy is . . . mingling . . . with another guy who is from San Francisco. Hmm . . .

"Pretty famous place for butterflies," he continued.

"Yes."

As he melded back into the crowd, he left with, "Too bad about Xerces" like an afterthought under his breath.

"Right," I said to . . . no one.

A real wife cornered me about how much fun it is to eat clam chowder out of a bread bowl down on Fisherman's Wharf. Something, she said, the grandkids and she did religiously every time they visited San Francisco. Her husband came up and joined us. "Hey, isn't that where that extinct butterfly is from? What's it called?"

"Xerces. The Xerces Blue."

"Right. Saw a whole drawer of them at a university once."

In the early years of this passion for me one truth became clear: we San Franciscans are more famous for what no longer flies here than for what still does. The Xerces Blue (*Glaucopsyche xerces*) became the first well-known butterfly (the Sthenele Satyr—*Cercyonis sthenele sthenele*—actually

Xerces Blue

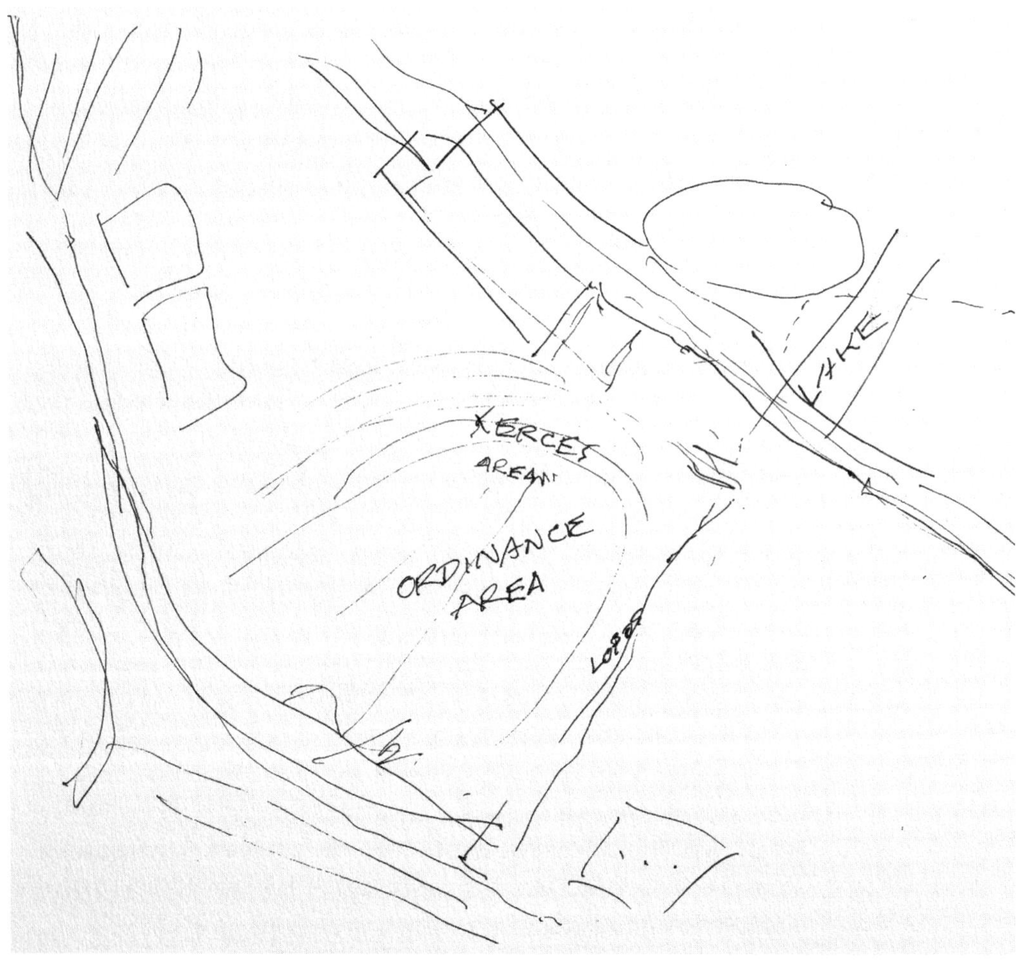
XERCES
AREA
ORDINANCE
AREA
LAKE

disappeared before Xerces in San Francisco, but Xerces seems to have obtained this dubious honor as being the first) to be lost from the planet Earth due to human development. As of March 23, 1943, one could no longer see this creature in flight, and its extinction has cast a constant pall over San Franciscan lepidopterists.

It all started with a Frenchman: Pierre Lorquin, who came to California during the 1840s to find gold in them thar hills. He never did find gold, but lucky for us he brought his butterfly net. He collected approximately thirty-five new butterflies for academic science. Orange Sulphurs, Green Hairstreaks, and Mylitta Crescents were either shipped to France or they returned with him at the end of his journey—a journey we know little about. He passed them to his friend Jean Baptise Alphonse Boisduval at the Musée de Paris. Boisduval, a naturalist and physician, described one of these in 1852 and named it after the Persian king, Xerxes. *Xerces* is the French spelling. The male is colored in cerulean blue, matching a clear sky day in San Francisco; the female is a slate brown gray from above. One of the forms of the boy has a Jackson Pollock smattering of white polka dots below, and the girl, ventrally, looks a great deal like its closest living relative of today—the Silvery Blue (*Glaucopsyche lygdamus*). The Xerces Blues came in at least three forms.

Collecting was huge in the Victorian era, and it was primarily a female hobby. Many folks held high on their wish lists "the polka-dotted butterfly from San Francisco." The expansion of post–gold rush San Francisco across the sand dune ecosystems to the west of downtown destroyed all but a few contiguous habitats for the butterfly. Horse and livestock grazing radically changed the terrain as well.

In the 1980s, Ed Ross, the last person to collect the Xerces Blue, jotted down a quick sketch of where he took it. The drawing was passed to me from Ruth Gravanis, and I will be donating it to the Presidio Trust's archives.

Known from colonies within the sand dunes, the Xerces Blue flew from near Twin Peaks to North Beach and from the Presidio southward to Lake Merced. It hosted (laid its eggs) on deerweed (*Acmispon glaber*) and yellow bush lupine (*Lupinus chamissonis*) and only flew between March and May. When San Franciscan lepidopterist Hermann Behr wrote a letter to his friend in 1875, the go-to quote for any article on Xerces was created:

> G. xerces is now extinct as to the downtown San Franciscan neighborhood. The locality where it used to be found is converted into building lots and between German chickens and Irish hogs no insect can exist. (Behr.1875).

The first sounding to the public of this pending ecological disaster was from F. X. Williams in an article he wrote in 1910, "The Butterflies of San Francisco," for the *Entomological News*:

> Very little is left of the former large areas of Ceanothus which at present is only found along the north and northwestern cliffs, and there in small patches. The other shrubs are still plentiful. As a consequence the insects are disappearing along with the destruction of their food, and the hunting grounds of the entomologist, though probably at no time very rich in this vicinity, becoming more and more restricted, and are soon destined to become a thing of the past. (Williams.1910).

By 1919, Xerces was restricted primarily to a single locality near Lobos Creek and its dunes within the Presidio. It might have been the tipping point where a species' numbers are so low it cannot rebound. But then I found an interesting tidbit while researching this piece. A guy by the name of R. G. Wind, a collector in the 1930s, tried unsuccessfully to relocate some Xerces to more suitable sites (Shapiro & Manolis.2007.149). Considering this is a huge part of what I and other butterfly conservationists do today, this window into early conservation attempts blew my mind. Yes, it failed but . . . he tried. I always say in my talks I just want to be a part of a group of people who try. A noble legacy, this R. G. Wind guy.

On March 23, 1943, Ed Ross collected the last few Xerces on the wing "near the old Veterans hospital," the story goes. When a reporter brought him out to the site years later, near the old military hospital in the Presidio, he alluded to the shadow over him and that, at the age of eighty-six, he hated being known for this inauspicious deed. He never could have imagined they would be the last seen alive.

When I surveyed all the butterflies of San Francisco in 2007, I came across a sympatric species of Xerces that thankfully still flies here to this day. The Coastal Green Hairstreak (*Callophrys viridis*) inhabits the same rocky outcrops and coastal cliffs in isolated pockets of the city much like Xerces did toward its end. The difference this time is that a whole lot of people are trying to help the Coastal Green from blinking out (see **The Green Hairstreak Project**). It's been embraced as a unique creature by the neighbors in one of its last holdouts in the Upper Sunset District of San Francisco and has been wonderfully marshaled through the years by Nature in the City, an organization led by Amber Hasselbring.

I certainly do not hold those Victorians and Edwardians accountable for Xerces's disappearance—it was another time that truly predated the idea of conservation, though rumblings were occurring. I, for one, am glad we still have Xerces to see in drawers in museum collections. It was a beautiful butterfly. But what San Franciscans are doing now to protect one's experience to see Mission Blues, Variable Checkerspots, and Green Hairstreaks ultimately should transcend the shadow of losing Xerces.

Now at a Lepidopterists' Society meeting when they see my name tag they say, "Oh, San Francisco? I hear you are doing wonderful things out there."

We are . . . because of Xerces, not in spite of it.

If you would like to become involved in butterfly conservation, contact Nature in the City in San Francisco at natureinthecity.org; *the Xerces Society in Portland, OR, at* xerces.org; *or the North American Butterfly Association in New Jersey at* naba.org. *These organizations appreciate your time and donations.*

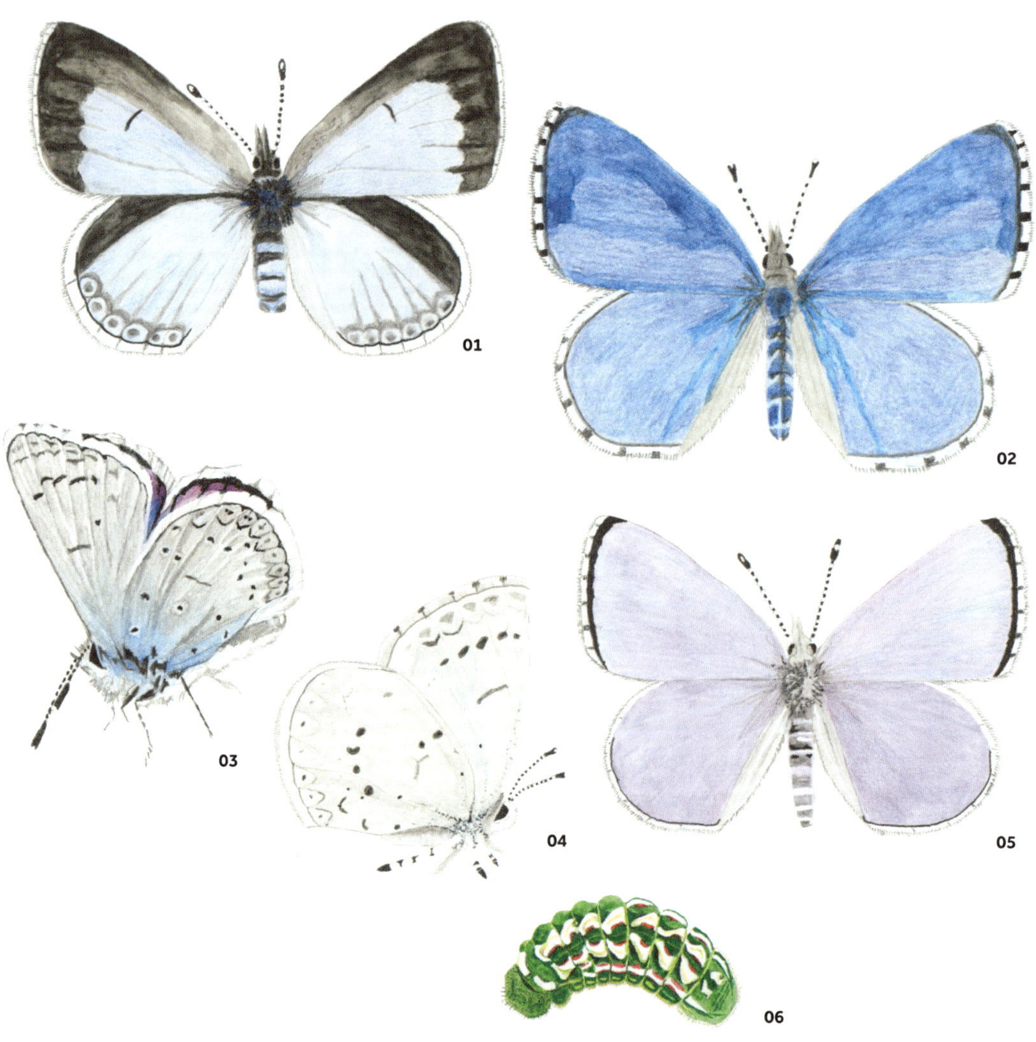

01
02
03
04
05
06

ECHO AZURE

W. H. Edwards 1864

Celastrina echo
also known as **SPRING AZURE**
also known as **ECHO BLUE**

An easy one to learn when first starting out because it flies at about shoulder height or high up in the canopy of buckeye trees. If you see males patrolling blossoms up in the tree looking for mates, you've got an Echo—one doesn't even need to see them up close. Let the behavior be your guide. All other blues, for the most part, are on the ground no higher than your knee.

Habitat: Found everywhere except cultivated farms (Garth & Tilden. 1986.151).
Host Plants: The Echo Azure is one of the few butterflies to lay its eggs on a tree—California buckeye (*Aesculus californica*), also California lilac (*Ceanothus* spp.).
Life Phases: Larvae feed quickly and develop in approximately twelve days (James & Nunnallee.2011.190). Pupa light tan with brown mottling. Overwinter as pupae. Female adults have strong black borders above, while males can come in sky blue or lavender forms. Multivoltine: March–July.
Counties They Fly In: All
Great Places to See Them: A strong mud-puddler, they can be seen in numbers after a spring rain along the trails at Mount Diablo State Park (Contra Costa County) or in the Marin Headlands (Marin County) over the coastal scrub.

01: dorsal ♀
02: dorsal ♂, spring
03: ventral, in situ
04: ventral
05: dorsal ♂, summer
06: mature larva

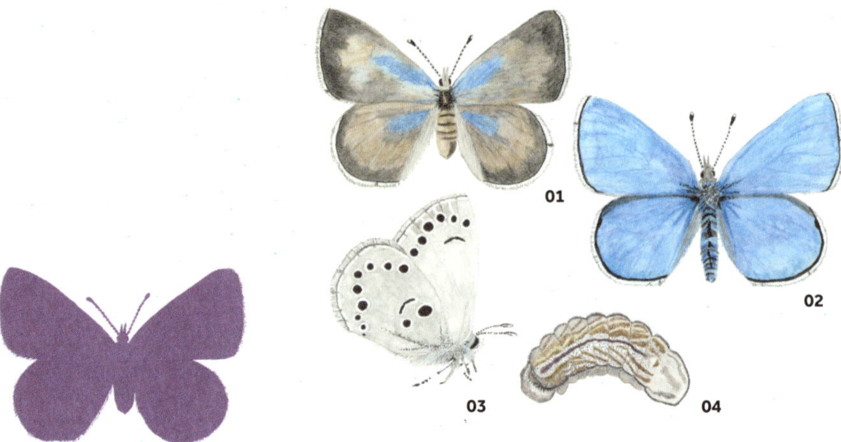

SILVERY BLUE

Tilden 1974

Glaucopsyche lygdamus

Silvery Blues are the closest living relative to the now extinct Xerces Blue (*Glaucopsyche xerces*). While they are present throughout the Bay Area, sadly, when I surveyed the butterflies of San Francisco in 2007 and again in 2009, I could not find this species within the city limits. The last confirmed San Francisco Silvery in the historic record was seen in the seventies on Billy Goat Hill. The main difference between this species and the endangered Mission Blue (*Icaricia icarioides missionensis*) is that Silveries have one row of black dots along the ventral margins on both wings and Mission Blues have two rows. Both are patrolling butterflies.

Habitat: Moist, grassy places; coastal scrub

Host Plants: Many in the pea family (Fabaceae), including vetches (*Vicia* spp.), lotuses (*Acmispon* spp.), and California golden banner (*Thermopsis californica*).

Life Phases: Larvae are tended by ants: a commensal relationship where the ants protect the larvae from other predators on the plant and the ants receive a sweet droplet of honeydew from the back end of the larvae. Both benefit. Pupa brown with dark mottling. Overwinters as pupa. Univoltine: February–June.

Counties They Fly In: All except San Francisco

Great Places to See Them: Mitchell Canyon Trail, Mount Diablo State Park (Contra Costa County); Briones Regional Park Overlook Staging Area (Contra Costa County)—hike north from the parking lot and out to Little John Cove. I've seen them there along the shaded trail. You can also find them at Garrapata State Park (Monterey County).

01: dorsal ♀
02: dorsal ♂
03: ventral
04: mature larva

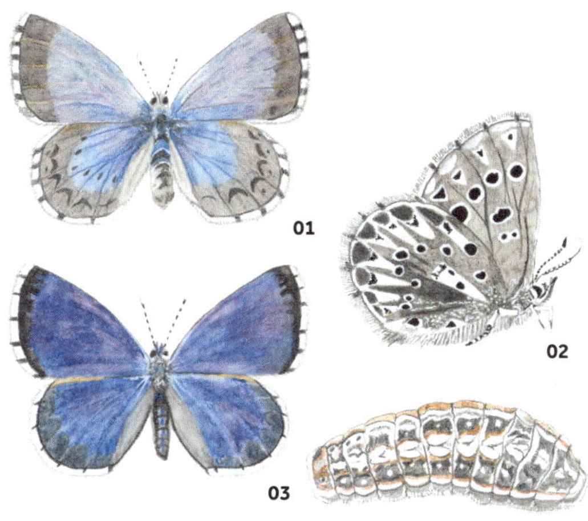

ARROWHEAD BLUE

Boisduval 1852

Glaucopsyche piasus

This is our largest blue. No other butterfly has the pattern of arrowheads that makes this instantly identifiable. However, it can be easy to overlook if it's flying with Boisduval's (*Icaricia icarioides icarioides*). Got to learn that underside if you really want to know what you're seeing.

Habitat: Coastal scrub
Host Plants: Silvery lupine (*Lupinus albifrons*), milk vetch (*Astragalus* spp.)
Life Phases: Larvae tended by ants. Caterpillars, which are small and sluglike, come in an array of colors. Pupa tan with yellow, wing area bluish green. Overwinter as pupae. Adult females are slate blue or gray above while males are royal blue. Named for the unique rows of arrowheads on the ventral hind wing. Males patrol all day in search of mates. Univoltine: March–June.
Counties They Fly In: Monterey. Extinct now in Santa Cruz and San Francisco Counties.
Great Places to See Them: Reported from Chews Ridge (Monterey County) in the spring. There's a recent report of a sighting on Cone Peak in Monterey County. This is a very isolated part of the state with no easy access. If it is there, it's local and rare. Further exploration is necessary. Sounds like an adventure.

01: dorsal ♀
02: ventral
03: dorsal ♂
04: mature larva

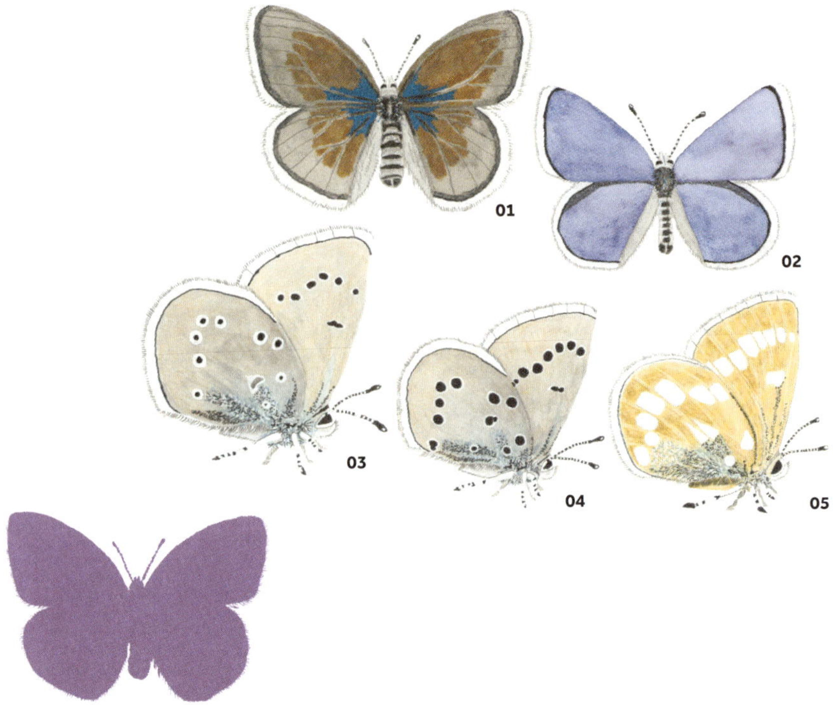

XERCES BLUE

Boisduval 1852

Glaucopsyche xerces

STATUS: EXTINCT

Adults came in three forms: one with white polka dots (f. xerces) and two others (f. polyphemus and f. antiactis) that looked like the ventral side of a Silvery Blue (*Glaucopsyche lygdamus*) of today, so much so that Xerces is considered by many to have been a subspecies of Silveries and not given full species status. A conservation group whose aim is the protection of threatened invertebrates, the Xerces Society, was formed by Robert Michael Pyle and named in honor of this lost butterfly.

Habitat: Coastal dunes
Host Plants: Deerweed (*Acmispon glaber*), bush lupine (*Lupinus arboreus*)
Life Phases: Larvae were tended by ants and came in a variety of colors. Pupa brown like that of Silvery Blue. Univoltine: March–May.
Counties It Flew In: San Francisco and San Mateo
Great Places to See Them: Drawers in the collections of the California Academy of Sciences (San Francisco County); the Essig Museum of Entomology, UC Berkeley (Alameda County); and the Bohart Museum of Entomology, UC Davis (Yolo County).

01: dorsal ♀
02: dorsal ♂
03: ventral, form polyphemus
04: ventral, form antiactis
05: ventral, form xerces

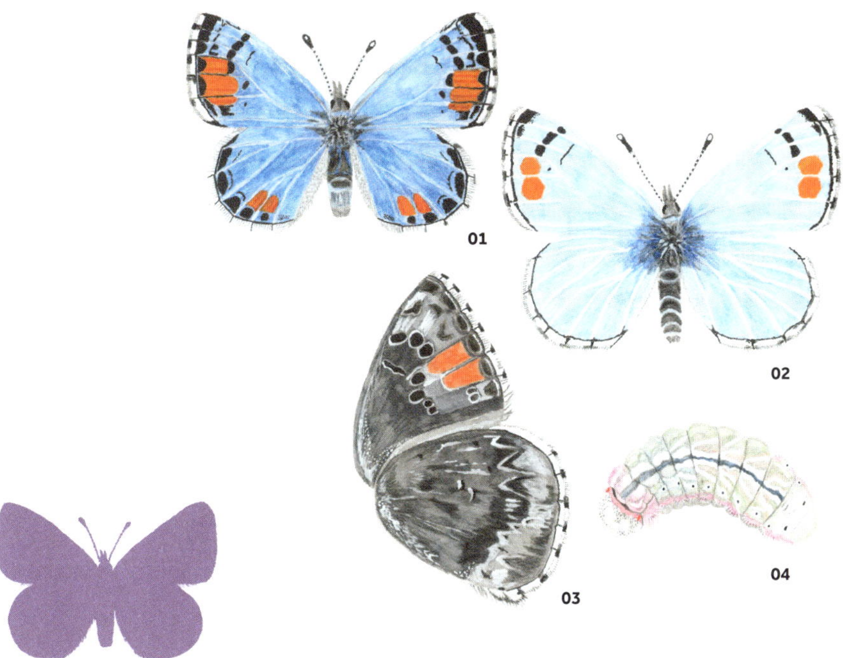

SONORAN BLUE

C & R Felder 1865

Philotes sonorensis

Considered by many to be the Most Beautiful Butterfly in the United States, no other Lepidoptera looks like it or flies like it. With its multiple wing beats per minute, the viewing of this butterfly should stop you in your tracks. For many years going to view the Alum Rock population has kicked off most of my butterfly seasons—it flies earlier than other butterflies.

Habitat: Rocky outcrops (always in tandem with its host plant)
Host Plant: Liveforever (*Dudleya cymosa*)
Life Phases: Early larvae eat within the succulent's leaves (Steiner.1980.100) and crawl into the leaf litter below the plant to pupate. Pupa green or olive green. The female adult has red spots on both her forewings and hind wings. The male has red spots on only his forewings. Both have a field of metallic blue. Elaborate dark spots in a dark field from below. Univoltine: January–April.
Counties They Fly In: Alameda, Monterey, San Benito, and Santa Clara
Great Places to See Them: Sunol Regional Wilderness (Alameda County) up along the ridgelines. In Alum Rock Park, San Jose (Santa Clara County), drive to the fourth and final parking lot, walk out the Alum Rock Falls Trail. The host plant should appear where the trail hugs the rocky cliffs. Pinnacles National Park (San Benito County)—any of the high rocky trails, mainly the Balconies.

01: dorsal ♀
02: dorsal ♂
03: ventral
04: mature larva

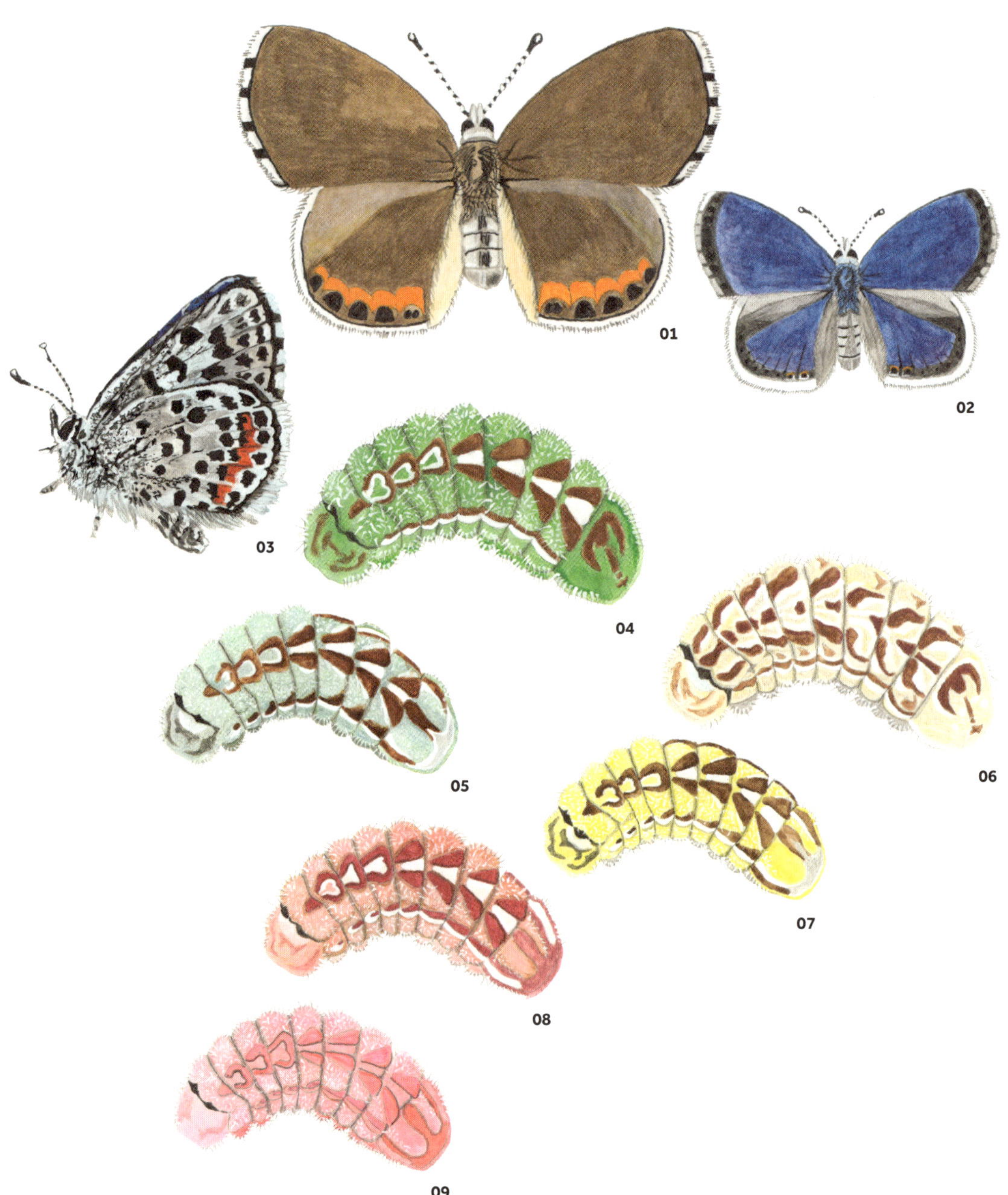

01
02
03
04
05
06
07
08
09

SAN BERNARDINO BLUE

Behr 1867

Euphilotes bernardino

The genus *Euphilotes* used to be divided into two broad groups, the Square-spotted and the Dotted Blues. Paul Opler, who studied this group in detail, often pointed out that if these tiny butterflies were the size of Fritillaries, we would have split them into many different species years ago. As science progresses, his vision is materializing. Where the San Bernardino Blue was once considered a subspecies of the Square-spotted Blue, it has now been given full species status. It seems to prefer dry washes and lower canyons.

Habitat: Coastal scrub, inland chaparral
Host Plant: California buckwheat (*Eriogonum fasciculatum*)
Life Phases: I've madly painted all the variations the larvae can come in: green to yellow to pink to rose. It has to do with camouflage, my friends, perfectly fitting in after each molt the area the larva is feeding on. Pupa is pale or orangish brown. Female adults are brown with orange auroras. Males royal blue above with dark margins on forewings. It flies quickly and erratically around the buckwheat shrubs. Other blues fly slowly and steadily. Both sexes exhibit angular and chunky black spots from below. (Important because it'll help you key it apart from Acmon Blues that might be nectaring on the plant as well.)
Counties They Fly In: Monterey and San Benito
Great Places to See Them: By far the best spot is Pinnacles National Park (San Benito County). Enter the park through the East Entrance and drive to the last parking lot, by the visitor's center. The host is actually present on both sides of the park—it's a small bush, knee-to-waist with balls of rusty white flowers. It's found all over the park in sunny, dry stream beds and on lower hill slopes. *Euphilotes bernardino* will be the most abundant and smallest blue here. Look for the chunky black spots from below.

01: dorsal ♀
02: dorsal ♂
03: ventral, in situ
04: larva, variation one
05: larva, variation two
06: larva, variation three
07: larva, variation four
08: larva, variation five
09: larva, variation six

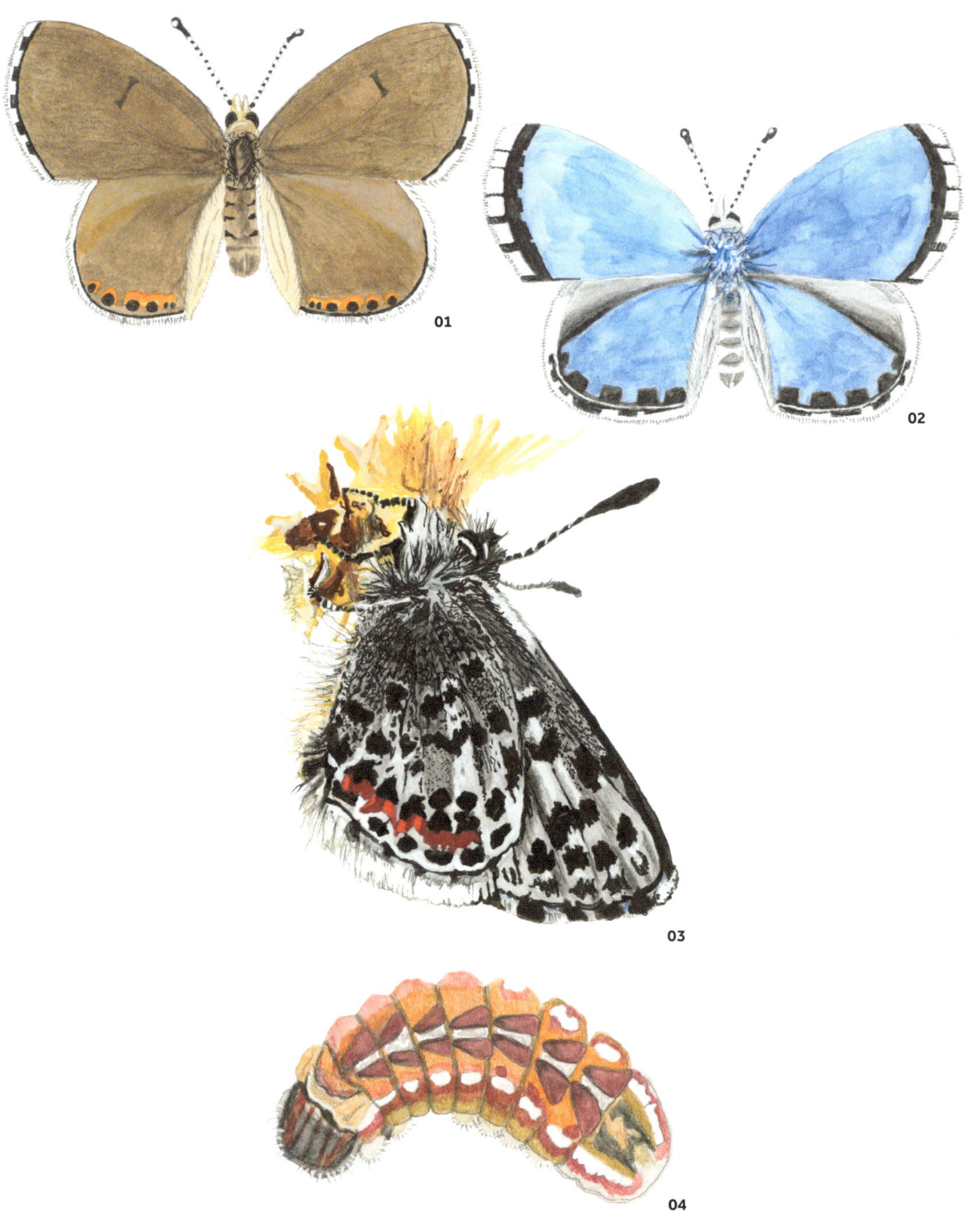
01
02
03
04

SMITH'S BLUE

Mattoni 1955

Euphilotes enoptes smithi
including the Marina Blue (*Euphilotes enoptes arenacola*)
STATUS: ENDANGERED

A butterfly a tad smaller than a dime, Smith's Blue was named to the U.S. Federal Endangered Species List on June 1, 1976. At the time, its range was limited to a few locales, but further fieldwork revealed it was much more widespread. The Marina Blue (*Euphilotes enoptes arenacola*), which is limited only to the Marina Dunes (Monterey County), was what the USFWS initially intended to list because that is the one that is in most peril. Two undergraduates from UC Berkeley collected this butterfly along the Big Sur coast in 1948. One of them, Claude Smith, met an untimely death not long after that. Rudi Mattoni, the lepidopterist who brought this species to science, named the butterfly after his friend.

Habitat: Coastal dunes, coastal scrub
Host Plants: Coast buckwheat (*Eriogonum latifolium*), dune buckwheat (*Eriogonum parvifolium*)
Life Phases: The larvae are quite the beauties and come in an array of oranges and russet reds. Pupa brown with slightly yellow mottling. Pupation occurs in the flower buds or sandy leaf litter below. Females are brown dorsally above with a band of orange spots along the bottom of the hind wing. Males are sky blue with wide, checkered black borders and fringe along the margins of its forewing. Univoltine: June–September.
Counties They Fly In: Monterey
Great Places to See Them: One of the best reasons to attend Chris Tenney's Monterey Butterfly Count is the chance to be sent to a place this special butterfly flies. Garrapata State Park (see **Best Butterfly Walks** for directions) (Monterey County). The dunes at Sand City (Monterey County).

01: dorsal ♀
02: dorsal ♂
03: ventral, in situ
04: mature larva

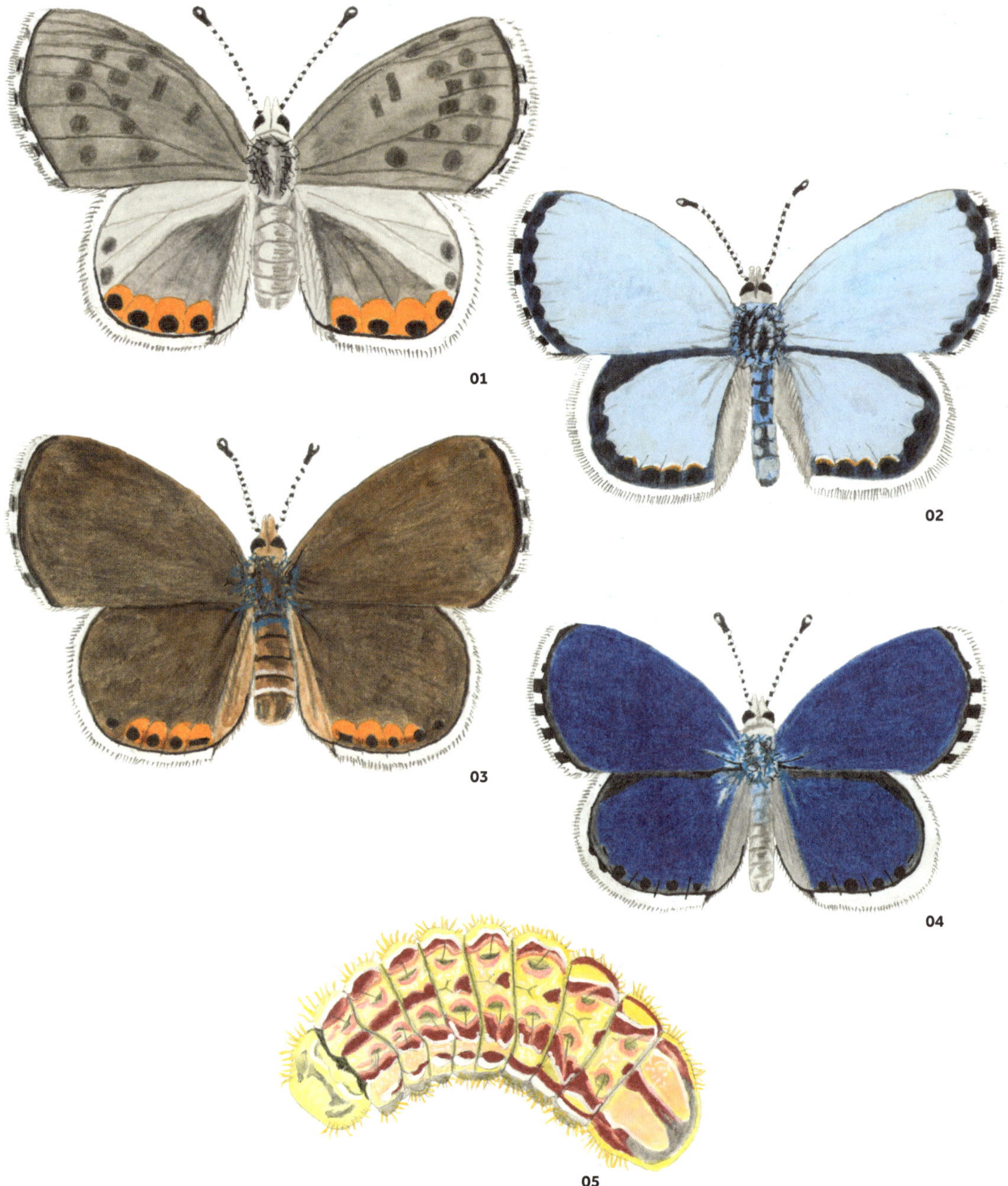
01
02
03
04
05

PACIFIC DOTTED BLUE

Boisduval 1852

Euphilotes enoptes
including other subspecies: *Euphilotes enoptes tildeni* and *Euphilotes enoptes bayensis*

The next two butterflies are very difficult to key apart. I've found that looking for the plant first with this genus leads to finding the butterfly hovering about it. "The genus *Euphilotes* is among our most confusing butterflies" (Kaufman & Brock.2003.138).

Habitat: *E.e. tildeni*: oak woodlands, chaparral; *E.e. bayensis*: coastal scrub, grasslands
Host Plant: Naked-stemmed buckwheat (*Eriogonum nudum*); *E. e. bayensis* also coast buckwheat (*Eriogonun latifolium*)
Life Phases: Eggs laid on flower buds. Larvae can remain overwintering up to nine months (Scott.1986.404). Pupa is see-through yellow. Overwinters as pupa. Adults lack the ventral metallic border spots Acmon Blues have—this is true for all *Euphilotes*. Females of *E.e. tildeni* are slate gray with pronounced orange crescent-shaped lunules below. Males are a subdued blue. The *E.e. bayensis* adult female is uniformly sable brown with a row of orange dots along her forewing margin. The dark royal blue of the male is accompanied by white borders and checkered fringe on the dorsal and ventral forewing. The black borders on the forewing are thinner than *E. e. tildeni*. Univoltine: June–September.
Counties They Fly In: *E.e. tildeni*: Santa Clara; *E.e. bayensis*: Contra Costa, Marin, Solano, and Sonoma
Great Places to See Them: *E. e. tildeni*: Mount Hamilton (Santa Clara County). *E.e. bayensis* is more of a coastal butterfly than Tilden's. Many spots are clouded in the fog zone. Kent Pumphouse Road, behind Alpine Dam (Marin County) (see **Best Butterfly Walks** for directions). Walk behind the dam along the road for about a mile until the cliff face comes perpendicular to you. You should see the Dotted Blues' host buckwheat there. Miller/Knox Regional Shoreline (in the old quarry), Point Richmond (Contra Costa County); Ring Mountain (Marin County).

01: dorsal ♀, Tilden's Blue
02: dorsal ♂, Tilden's Blue
03: dorsal ♀, Bay Region
04: dorsal ♂, Bay Region
05: mature larva

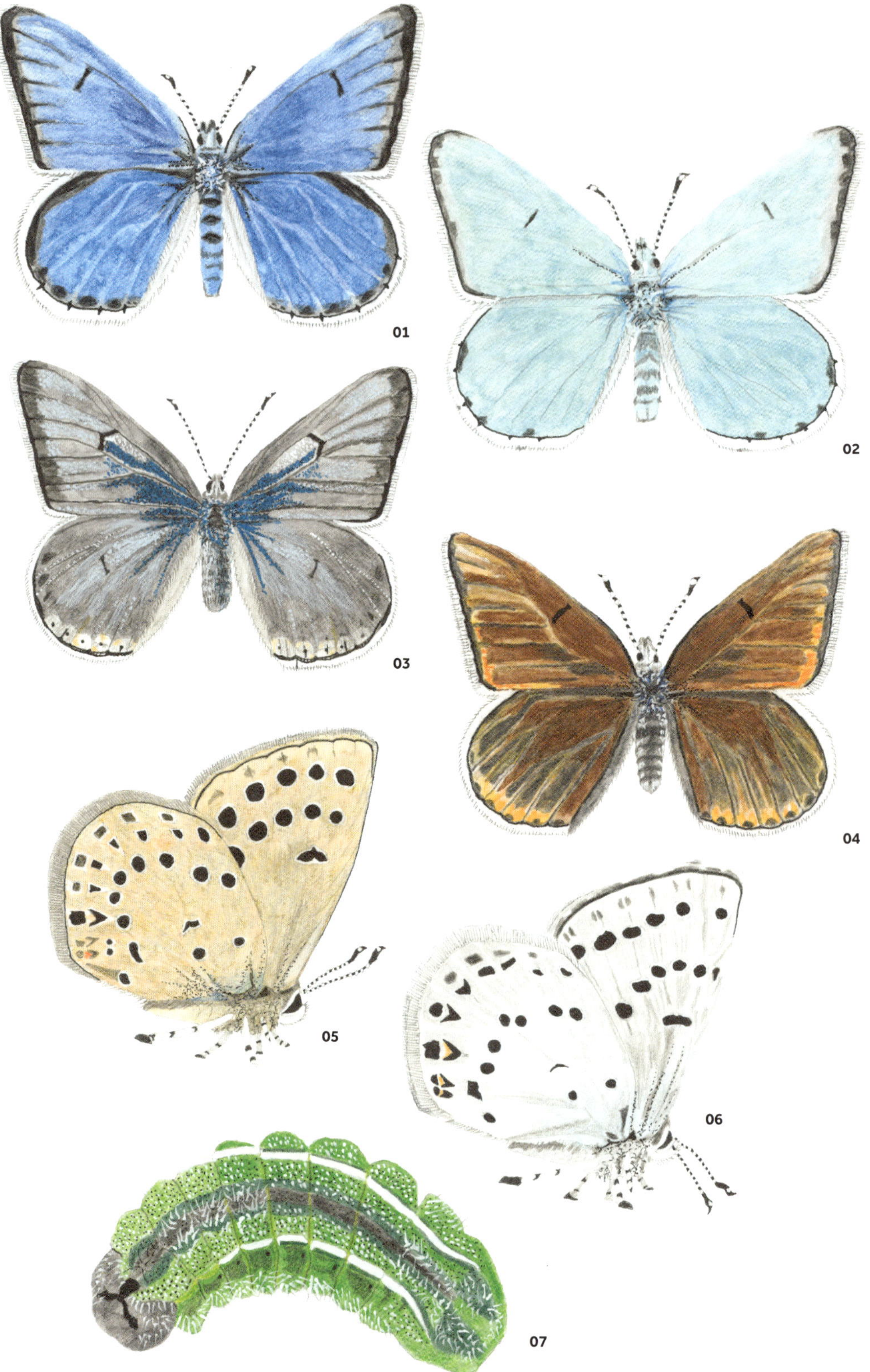
01
02
03
04
05
06
07

GREENISH BLUE

Boisduval 1852

Icaricia saepiolus

Icaricia saepiolus can be mixed up with Boisduval's from below. This butterfly is noticeably smaller. Both within a population and among populations, the adults can vary widely in how they look. A dorsal disc spot on the forewing helps key this creature out. They fly very close to the ground.

Habitat: Moist meadows, forest glades, marshes, grasslands
Host Plants: White clover (*Trifolium repens*), or bird's-foot trefoil (*Lotus corniculatus*) when the clover dries up. Males stay close to the host all day (Scott.1986.409).
Life Phases: Larvae overwinter and resume feeding in spring. Pupa has a gray field with black spots. Female adults can come in a blue or brown form. Males come in a light or dark blue form. Can have a coppery field on the ventral side. I find using this helps key them away from others: the bordered edge of the forewings is almost perpendicular from its bottom corner to the tip of the apex—much straighter than most blues.
Counties They Fly In: Marin, San Mateo, and Sonoma
Great Places to See Them: Good question. The historic record indicates a sighting in each of the counties listed. A solid recent sighting was when Donna Pomeroy, an avid iNaturalist user, identified one a few years back along Highway 35 in San Mateo County. I believe the Greenish Blue is going the way of the two Parnassians, the Pacific Fritillary and other high-elevation species that thrived when the coast was colder. It's too warm here for these guys anymore. Perhaps folks exploring Sonoma County a little more thoroughly might strike gold.

01: dorsal ♀, blue form
02: dorsal ♀, brown form
03: dorsal ♂, dark blue form
04: dorsal ♂, light blue form
05: ventral ♀
06: ventral ♂
07: mature larva

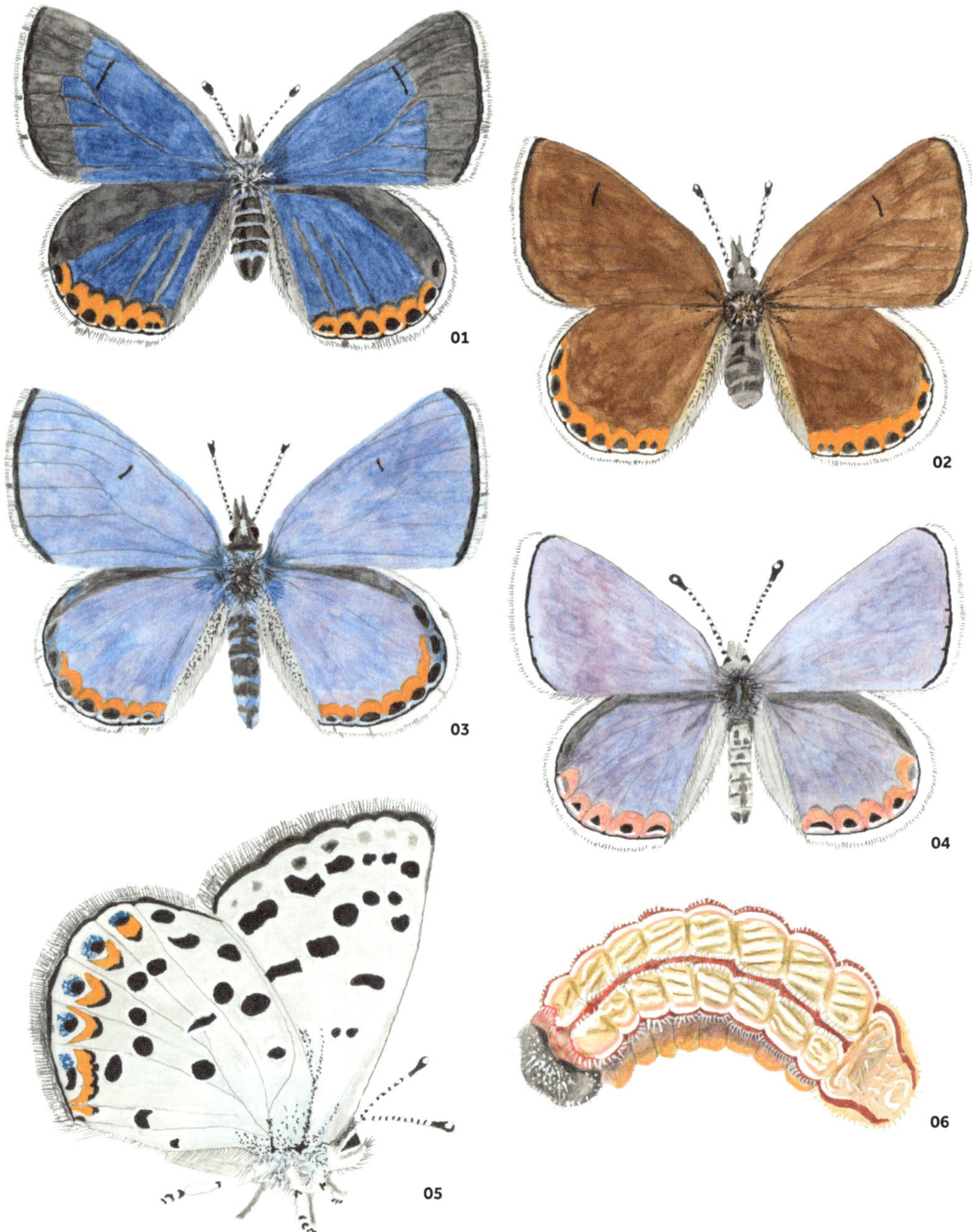
01
02
03
04
05
06

ACMON BLUE

Westwood & Hewitson 1852

Icaricia acmon

Can easily be the first blue you learn because of the orange aurora (band) at the base of the upper hind wings on both sexes. Most of the tending by ants is within the native ant genus *Formica,* though the non-native Argentine ant is known to tend as well.

Habitat: An array of communities, both disturbed and native environments. Not usually found in shady enclaves.

Host Plants: Acmon Blue parties on the pea family: bird's-foot trefoil (*Lotus corniculatus*), deerweed (*Acmispon glaber*), Spanish clover (*Lotus purshianus*), clover (*Trifolium* spp.), lupines (*Lupinus* spp.), and buckwheats (*Eriogonum* spp.)

Life Phases: Larvae are tended by ants. Pupa tannish with a greenish middle. The adult female comes in two forms: a dark blue version in the early spring (f. cottlei) and a brown wing in the summer. The males range from sky blue in the spring to an almost pinkish lavender wing later in the year. Multivoltine: almost year-round.

Counties They Fly In: All

Great Places to See Them: Take any one place mentioned in the book up until now, with perhaps the exception of a redwood forest, and Acmon Blues will be there. I find them readily in vacant lots because of the non-native trefoil host.

01: dorsal ♀, early spring
02: dorsal ♀, summer
03: dorsal ♂, spring
04: dorsal ♂, summer
05: ventral
06: mature larva

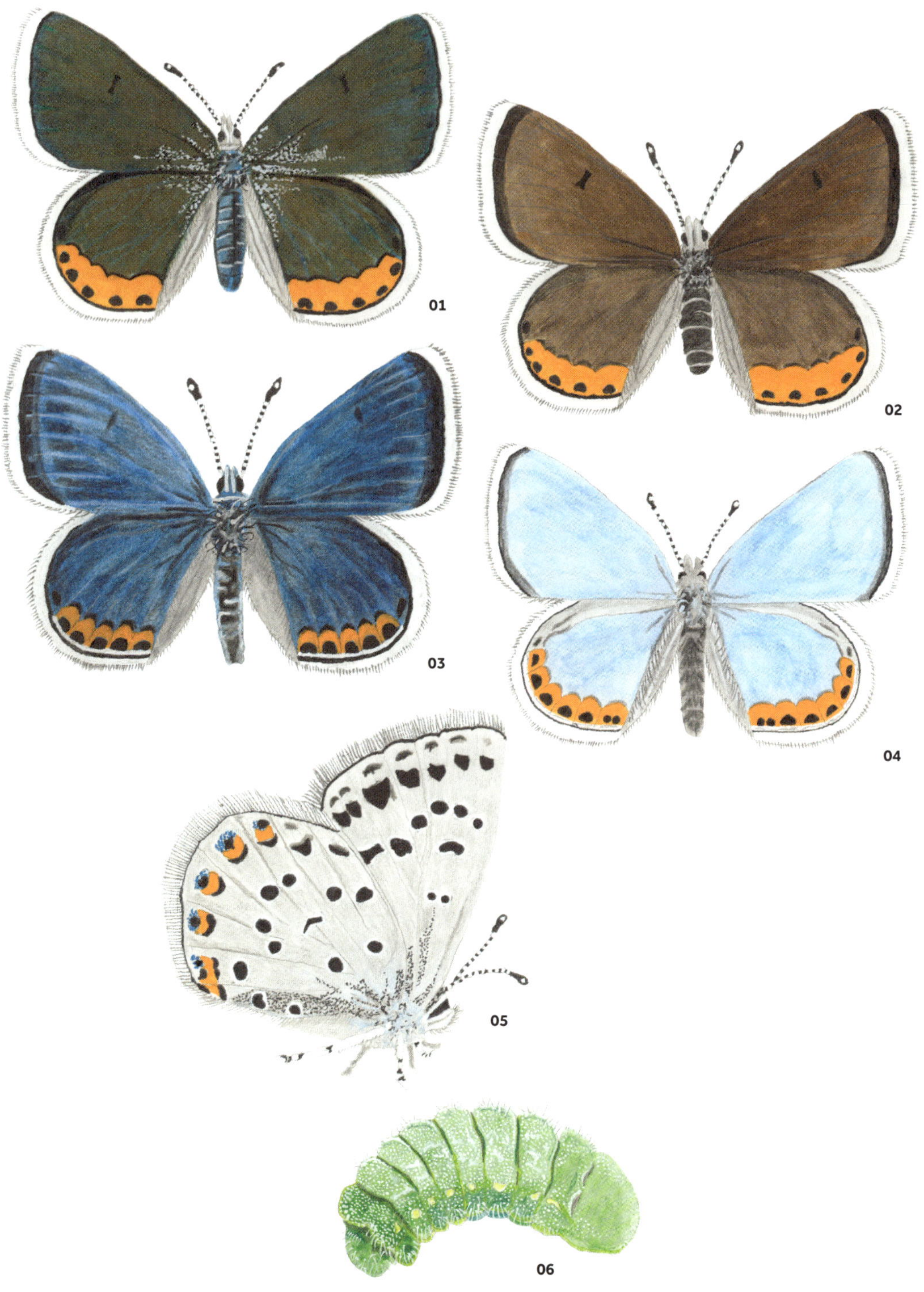
01
02
03
04
05
06

CLEMENCE'S BLUE

Icaricia monticola

Clemence 1909

The name "Clemence's Blue" has been bandied around of late to replace a subspecies of Lupine Blue (*Icaricia lupinus monticola*) as the overall common name. Strangely enough, this butterfly had nothing to do with lupines, so a name change seemed appropriate. They are known to hybridize with Acmons in places beyond the scope of this book—clearly, they have no patience with our desire to keep them in just one category. Clemence's are much rarer than Acmon, and only really identifiable in their summer form. Folks, this one took me years to even broach. Allow yourself to be wrong most of the time—it's a big bag of crazy.

Habitats: Sandy-gravelly slopes, alluvial fields, and flood plains.
Host Plant: California Buckwheat (*Eriogonum fasciculatum*)
Life Phases: Larvae coloration varies according to host plant species and whether leaves of the flowers are eaten (James & Nunnallee.201.223). Pupa is similar to Acmon. Adult females come in two forms: spring, a blue-green gray; and summer, brown (the latter being virtually identical to the summer form of the Acmon Blue female). The male is dark, cerulean blue in the spring and comes in two forms in the summer: a light lavender above or a light blue above. Univoltine: April–September.
Counties They Fly In: Alameda, Monterey, San Benito, Santa Clara, and Santa Cruz
Great Places to See Them: Pinnacles National Park (San Benito County). The two species —Acmon and Clemence's—are sympatric and synchronic here (they fly in the same place at the same time) and it's an excellent place to practice pulling your hair out over them. Enjoy them (see **Appendix B, Tableau 12**). I believe they are also known to fly on Mines Road (Alameda County).

01: dorsal ♀, blue-green form
02: dorsal ♀, summer form
03: dorsal ♂
04: dorsal ♂, form monticola
05: ventral
06: mature larva

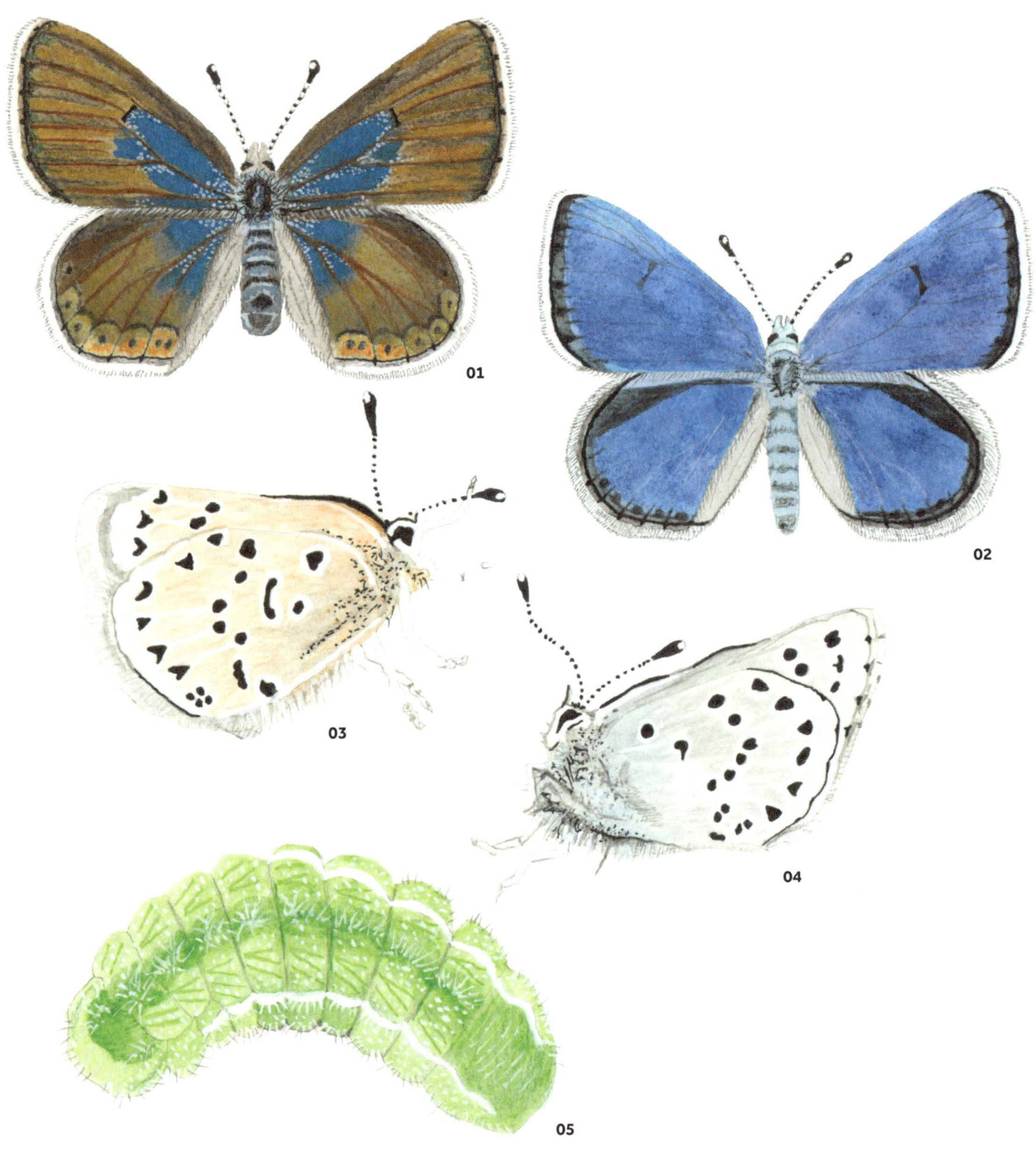
01
02
03
04
05

MISSION BLUE

Hovanitz 1937

Icaricia icarioides missionensis

STATUS: ENDANGERED

I've worked on this butterfly for ten years in all three counties it flies in. It was named after the Mission District of San Francisco when that neighborhood included the peaks. Twin Peaks is the type locality for this insect.

Habitat: Open grasslands, edges of coastal scrub
Host Plants: Silvery lupine (*Lupinus albifrons*), summer lupine (*Lupinus formosus*), varied-colored lupine (*Lupinus variicolor*), bi-colored lupine (*Lupinus bicolor*), bush lupine (*Lupinus chamissonis*)
Life Phases: See entry for Boisduval's Blue. Larvae overwinter in among the leaf litter below the lupine, then crawl back up onto the new growth of the next season. Larvae are ant tended. Pupa reddish brown with spots of green. Females below have a watery copper field while the male's is chalky white. Adults live six to ten days. Univoltine: March–May. Bivoltine on San Bruno Mountain, the first using *L. albifrons* as its host, the second flight using *L. formosus*.
Counties They Fly In: Marin, San Francisco, and San Mateo
Great Places to See Them: The grasslands in the immediate coast are some of the most fragile ecosystems we have. Please stay on the trail in these locations and use your binoculars. At the Old Rifle Range in the Marin Headlands (Marin County), walk along the path on the left. One could catch a glimpse of a patrolling male up on the cliff above. It's where I first saw it. At Twin Peaks (San Francisco County), on east peak, along the northern edge of the road, the odds are with you in the month of May to see it. On San Bruno Mountain (San Mateo County), take the Ridge Trail from the summit parking lot out about three-quarters of the entire trail (it's a long hike; bring lunch and water). Blues should start appearing along the trail patrolling over the lupines.

01: dorsal ♀
02: dorsal ♂
03: ventral ♀
04: ventral ♂
05: mature larva

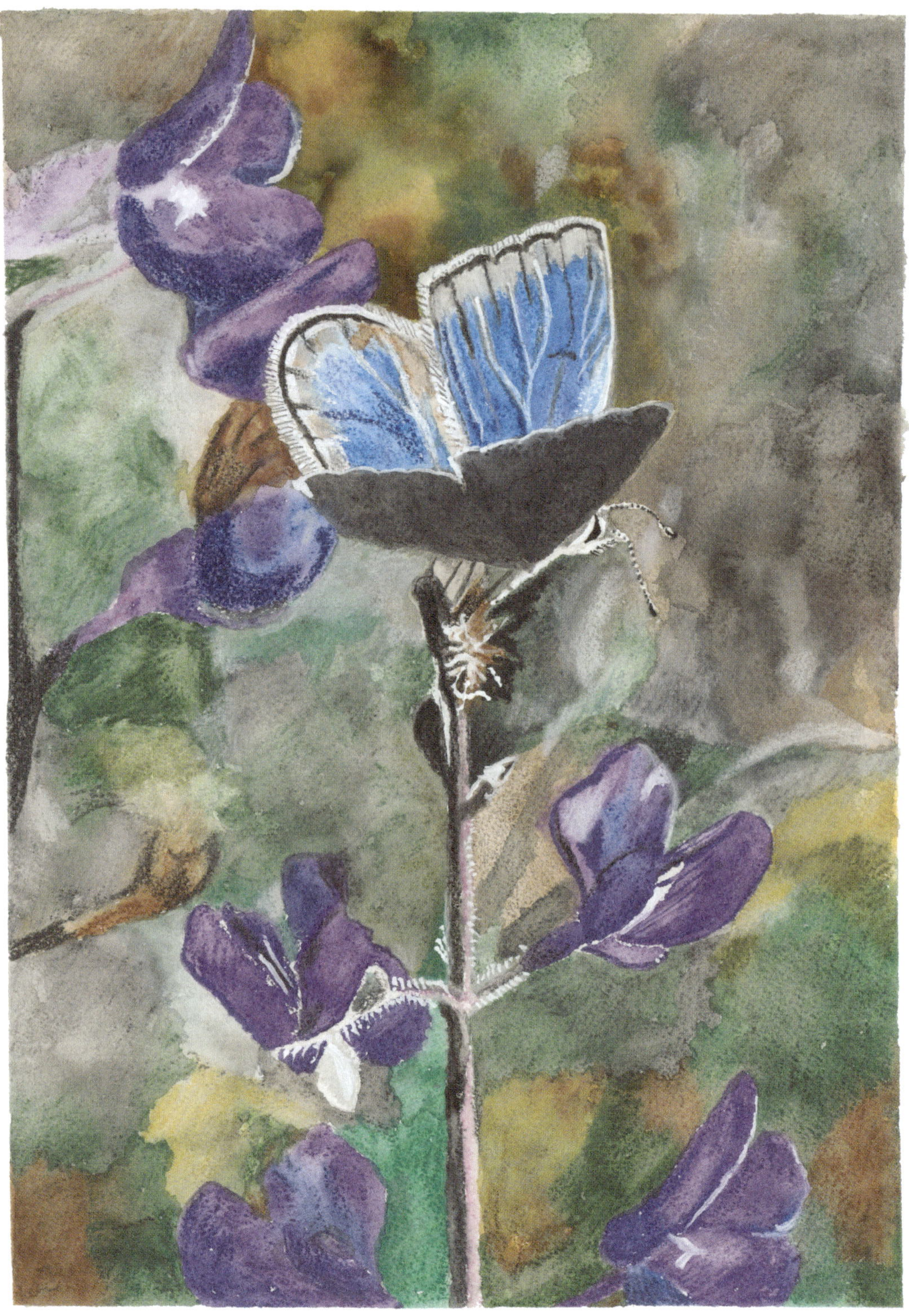

ON A MISSION

Each morning for the last eight years between April and June I've been picked up by Ranit Cohen on the corner of 19th Avenue and Judah in San Francisco because, frankly, I have no car. Ranit works for Coast Ridge Ecology, and I listen in daily amazement to her stories of how she corrals three kids to get them off to school each day—god bless working women. We usually have a good idea how the day will go when we turn north onto the Golden Gate Bridge and see our destination through the marine layer. (Someone barked at me once . . . I think it was Jerry Powell in fact . . . for calling it "fog" when "fog" is created over land. What storms through the San Francisco Bay gate each day like an ephemeral iceberg, then retreats through the famous strait is a "marine layer" formed out over water.)

The Marin Headlands is an extraordinary piece of land saved for the public when the Golden Gate National Recreation Area (GGNRA) was formed back in the seventies. Though much of the headlands is closed to the general public, the place holds such intriguingly named geography as Wolfback Ridge, Hawk Hill, and Slacker Ridge as well as a slew of military batteries, some dating back to the Civil War. We comb the place annually in search of a small, blue butterfly called the Mission Blue (*Icaricia icarioides missionensis*): the male is sky blue from above, and the female is slate gray or brown with some blue toward her thorax. It was named to the federal Endangered Species List in June of 1976 by Paul Opler, a lepidopterist who grew up in Alameda and was chosen to add invertebrates to the list three years after it was created. The Mission Blue enjoys a large tract of land where it can spread out. Much of the work is needle-in-haystack journeying in among precipitous grades of crumbing Franciscan chert rock, watching your foot at every step. It's exhausting some days, but my friend and boss

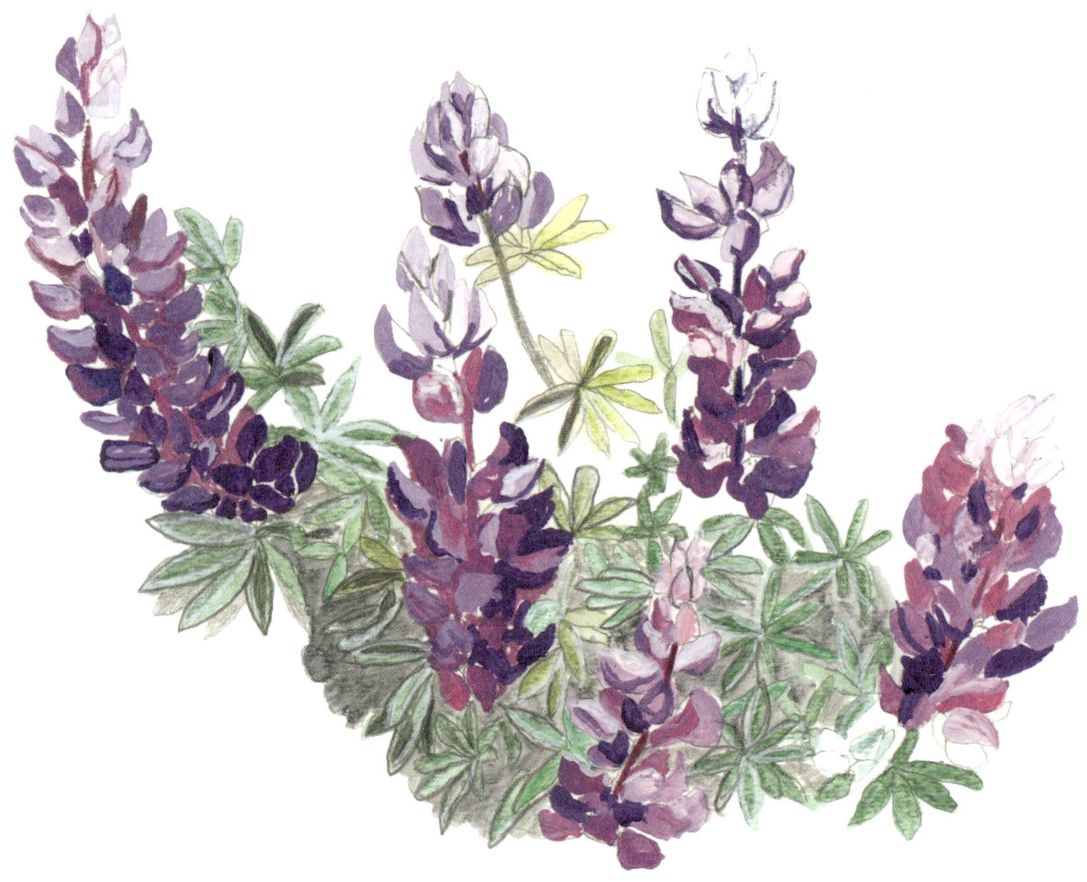

Silvery Lupine (*Lupinus albifrons*)

Ranit has the stamina of twenty, and she sets a mean pace if the weather cooperates with us. She might be the most extraordinary field observer I've ever worked with. Like serious basset hounds doggedly focused—but we laugh in between and kvetch when we don't see them.

* * *

Named after the Mission District of San Francisco when Twin Peaks (its type locality) was considered part of that neighborhood, this is a butterfly I have a long history with. After surveying the county in 2007 for all the butterflies that remained in San Francisco, I put together a PowerPoint presentation with my big announcement as a finale: that the Mission Blue no longer flew on Twin Peaks and we should take the current signs down acknowledging its presence. I'd volunteered enough days back then with the folks who monitored the butterfly there to know that no one from the staff of what's now called the Natural Resources Division of San Francisco Recreation and Parks had seen the butterfly in years and that their methodology of egg monitoring was questionable. I remember one blue streak flying by—"There's one!" a staff member yelled out. It was an Acmon Blue (*Icaricia acmon*). Their cumulative data revealed that nothing could possibly be flying with such low numbers. It really wasn't their fault, as the staff are all exceptional botanists, just not really, at the time, up on their Lepidoptera.

A wiser mind stepped in. Lisa Wayne, who was the executive director of the Natural Resources Division, thought we could do better than that sort of announcement. "Why don't we have a meeting with all the parties involved?" The U.S. Fish and Wildlife Service, her department, et cetera, and see what we could do. Who the hell was I but a loud, braying concerned citizen and bug nerd with no degrees behind his name? Nevertheless, Lisa listened to me, genuinely acknowledged I was on to something, and gave me a place at that meeting even though she really didn't have to do so. My first foray into the politics of butterflies.

> Declining species may require intervention at several levels. At the first, local populations may decline or disappear, while at the second, a species may decline throughout much of its range. In the first instance, local citizens should call these changes to the attention of community leaders and conservation groups; while rangewide changes should be made of national, state or provincial concern. Species or subspecies with naturally small ranges or narrow habitat preferences should be watched especially closely.
>
> (Opler/Wright 1999.85)

It was about this same time that some strange fate stepped in. I was out at the Antioch Dunes National Wildlife Refuge helping to monitor another endangered butterfly—Lange's Metalmark (*Apodemia mormo langei*). Volunteers spread out in a long line with hand counters and then tabulated the butterfly as they walked forward on transect lines. After standing next to this guy all day long he introduced himself. "Hi, my name is Dave Kelly from the U.S. Fish and Wildlife Service. I monitor all the endangered butterflies for them, and I heard you say that you were from San Francisco. You wouldn't have any idea how the Mission Blue is doing there, would you?"

A meeting was set up with Dave Kelly, Stu Weiss from Creekside Science, Lisa, and myself. I'd had a chance to review the restoration plan written in 1976 for the butterfly when it was first listed. It said the Twin Peaks location should never let the Mission Blue population blink out because it might be needed to bolster the remaining population on San Bruno Mountain in San Mateo County. (The opposite occurred.) It also said that the Twin Peaks population might need to be augmented in the future, which translated to needing more butterflies transferred to it. This is a grassland/coastal prairie species, and the natural corridors had been closed down decades earlier with all the development around Twin Peaks, destroying contiguous grasslands, and now its closest population was isolated on San Bruno Mountain to the south. The butterfly has proven itself not to fly over urbanized streets. If she bolts away from the Twin Peaks refuge, the odds of finding her host plant are slim to none.

Twenty-two Mission Blues were moved that first year (ten males, twelve females) from San Bruno Mountain to Twin Peaks. Individual Tupperware containers were used in their transport. It was the first time I was on a permit to catch and handle (as little as possible) an endangered species. I got there early on press day to make sure I found one. High-fives all around to everyone involved in this relocation. Subsequent years have included further augmentation. The Twin Peaks population to date is thriving and in the hands of the Natural Resources Division staff Christopher Campbell, Dylan Hayes, Parke Custis Lewis-Deweese, and others. The butterfly flies on.

The historic rains of 2023 made the slopes of the Marin Headlands explode with wildflower color: lipstick reds of the paintbrushes, the indigo purple of larkspur, and carpets of white Douglas' stitchwort in between the sea of green, and most important to this story, silvery lupine (*Lupinus albifrons*)—the Mission Blue's host plant, the plant she lays her eggs on. Since butterflies co-evolved with flowering plants, a great indicator as to when they might begin their seasonal flight is when the lupine shoots up its grape-soda-colored blossoms. Males patrol between the lupine all day in search of a mate. Females, for the most part, sit directly on the plant. When surveying for the Mission Blue, the headlands are divided into sixty quadrants (each about a half mile across). We try to visit every lupine in our assigned quadrant, marking presence or absence and male or female. If we don't see a Mission Blue, we return to that quadrant at least two more times. We don't always get to all the quadrants (I think our record was thirty-three in a season), but after all these years we have a pretty good idea where we are going to see them. Like any metapopulation—a large population made up of many smaller ones—numbers ebb and flow from year to year. We are in contact with Jeremiah Jolley and Bill Merkle of the National Park Service constantly throughout the survey.

The butterfly has plenty of land here, so it can retreat to certain smaller areas in bad years and spread out widely in good ones. Our numbers go up greatly when we see more females on the wing, and this also indicates the season is coming to an end. A season of surveying can go from late March to early June, throughout the period of the butterfly's annual springtime flight.

Recently, Coast Ridge Ecology added a population in the oak woodlands up above Tennessee Valley. This isolated group uses summer lupine (*Lupinus formosus*) as its host, and the butterflies are much smaller than in other populations—more like a dime as opposed to the quarter size we regularly see. This oak woodlands group is much more dimorphic from below as well, with females having a coppery wash ventrally and the males a chalky blue. There is also a size dimorphism between this population and the Greater Marin Headlands—more dime than nickel. For this community we walk a timed, linear transect with butterflies only counted from seeing what we see on this course and just a little beyond. We note male or female here as well.

It was a rough year for the butterflies in 2023, with many butterflies in California being inundated with heavy rains and fungal growth at the larval stage, as this is a species that overwinters in the larval stage below the lupine in the leaf litter. Overall numbers seemed low as well, but that is a perfectly normal fluctuation for a metapopulation such as this.

A day's work is exhausting with the relentless hiking, but amazingly rewarding. Giving one's time to another creature who needs it is invigorating. Foghorns on the famous bridge sound as our shift comes to an end. The butterfly smartly just hunkers down in inclement, howling weather, and that makes it tough to find. We join the other commuters and tourists crossing back over the Golden Gate Bridge, most of whom are entirely unaware of the extraordinary gossamer-winged beauty that floats in the swales of these headlands.

I now have been privileged to work on this butterfly in three counties. I've helped with a survey for it on Sweeney and Milagra Ridges in San Mateo and was part of the relocation team from San Bruno Mountain to Twin Peaks in San Francisco. I have sung alone on a Broadway stage, but this job transcends all other experiences. There is a completeness in me here.

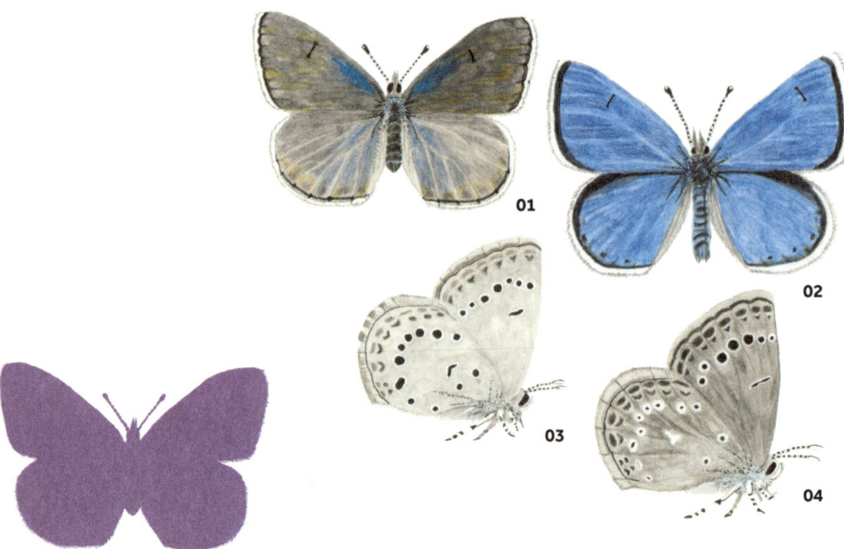

BOISDUVAL'S BLUE

Behr 1852

Icaricia icariodes

Includes subspecies: *Icaricia icariodes icariodes*, *Icaricia icariodes pardalis*, and *Icaricia icariodes parapheres*

With all the subspecies around in the Greater Bay Area, it's difficult sometimes to find pure Boisduval's. *I. i. pardalis* has similar markings to the Mission Blue (*Icaricia icariodes missionensis*) but a darker field from below. In the Marin Headlands, the two fly together, and it makes distinguishing them somewhat difficult. The female *I. i. pardalis* has a color range from the dark slate gray of the Mission Blue in the Marin Headlands to the solid light gray of Pardalis in Point Richmond. I was hired with Ranit Cohen of Coast Ridge Ecology a few years back to inventory the butterflies of Ring Mountain in Marin. We didn't come across *I. i. pardalis* there, but we found it next door at Old Saint Hilary's open space. Check along the creek at the base of the slope. I thought at first it was a Gray Hairstreak . . . one without tails.

The Point Reyes Blue (*I. i. parapheres*) was created from a reclassification of the now extinct Pheres Blue (*Icaricia icariodes pheres*) by lepidopterists Jon Emmel, Tom Emmel, and Rudi Mattoni in the 1980s.

Counties They Fly In:

I. i. icariodes: Monterey

I. i. pardalis: Alameda, Contra Costa, Marin, and Santa Clara

I. i. parapheres: Marin and Sonoma

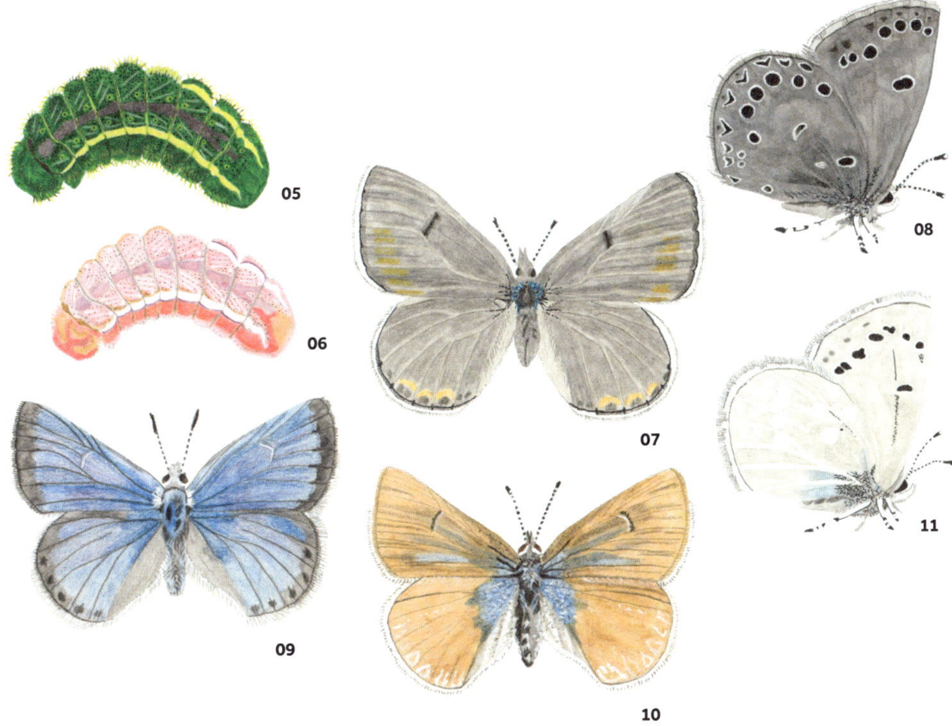

01: dorsal ♀
02: dorsal ♂
03: ventral ♀, light variation
04: ventral ♂, dark variation
05: mature larva
06: mature larva, red morph
07: dorsal ♀, Pardalis Blue
08: ventral, Pardalis Blue
09: dorsal ♀, Point Reyes Blue
10: dorsal ♂, Point Reyes Blue
11: ventral, Point Reyes Blue

Habitat: Grassland, coastal scrub, coastal dunes

Host Plants: Various lupines (*Lupinus* spp.), silver bush lupine *(Lupinus albifrons*), summer lupine (*Lupinus formosus*), bush lupine (*Lupinus chamissonis*), bi-colored lupine (*Lupinus bicolor*), varied colored lupine (*Lupinus varicolor*), coastal bush lupine (*Lupinus arborealis*)

Life Phases: Eggs look like deflated Ping Pong balls. Females lay a single egg three-quarters of the way out from the base of the lupine lancelet. Pupa is green. Adult males are sky blue. Adult females look different among subspecies. Female Pardalis is solidly gray (like a Gray Hairstreak without the tails). Female Point Reyes Blue is coppery gray above with blue toward the thorax. The underside spots of the hind wing of the Point Reyes Blue are white (like Xerces was) without any trace of black.

Great Places to See Them: Boisduval's Blue: Chews Ridge (Monterey County) The species that flies here is clearly *Icaricia icariodes.* What the subspecies here is yet to be determined. *I. i. pardalis*: Ring Mountain (Marin County); Saint Hilary's Open Space (Marin County); Miller Knox Regional Park (Contra Costa County). *I. i. parapheres*: Bull Point Trail, Point Reyes National Seashore (Marin County)—look in the yellow bush lupine and around the buckwheat in the coastal dunes close to the parking lot.

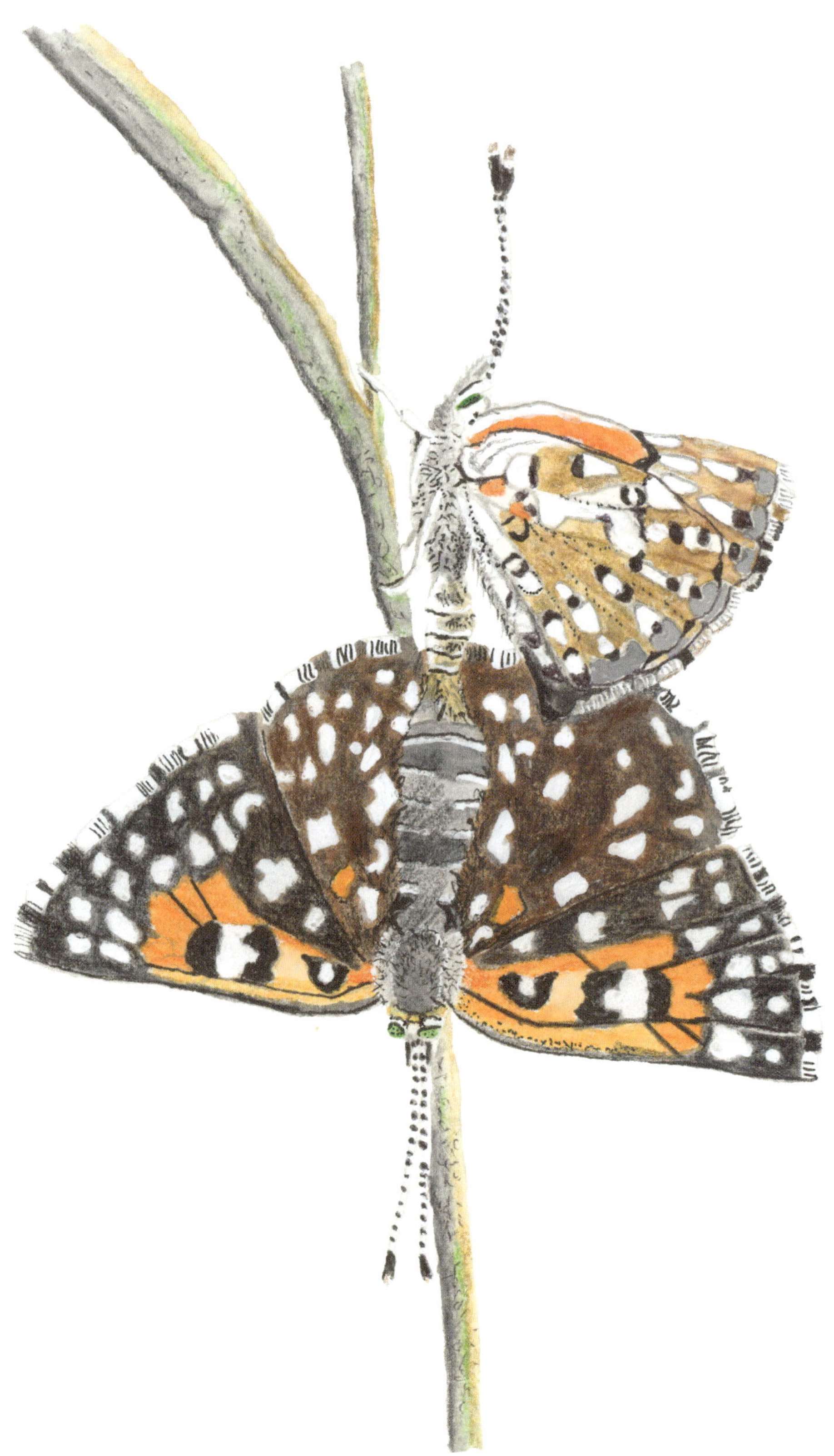

The Metalmarks (*Riodinidae*)

With over a thousand species described worldwide, the metalmarks are represented by twenty-one species in North America.

The butterflies have short wings, are small to medium in length, have long antennae, and are named for the presence of metallic colored thread-like marks on some of their wings. Most of them don't have these markings, and it makes one wonder if this is not a confusing label. There is a characteristic manner in which metalmarks hold their wings when perched. Instead of closed and up over their heads they hold their wings out laterally, flat against the substrate, like a geometrid moth.

Metalmarks used to enjoy being in their own family, then they were lumped back into the Lycaenidae, and now again have been given their own family. We have two here, with the highly endangered Lange's Metalmark being one of them.

Mormon Metalmarks mating

LANGE'S METALMARK

Comstock 1938

Apodemia mormo langei

STATUS: ENDANGERED

In 1997 I was part of a group of volunteers that walked the Antioch Dunes refuge counting this butterfly, for whose survival the land had been set aside in the first place. With clickers in hand, we fanned out in a long line and marched forward through the profuse patches of the butterfly's buckwheat host plant, making a click every time we saw the butterfly. In the end we had 2,246 Lange's for that day. I returned twenty years later to show my friend Cat Chang this remarkable creature. A sizable group of volunteers that day in 2017 had a grand total of one butterfly for their efforts. One butterfly. The cataclysmic collapse had begun in 2000 with a downward trend every year after that, with only 45 seen in 2006. A captive breeding program set up in the subsequent years yielded few results. Though not officially announced yet to the public by the U.S. Fish and Wildlife Service, I believe this butterfly is now extinct.

Ultimately I am so very sad for all of you who will never see a Lange's Metalmark in flight. About the size of a nickel, it had green eyes and blood-orange forewings. It truly was something. A marvelous beauty that not only was silently going but is now irrevocably gone.

01: dorsal, in situ
02: ventral
03: mature larva

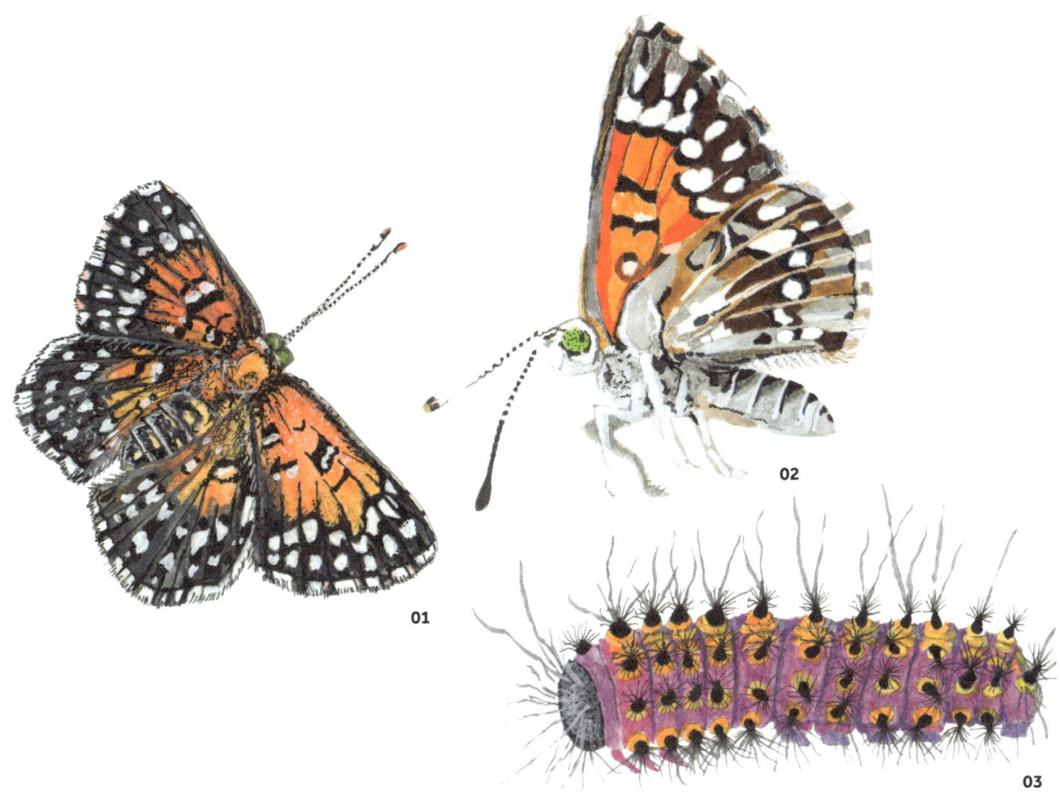

Habitat: Sand dunes

Host Plant: Antioch Dunes buckwheat (*Eriogonum nudum* var. *psychicola*)

Life Phases: Eggs are laid singularly on the host. Pupation occurs just below the leaf litter. Adults are more red orange, with color extending toward the thorax of the hind wing, than other Mormons—though some on the refuge looked like pure Mormons (Powell.pers comm.2010). Univoltine: July–September.

Counties They Fly In: Contra Costa

Great Places to See Them: The only place to see them is the Antioch Dunes National Wildlife Refuge in Antioch (Contra Costa County). The dunes are a disjunct, relict piece of what was once a giant sand dune ecosystem that connected it to the Mojave Desert. The refuge was established in the 1980s when the butterfly was federally listed as endangered. It was the first piece of land set aside in the United States for the protection of an insect. The only way to enter the locked refuge is by invitation.

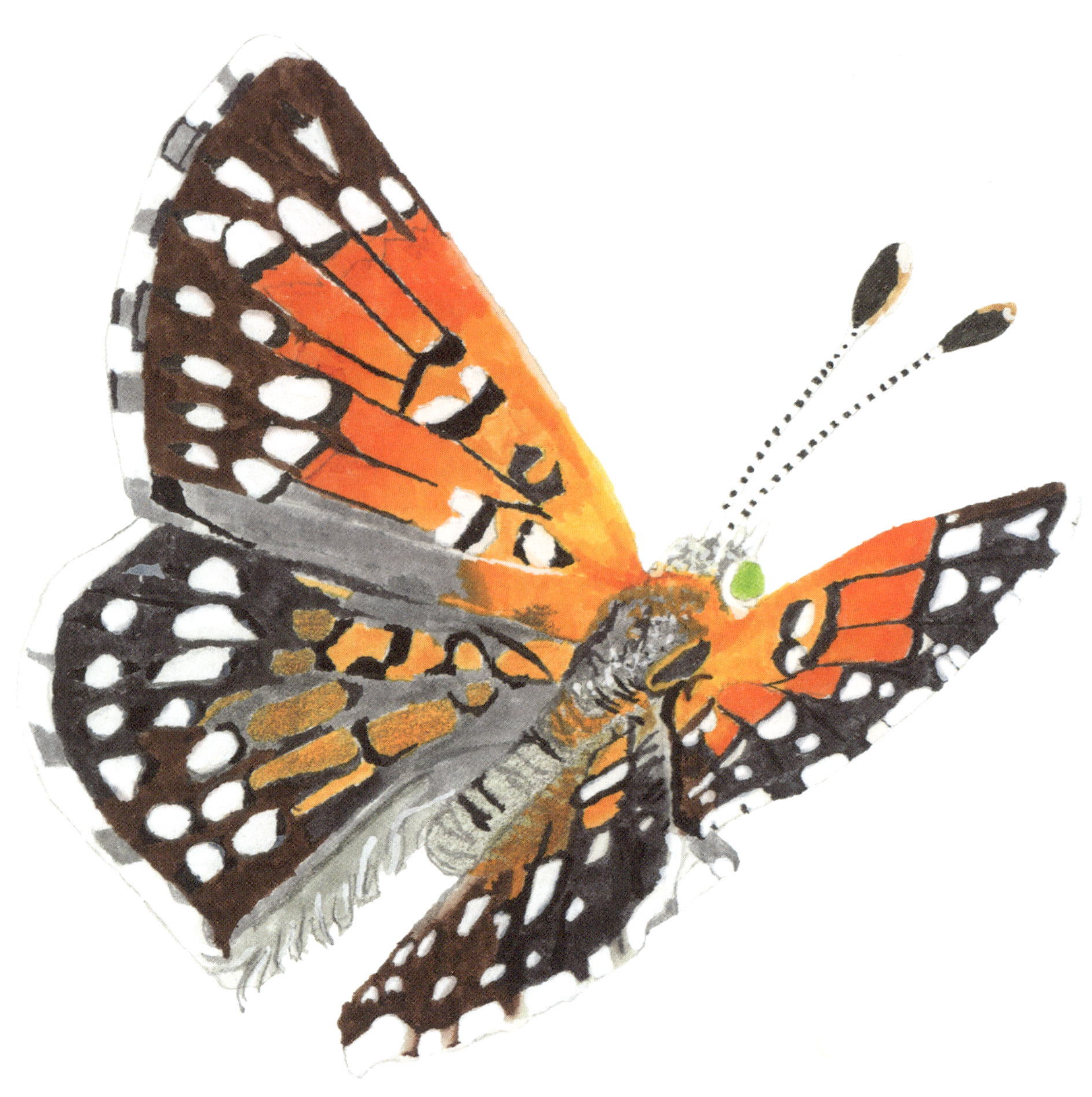

THE HOPE BUTTERFLY

A number of factors played into this catastrophe: it's an isolated population with no new genetics coming in; the refuge had become compromised by the influx of non-native plants, and along with them came a new set of predators that decimated the butterfly at the larval level; sand mining of the dunes continued until recently despite the area's refuge status. Stuart Weiss had an idea to release Mormon Metalmarks into the place to try and capture the last of the genetics of Langes. In 2023 I was informed numbers had dropped so low that the army of volunteers was no longer invited in to help with the survey. Then in November of 2023 I was with a group of botanists in the refuge making a rare-plant inventory along the water's edge. (No chance of seeing the butterfly that day as its flight for the season was finished.) Word on the street that day was that the 2022 survey revealed one butterfly, and the last inventory in 2023 had . . . no Lange's Metalmarks.

I was told to "have hope" that day, that it might rebound on its own. "It might have retreated to a smaller sliver of the refuge we've not seen." But it's a small place, and isn't that what an inventory is for? I just stared at this person incredulously. A butterfly population cannot rebound from numbers like one and none. When does hope cross the line into false hope?

I was considering the crashing trend since 2000—where numbers seemed to have gotten worse each year—while I stood there listening. *Lange's is gone,* I thought at that moment. *It's gone.* Just announce that the butterfly (an *endangered* butterfly at that) is now gone. Extinct. No longer flies at the refuge or anywhere else on the planet. Human management went horribly wrong somewhere here, and the alarm should have been sounded earlier. No one wants to have been in charge when a protected creature goes missing, but playing this information so close to the chest isn't good either. Red flags waving earlier on might have brought forth that someone with the right idea that could have saved it.

Attempts with the captive breeding were in earnest, but one needs a massive influx of butterflies and a commitment by all parties involved over a period of years for this type of reintroduction to be successful. Alexander Broom wrote a blog posting for Save Mount Diablo titled "Is Antioch's Lange's Metalmark Butterfly Extinct?" just one month prior to my visit (Broom.2023. blog). He walked a fine line in coming to any conclusion about the situation. I might be the first to make the leap and say, yes, it's gone. Even if there is a hiccup of a few found in the next few years, the trend is set in cement that this place has become irrevocably compromised. The U.S. Fish and Wildlife Service has taken a "wait and see" approach, but I wonder how long that will go on. And where is the Xerces Society? I wonder if they are even aware of the situation. I will gladly wear the mud on my face if proven wrong and they suddenly appear again in numbers, but rising from the dead should not be any part of a recovery plan. "Fingers crossed" is not conservation.

When writing about the Xerces Blue, I mentioned its extinction had become a banner for conservation and how we must never allow this to happen again. I'm afraid that despite the efforts of a few to save this butterfly from the precipice, it has happened again and in my own lifetime. It flew out there forever in the sand dunes, and for most of that it was blithely unaware of humans. Then a Jepson mining factory went in next door, and the place was cyclone-fenced off. The dust from the factory on the host plants didn't help matters.

We "managed" it and then a tsunami of non-native plants swept in and radically altered this pristine, unique ecosystem. People thought everything was all right because there was a sign saying the dunes were a protected sanctuary for the butterfly. The catch-22 of any fenced-off area is the silent implosion from within and the distracting anesthesia of hope. Keep up the signs of its presence and its picture on kiosks and you really . . . don't even need the actual butterfly anymore. It's relegated to a phantom.

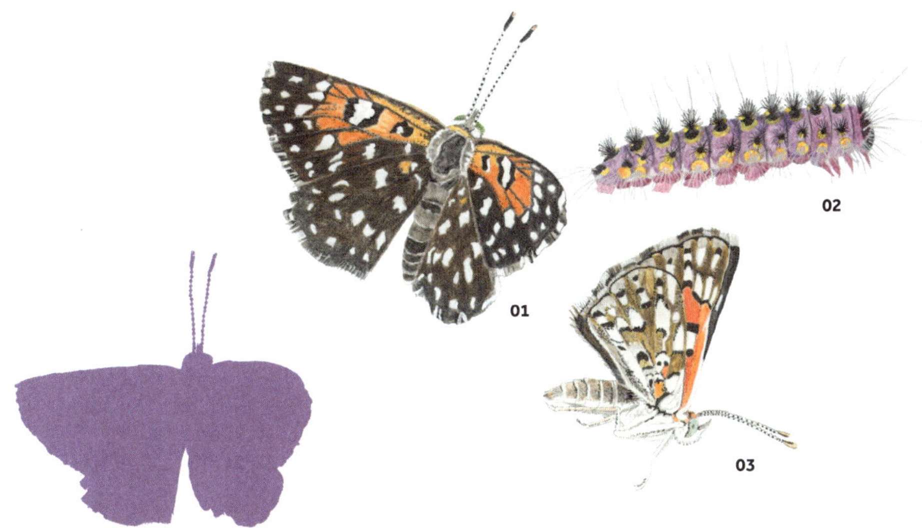

MORMON METALMARK

Felder & Felder 1859

Apodemia mormo mormo

With no actual "metal marks" on the metalmark, this butterfly is the most widespread *Apodemia* in North America (Howe.1975.268). The flight is quick and shimmering. This is the nominate subspecies the entire set of Mormon Metalmarks branches from. When I first started this mania a few decades back, it seemed like a new population of this butterfly had been discovered monthly in some remote part of California and given full species status; no other butterfly has been split up by so many. It's clearly a complex with intense geographic variation. Lucky for us, almost everything within the scope of this book is *Apodemia mormo mormo* . . . except for Lange's.

Habitat: Rocky slopes, chaparral, oak woodlands
Host Plants: Several wild buckwheat (*Eriogonum* spp.), including coast buckwheat (*E. latifolium*), Wright's buckwheat (*E. wrightii*), California buckwheat (*E. parvifolium*), naked-stemmed buckwheat (*E. nudum*)
Life Phases: Larvae are purple with long hairs protruding from the body. Overwinter as larvae. Pupa brown and furry. Adults do not travel far from where they are born. Univoltine: August.
Counties They Fly In: Contra Costa, Monterey, Napa, San Benito, Santa Clara, Santa Cruz, and Solano
Great Places to See Them: Sierra Azul Open Space Preserve (Santa Clara County)—on the trail that goes toward Mount Umunhum; Del Puerto Canyon (Santa Clara County); Pinnacles National Park (San Benito County); Sonoma Coast State Park (Sonoma County); Marina State Beach, Marina (Monterey County)

01: dorsal, in situ
02: mature larva
03: ventral

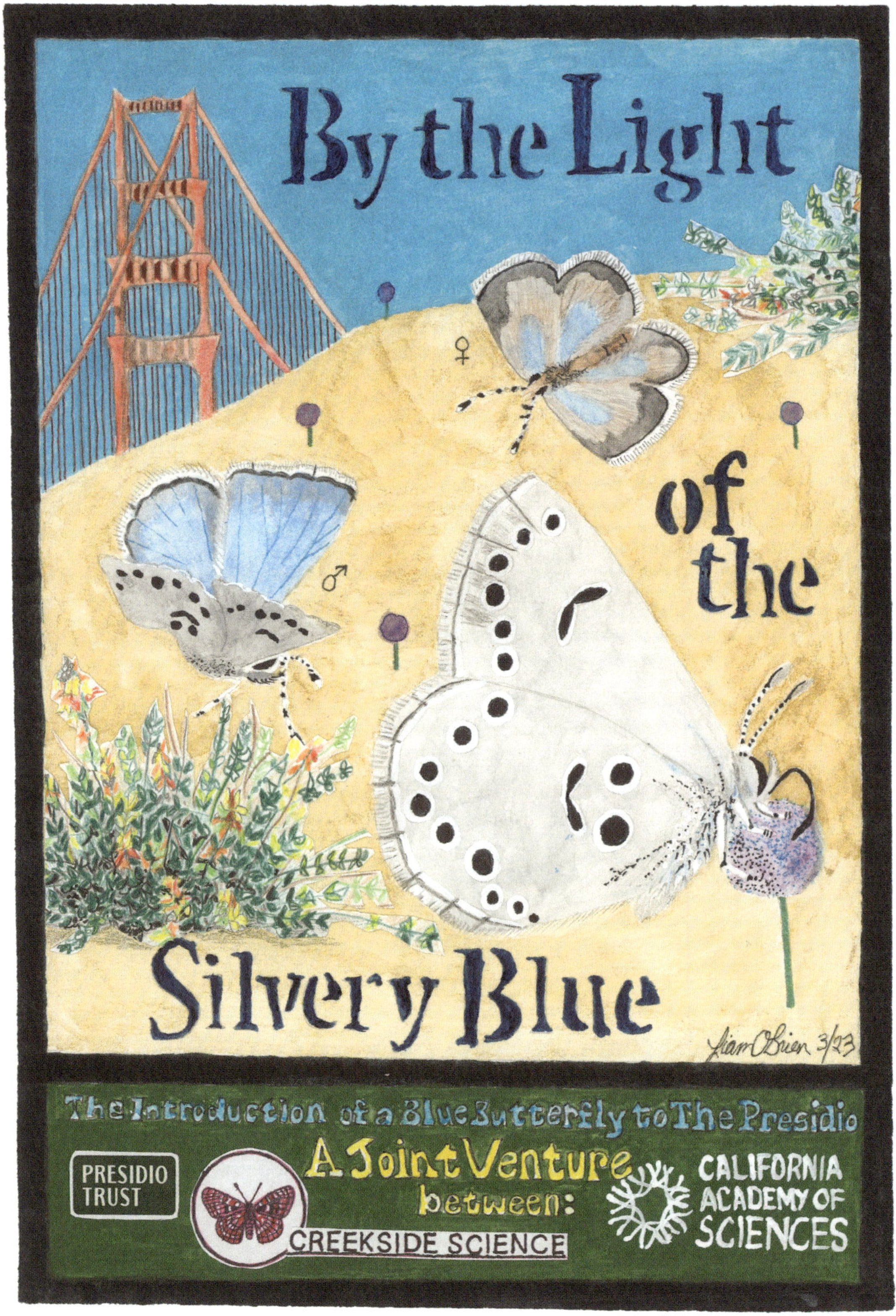

By the Light of the Silvery Blue

Liam O'Brien 3/23

The Introduction of a Blue Butterfly to The Presidio

A Joint Venture between:

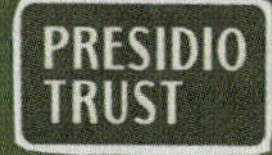

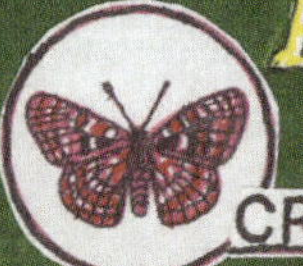

CALIFORNIA ACADEMY OF SCIENCES

XERCES RISING (SORT OF . . .)

A decade or so back I was taken to lunch by Ryan Phelan, who, with her husband Stewart Brand, created a company called Revive & Restore. Inspired by the headway made in the cloning of the woolly mammoth, they are the folks behind the possible cloning of the passenger pigeon, and she wondered what I thought of bringing back another creature they had their eyes fixed on: the Xerces Blue, the first butterfly to disappear as a result of human development (see **In the Shadow of Xerces**). Stunned (and wondering why she would have contacted me), I think I said, "Is that even possible?" The little I knew of it was that it was far easier to clone a sheep or a dog (I knew Barbra Streisand lived with two cloned dogs) than an insect. I remember hearing at the time that we barely had enough sequencing technology to figure out two legs of a butterfly, let alone the whole creature. It was a fascinating lunch, but I didn't hear much about it after that. I think she must have quickly realized I'm not a science guy.

Then at a meeting in 2020 I had been asked to attend, it quickly became clear just how much the project had ramped up. The project had been broken down into two phases: the first would be introducing Xerces's closest living relative, the Silvery Blue (*Glaucopsyche lygdamus*) into portions of dunes within the Presidio that the Presidio Trust has managed and restored. Silvery Blues are a patrolling butterfly, males and females flying in between the host plants (in this case, deerweed [*Acmispon glaber*] in search of one another just like the Xerces did. Stu Weiss from Creekside Science drove that portion of the discussion along with Lew Stringer from the Presidio Trust. It was important that the Silvery be established as an indicator species and fill in the missing component to the ecosystem that Xerces had. Ben Novak, the lead scientist for Revive & Restore, and Durrell Kaplan from the California Academy of Sciences, who has the daunting task of extracting

A poster I created for the introduction of the Silvery Blues to the Presidio the genome from Xerces specimens at the academy (a procedure the entire project hinges on) were in attendance as well. Ryan Phelan and Stewart Brand of course were there too.

The Academy holds over four hundred specimens of Xerces in its collection (multiply that by the many university collections throughout the country and one gets a sense of what an impact collecting had on this butterfly—it was highly sought after.) Sequencing will occur on the Silvery Blues as well to relocate just the right individuals from a coastal colony and also to see if they are in any way relatives of one another—Silveries and Xerces. Many people have argued that Xerces was not a separate species but a form of the Silvery Blue, that it came in three forms, two of which look incredibly like the Silveries of today. The latest belief is that it is indeed its own species. Kaplan and the California Academy of Sciences received a grant of $18,000 from Revive & Restore to tackle isolating the genomes. As much as the project centers on this, it is not even known if cloning is possible yet. Much of this is way over the head of any average layperson, including myself. But I gotta say—it was all very exciting stuff to listen to.

In the years since that first lunch in the late 2010s, the Presidio Trust has done an extraordinary job in coastal dune restoration that has gone on with little fanfare. Mountains of sand, much of it removed to make way for the new tunnels along Presidio Parkway, have been placed along Wedemeyer Street, towering high like the dunes that used to roll through the western part of the city. Native plants like dune gilia, coast buckwheat, sea rocket, Indian paintbrush, and deerweed anchor this renewed piece of San Francisco. The Xerces Blue is a part of that still incomplete dune ecosystem. I asked Lew Stringer, the associate director of nature resources for the Presidio Trust, to comment on this portion of the project for me:

> Once the fourth largest dune system in California, San Francisco's dunes have largely given way to urban development. However, since the Presidio became a national park in the 1990s, efforts have been made to bring back a piece of this natural heritage. The National Park Service and the Presidio Trust have successfully restored about a hundred acres of the original sand dune ecosystem. This has provided a much-needed refuge for the plants that were once on the verge of extinction, like the San Francisco *Lessingia*, which are now flourishing in these pockets of reclaimed wilderness.

This endeavor is not just about cloning a butterfly (though that will be an accomplishment in itself) but about making whole again something that was damaged. And in a fascinating irony, the area right behind the old veterans' hospital (now apartments) where the last Xerces was captured in 1946 will play a great deal into its return. If everything goes right, the scientists believe it could be as short as a few years before fully realized generations of Xerces are back on the wing after the initial cloning.

In the fall of 2001, Robert Michael Pyle put forth in a quarterly issue of *Wings* (a publication of the Xerces Society) the idea of "resurrection ecology" for the first time. His article specifically said that introducing "an organism closely related to an extinct type can result in the functional reconstruction of the animal thought to be lost" (Pyle.2001). He was listening to a lecture one night in England by T. G. Howarth on the demise of the Large Blue (*Phengaris arion*) when he came up with the idea for the Xerces Society, a conservation group formed for the protection of stressed invertebrates living on the edge.

Pyle makes no mention of cloning in his proposition because, quite frankly, it wasn't on anyone's radar at the time. He did, however, propose moving Xerces's closest-living relative, the Silvery, into newly restored habitat within the Presidio (back then he was referring to the area near Crissy Marsh along the bay). The current idea is to reintroduce them into a much larger area. Whether or not cloning works, we can try to offer a repaired ecosystem, with pollinators and plants helping each other thrive.

There is a long way to go on this project, but the wheels are turning. I humbly wonder if I'll be alive to see a pure Xerces on the wing. If all the powers that be can pull this off, it will be a game changer. We all have to be patient and let the scientists do what they need to do. Now is not the time for cynicism. We should be in full triage mode with our insects as they disappear at an alarming rate. The work on Xerces has got to be some of the most cutting-edge conservation going on in the nation. New ideas have sprung forth from San Francisco since the day it was founded. It gives me hope that this tool (cloning) will be invaluable for other declining creatures—maybe the Lange's Metalmark, which we've lost in my lifetime. Though as Lange's shows, without pouring real effort into protecting a thriving habitat, even cloning won't be enough to save our most vulnerable creatures. At the very least, a new butterfly, the Silvery Blue, will fly in our county to help stitch the tatters of our dunes back together.

The Lepidopterists' Society Open House was held at UC Davis in February of 2023. I have it on good authority from someone highly involved with the project that the genome sequencing for the Xerces Blue has been completed, and they have successfully isolated it. I confirmed this with Durrell Kaplan. He believes his data shows that Xerces was a distinct species and separate from Silvery Blues. That is a wonderful breakthrough. I reentered the project after interest by the Cal Academy in a poster I created regarding Silveries' introduction to the Presidio.

This poster will help inform the public of the ongoing efforts with the Silvery Blue. It was at a meeting for the poster that Durrell said in passing that "cloning had never been part of their intent" for this project. What? Did I get it wrong from the beginning here? It was more of Revive & Restore's dream of cloning Xerces, and I jumped to a conclusion that that was even possible when they funded research for Durrell's work. I'd even started making art around Xerces's return. We are still light years off from cloning a butterfly, and the implementation of Pyle's dream to restore the Xerces's closest living relative to the Presidio was the actual high bar that was set.

Acknowledgments

In the spring of 1996 I did my first butterfly count. It was called "South San Francisco" and was centered out of the Don Edwards San Francisco Bay National Wildlife Refuge in Alameda County. The count leader was John Steiner, an amicable guy who had authored a small pamphlet *Butterflies of the San Francisco Bay Region: A County Species List* that was for sale that day. This was a pivotal day in my journeyman years with butterflies. The purchase of that small, three-fold brochure became the Rosetta Stone for me for the next three decades as my desire to see all the ones on John's list grew.

Fast forward to 2010. I was contacted by Bill Shepard, who had been contacted by John Steiner. "Would you or Liam like all my research documents?" This was data for his master's thesis: "Bay Area Butterflies: The Distribution and Natural History of San Francisco Region Rhopalocera." Bill said he didn't want them anymore. "He decided just to get rid of a lot of stuff and wondered if you or I wanted it."

And for some reason . . . I said yes, I'll take them. Bill didn't want them. I think somewhere in the deep recesses of my brain I felt . . . protective? of this source material that had ultimately created that little pamphlet I cherished so much. "Great," Bill said. I'll drive them to your place sometime." The day Bill delivered to me the four boxes, they went immediately into the back of my closet, and there they sat until I entered into conversations with the folks at Heyday in 2019. Without this gold mine of research, writing this book would have been impossible. It is the cornerstone of this book—a source I returned to time and again. I am deeply grateful John saw something in me that valued what he had done. It has taken me into our butterflies in a way I never could have imagined. John Steiner passed away in 2017.

And then another fortuitous event occurred: I got an email from a guy named Tom Wysham right about when I was going to start this project. (The first three years focused on the paintings and the last two years the writing.) He wondered if I'd be interested in his collection, as he and his wife were moving to Washington to be closer to the grandkids. He'd collected two of every male and female species in Sonoma and Napa Counties. My mouth was permanently ajar when I saw them. The most exquisitely spread specimens I'd ever seen, even beyond museum quality and not a nick out of them nor a scale missing. Truly unbelievable. I'd decided previously that I was going to paint in the scientific-spread manner as a homage to the great illustrators that had gone before me: Maria Merian, W. J. Holland, William H. Howe, and Titian Peele. Even Audubon painted butterflies, but he did it on glass in reverse just to blow all of our minds. Tom, your gift was manna from heaven. I was overwhelmed by the scope of this book for most of its journey and your butterflies gave me a starting point—one butterfly at a time. Thank you.

Many authors have discussed that the most trying part of writing a book is isolation. From the day she approached me at a *Bay Nature* magazine gala to discuss the possibilities of a book, Marthine Satris from Heyday has been my base camp. This woman, a fabulous naturalist in her own right, was my den mother, wonderful mentor, champion, and de facto therapist. Her patience and sublime computer skills were everything at times. I'm still not sure I have the temperament or patience to be an author, but she talked me back from the ledge on many occasions. Riding shotgun to Marthine, Emmerich Anklam from Heyday. Thank you Emmerich, and everyone at Heyday: Diane Lee, Archie Ferguson, Victor Mingovits, Marlon Rigel, and Valerie Kamen for turning my paintings into this gorgeous book; Gayle Wattawa and Steve Wasserman; Kalie Caetano, Chris Carosi, Eve Sheehan, and Bradley Trumpfheller.

Jerry Powell, PhD, to whom I dedicate this book: I spoke at your memorial a while back, Jerry, and I really tried to get through it without crying. Didn't make it. It still makes me numb to think you won't be at all the butterfly counts in the season to come. Friend, mentor, and one of the only people I knew that actually ate Kentucky Fried Chicken. Jerry, little did I know that all those road trips and talks would add up to a pavement of golden knowledge. You were quiet but packed a mighty fine heart. Your memories are gifts that I still open from time to time. I was the fool you suffered gladly for. Lucky me. You picked a fine partner in Liz Randall. We all miss you, Jerry, terribly so.

Robert Michael Pyle, the godfather of American butterflying, was an early supporter of my conservation work. I met him and his now-departed wife Thea in 1999 in a butterfly class on Washington butterflies for the North

Cascades Institute. To be around Bob Pyle is a thing of beauty: brilliance in the subject at hand and an undivided gaze deep into your soul when he's talking to you. His library of books is a must for any serious, passionate lepidopterist. His words allow the reader to float without a safety net and remain transfixed in this world. Thanks for the factchecking, Bob.

To Paxton Gate, a quirky little store full of oddities in San Francisco where I bought my first butterfly net back when they were on Brady Street.

I've done all of this, folks, without the use of a car. I bummed rides from people. I drive. Just owning a car in the city of San Francisco is too much of a liability these days. There are two people, however, who not only took me to assorted locales but became friends along the way. Tony Iwane from iNaturalist is the king of the road trip, spontaneous in manifestation and duration. We think traveling hundreds and hundreds and hundreds of miles in two days is nothing if you have the right music playing. For your company, your curiosity and general joie de vivre, thank you, Tony. Cat Chang, a person I met through lichens, might be the most well-rounded naturalist I've ever been afield with. Smart, analytical, and with a killer sense of humor, Cat is pure joy for me to be around. She vetted all the host plants here. And I'd be remiss not to thank her husband, Jim Kaiser, who was glad there would be another body between Cat and the proverbial mountain lion he was convinced she would succumb to if she was out in nature alone. Tony and Cat have amazing brains, and we laugh our asses off. I love them completely.

The task of vetting all the lepidoptera fell on Paul Johnson, maybe the best lepidopterist I know. He's put up with my neurosis for many decades now, simultaneously laughing at my humor and being horrified at what comes pouring out of my mouth. We've fought like family but we seem to survive the estrangements. There was no one I knew skilled enough to dive into this pool. I'm so very lucky he agreed to help me. I'm so very lucky to know such an honorable man.

Thank you, David Loeb and Dan Rademacher from the early days of *Bay Nature* magazine: I walked into that office confident that my art belonged in their periodical, and thank God they agreed. My first big break. And a shout-out to the current executive director, Victoria Schlesinger. She gave me my first cover, which directly led to this book.

Lisa Wayne, former director of the Natural Resources division of San Francisco Rec & Parks for listening to this opinionated, upstart crow telling you how we might save this endangered butterfly on Twin Peaks in San Francisco. Extraordinary leadership this woman possesses.

To the late Bill Shepard whom I'd only see annually around count season and could talk about butterflies like presents on Christmas Day. I miss your joy, Bill. Your widow, Ginger Morris, continues your passion for butterflies.

To my brothers, Sean and Colin, whom I love tremendously. I'm sure you both reached the end of your patience with me prattling on about nothing but the book for the last five years. Your support has been wonderful. Colin, coming through with a new computer for me at a critical juncture was an act of generosity and understanding that I'm utterly humbled by. Thank you.

To my roommates, Bradley Jackson and Mike Rittenour, who've put up with all my quirks and peccadillos and eccentricities through this process.

Ken-ichi Ueda, for the use of his open sourced data on the marvelous iNaturalist website. Join iNaturalist, folks. It'll change your life for the better. Go be curious.

Thank you to my mother, June O'Brien, and to my aunt, Patricia Roth, some of the first people that encouraged my art as a boy. And my kindergarten teacher, Miss Pearl, who put my papier-maché grasshopper on display in the principal's office in 1968, and I remember thinking at the time at the ripe old age of six, "I'm on to something here."

To Art Shapiro, retired professor emeritus at UC Davis, who mentioned in passing in a letter to me once, "You are a fine illustrator, Mr. O'Brien."—one of those small yeses we gather along the way.

The cavalcade of thank you's continues: Patrick Kobernus and Ranit Cohen from Coast Ridge Ecology, Stu Weiss from Creekside Sciences, Mia Monroe, Barbara Deutsch, Ruth Gravanis, Jake Sigg, Jonathan Young, Damien Raffa, Peter Brastow, Amber Hasselbring, Matt Zlatunich, Bob Hall, Dave Bartholomew, Chris Tenney, Eric Simon, Leslie Flint, Donna Pomeroy, Rebecca Johnson, Alison Young, Lew Stringer, James Kruse, Opal Essence, Mary Ellen Hannibal, Erica Harris, J. J. Johnson, and the late Greg Lasley.

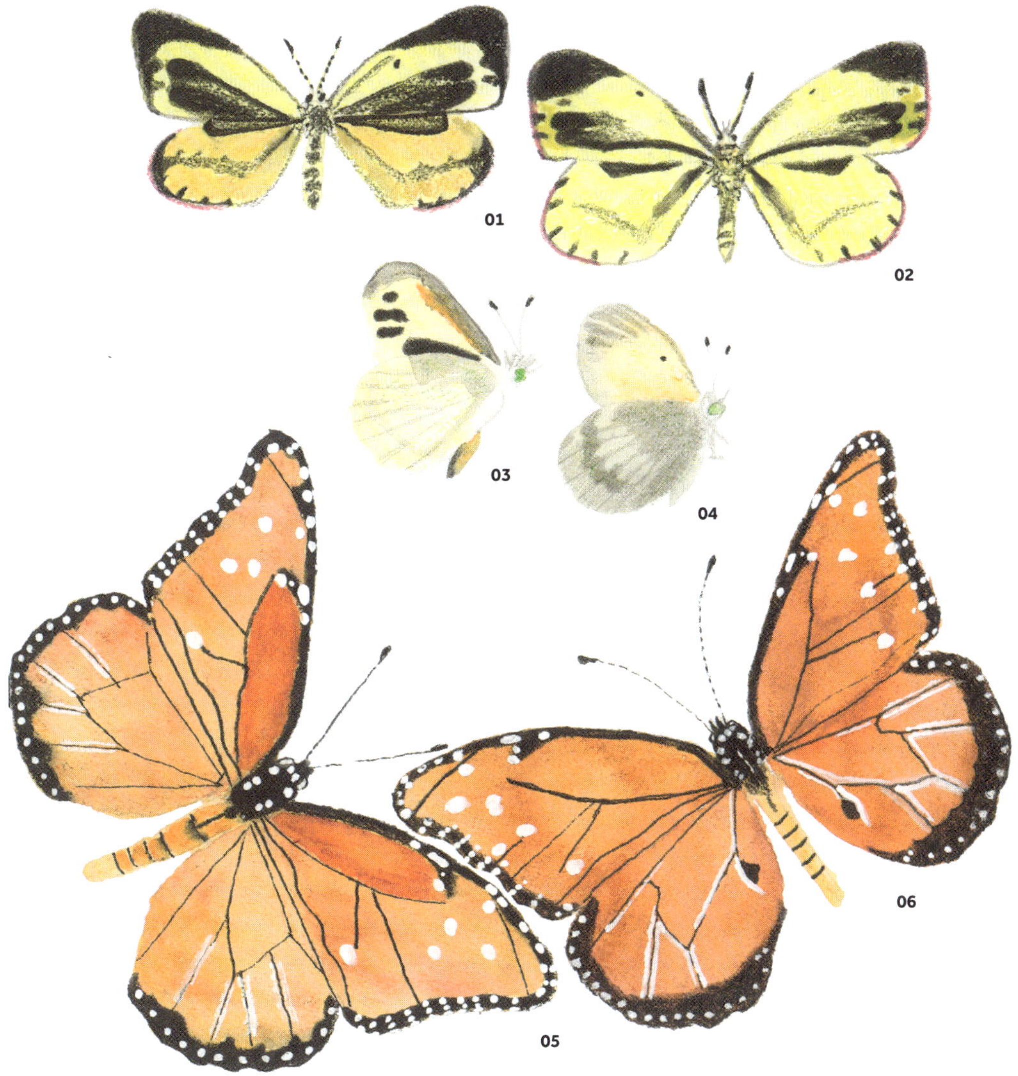

Appendix A: Strays

01: dorsal ♀, Dainty Sulphur
02: dorsal ♂, Dainty Sulphur
03: ventral, Dainty Sulphur, summer
04: ventral, Dainty Sulphur, dry season
05: dorsal ♀, the Queen
06: dorsal ♂, the Queen

Is it a vagrant that got blown off course? Is it a colonial butterfly pushing it's northern range? Did it escape your local butterfly house? The answer is probably yes and no. From time to time these anomalies appear in our area, some of them with enough regularity to predict we'll see them again. I'm so codependent I would carry enormous guilt if something landed before you in your journeyman or journeywoman years and I had not mentioned it. Here are two that you just might likely see:

The Dainty Sulphur (*Nathalis iole*). Pierids don't get any smaller than this one. A large push of them happens every once in a while up from the southern section of the state. They love disturbed habitats full of weeds, specifically sneezeweed (*Helenium autumnale*), cultivated marigold (*Tagetes*), and common chickweed (*Stellaria media*). They die back annually from frost conditions.

The Queen (*Danaus gilippus*). This one has been reported with enough sightings in San Francisco to make it to the list. We have a butterfly house in the Conservatory of Flowers in Golden Gate Park that is run from time to time but not since COVID, so the sighting of one of these in the Botanical Garden in 2023 was solid. We also have a permanent butterfly house at the California Academy of Sciences. Butterfly houses import many non-local, tropical species so I naturally assume these are escapees.

They'd been reported years back in the Lobos Dunes of the Presidio as well. I always think *Danaus gilippus* falls under "advanced butterflying" because it looks so much like a Monarch (*Danaus plexippus*) that is easily overlooked, and we aren't really looking for it within the scope of this book. Much more brown or dirty orange than the bright orange of a Monarch. Another one that blinks out in a frost, Queens seem to visit us randomly. They are milkweed-dependent butterflies for their egg-laying and use many species of *Asclepias* including the native narrowleaf milkweed (*Asclepias fascicularis*) but are probably using mainly the non-native tropical milkweed (*Asclepias curassavica*) to hopscotch north. It does not overwinter in our area.

Appendix B: Tableaux Keys

TABLEAU 1

Persius Duskywing vs. Pacuvius Duskywing Side by Side

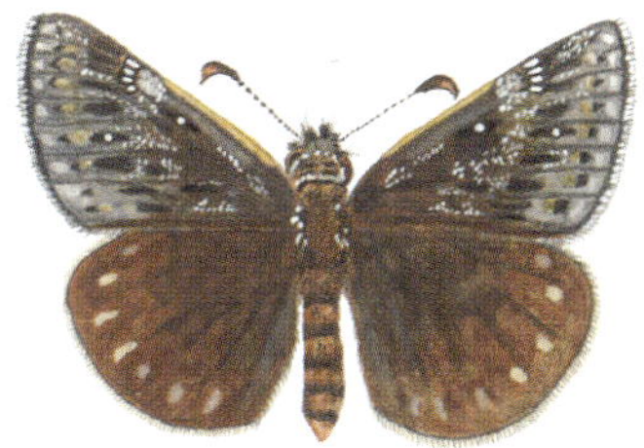

Persius Duskywing (*Erynnis persius)*
male, dorsal

- extra spot of forewing disc cell than Pacuvius
- black hairs on tibia

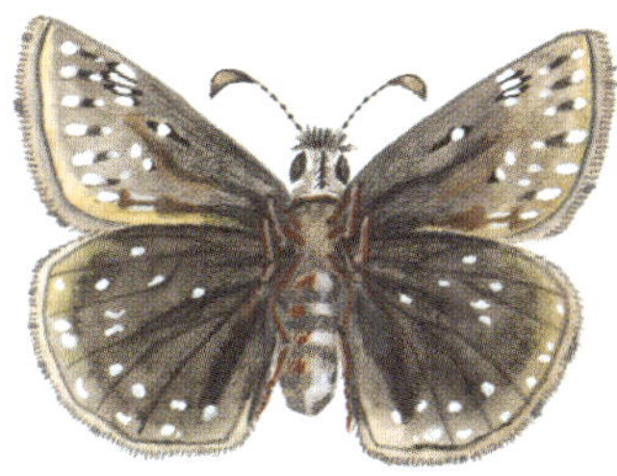

Persius Duskywing (*Erynnis persius*)
female, ventral

- heavily frosted dorsally
- hyaline spots dorsally
- black and white patches ventrally

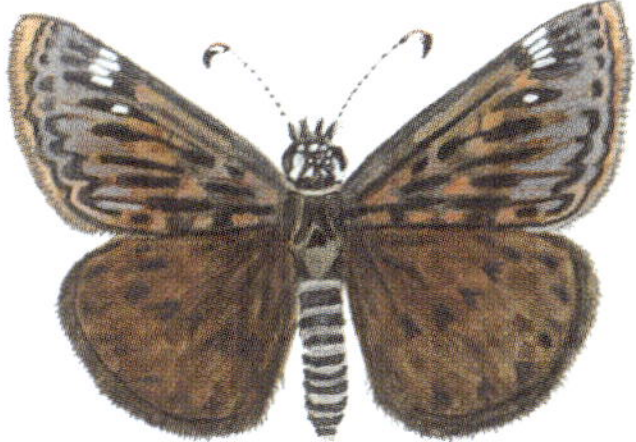

Pacuvius Duskywing (*Erynnis persius*)
male, dorsal

- prominent orange/brown spots above
- no extra spot on forewing disc cell like Persius
- gray body rings

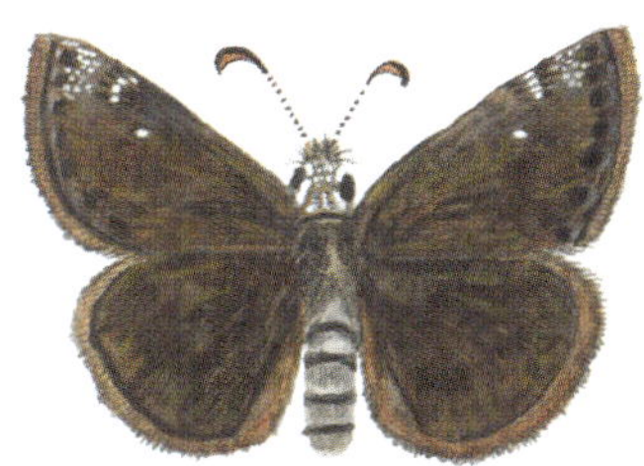

Pacuvius Duskywing (*Erynnis persius*)
both male and female, dorsal

- our darkest duskywing

TABLEAU 2

Dodge's vs. Tilden's Skipper

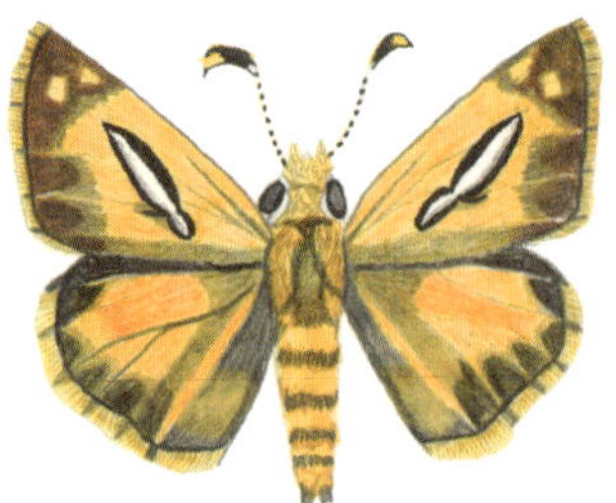

Dodge's Skipper (*Hesperia colorado dodge*)

- overall much darker than Tilden's
- can come both plain or incredibly patterned
- has long stigma with white felt within it
- more coastal (think Marin, Sonoma, and Mendocino)

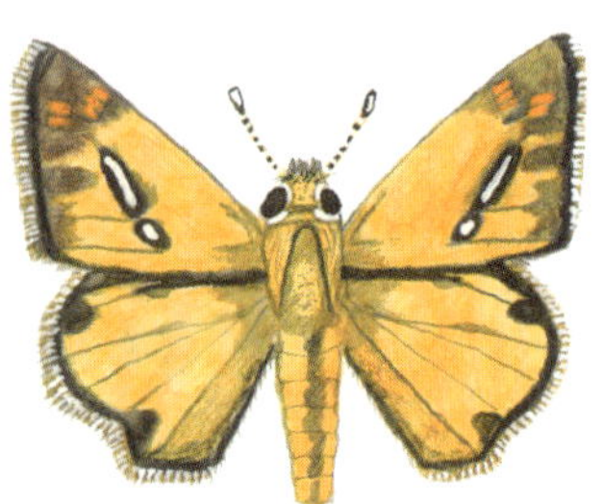

Tilden's Skipper (*Hesperia colorado tildeni*)

- overall much lighter than Dodge's
- can come both plain and incredibly patterned
- has long stigma with white felting
- more found in the interior (think Mount Hamilton Range/San Jose)

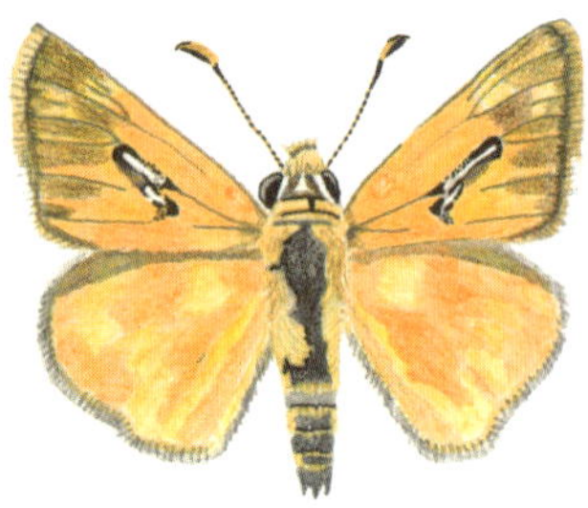

Both maybe confused with Lindsey's Skipper, *Hesperia lindseyi* (larger)

TABLEAU 3

Rural Skipper vs. Woodland Skipper Side by Side

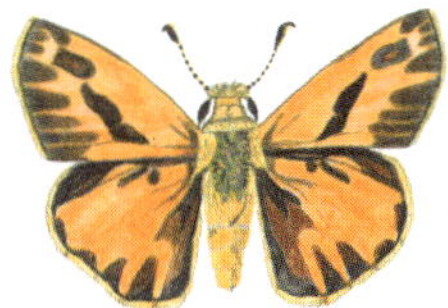

Woodland Skipper (*Ochlodes sylvanoides*)
male, dorsal

- prominent thin, black stigma
- heavily dark brown teething along margin

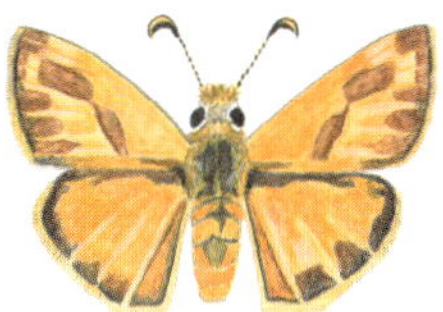

Woodland Skipper (*Ochlodes sylvanoides*)
female, dorsal

- brown patches instead of stigma
- light brown teething
- field light orange hind wing

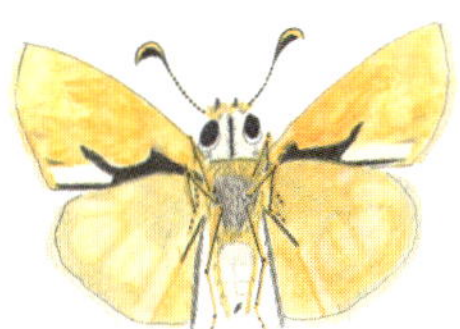

Woodland Skipper (*Ochlodes sylvanoides*)
male and female ventral, inland form

- just barely in view beige blocks

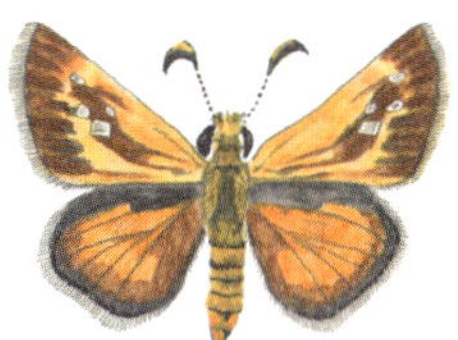

Woodland Skipper (*Ochlodes sylvanoides*)
male and female ventral, coastal form

- field forewing dark orange
- field hind wing purplish
- orange blocks on hind wing

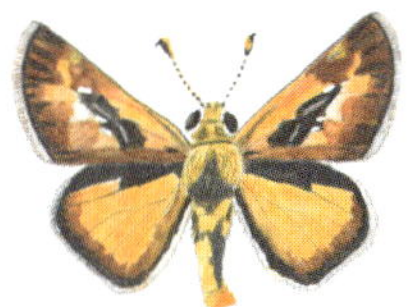

Rural Skipper (*Ochlodes agricola*)
male, dorsal

- uniformly orange on wings

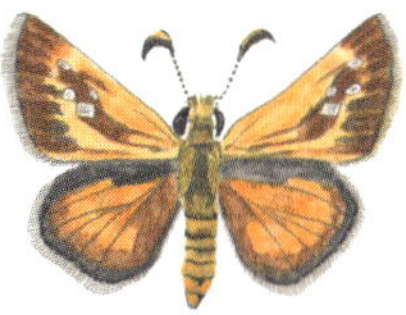

Rural Skipper (*Ochlodes agricola*)
female, dorsal

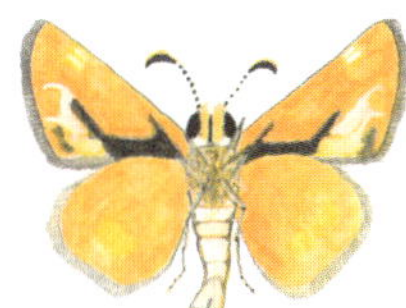

Rural Skipper (*Ochlodes agricola*)
male, ventral

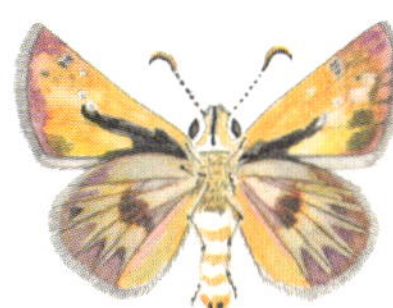

Rural Skipper (*Ochlodes agricola*)
female, ventral

- purplish on all four margin edges

TABLEAU 4

Differences between Our Two Most Common Yellow Swallowtails: The Anise and the Western Tiger in Flight

Let's say a yellow swallowtail shoots by you. Did you know that you can actually identify it without it ever needing to land? Get this little ditty into your head: "Black Shoulders vs. Black Stripes."

Anise Swallowtail (*Papilio zelicaon*)

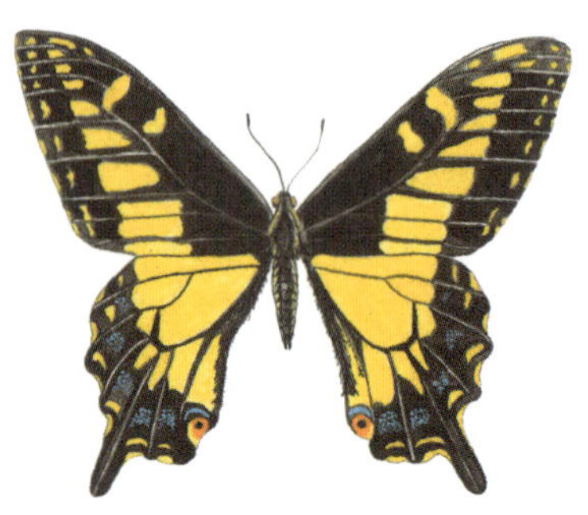

- black shoulders
- both have false heads on their tails

Western Tiger Swallowtail (*Papilio rutulus*)

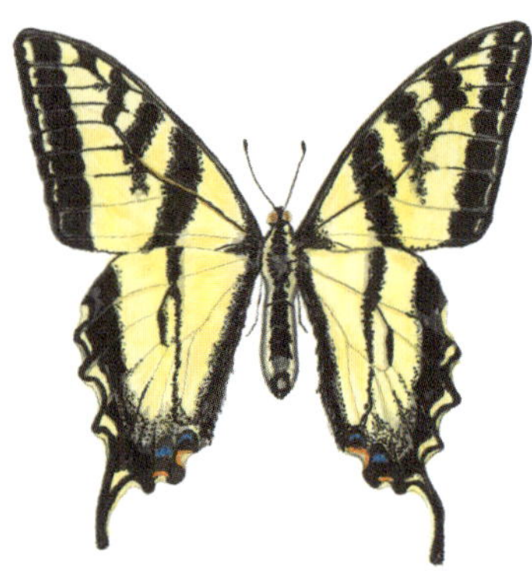

- black stripes
- both have false heads on their tails

Also, the Anise Swallowtail is always going to be the smaller of the two with a Western. Their flights are different as well. Anise Swallowtail may have many more rapid wing beats per minute than the Western. This enables one to really see the black shoulders on this creature. The Western Tiger is more like the Turkey Vulture with mainly gliding and occasional flapping.

More Advanced: Both the Two-tailed Tiger Swallowtail (*P. multicaudata)* and the female Pale Tiger (*P. eurymedon*) come in a muted yellow as well. These are lesser seen species than the Anise and the Western Tiger. The only two places I've known of all four flying together within the scope of this book is Mitchell Canyon, Mount Diablo State Park (Contra Costa County) and Mount Wanda at the John Muir Historical Home State Park in Martinez (Contra Costa County).

TABLEAU 5

Differences in Flight among the Cabbage White, the Large Marble, and the Orange Sulphur (Form Alba)

Congratulations! Your first few baby steps toward keying things on the wing. Did you know that you'll be able to identify a white butterfly without it ever having to land? It's all in the flight.

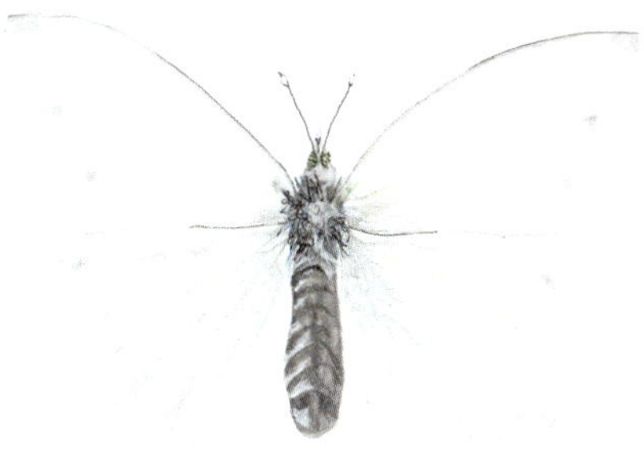

Cabbage White (*Pieris rapae*)

slow, ponderous flight that doesn't seem to have any purposeful direction

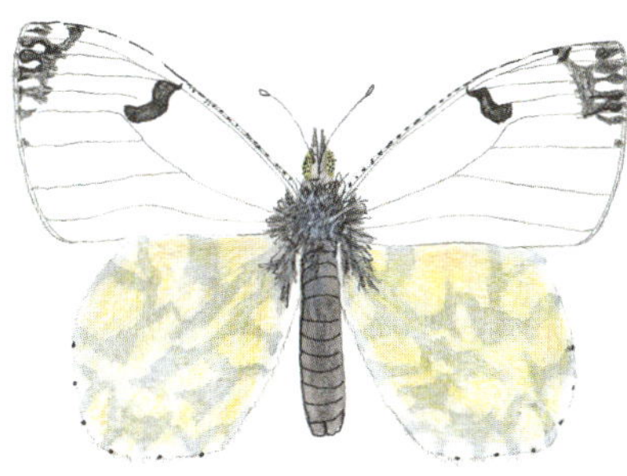

Large Marble (*Euchloe ausonides*)

quick, direct flight, as if it has something it is looking for (it does, some snacks and a mate)

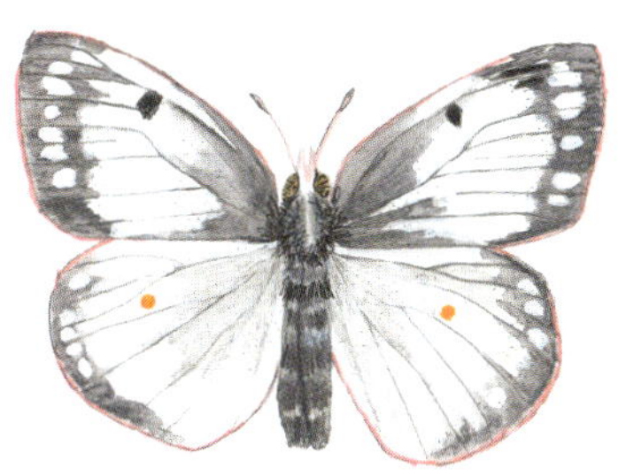

Orange Sulphur (*Colias eurytheme* f. alba)

white form female

sort of a combination of the two somewhat direct, but at times flight seems lazy; you should be able to see its black borders muted but visible in flight

TABLEAU 6

To Be Schooled in the Genus *Vanessa* and How to Sort the Ladies out in the Field

Take a breath. Perhaps one of the more exciting breakthroughs in your butterflying days is when you can look down at a Painted Lady (the group, not the species) and figure out which one of the three you're looking at. And when you have your first "quad" day—seeing all four in a walkabout. It took me about four years to master this, and it is quite doable.

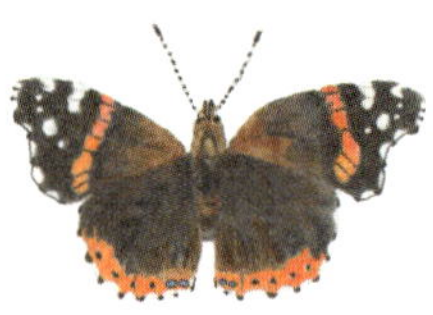

Red Admiral (*Vanessa atalanta*)

- this should be the easiest to key away from the others
- historically known to hybridize with West Coast Ladies

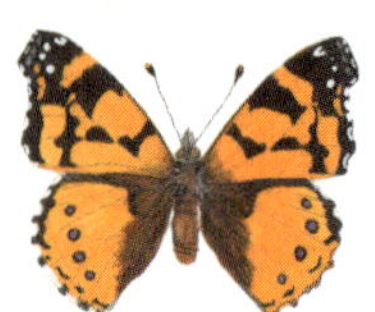

West Coast Painted Lady (*Vanessa annabella*)

- forewing tip extended and squared off (other two scalloped edge)
- coastal bar: orange (most orange in flight, smallest *Vanessa*)
- hind wing spots separate and blue/purple centers

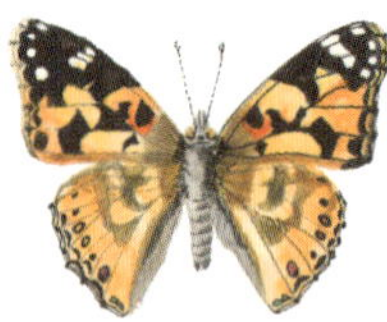

Painted Lady (*Vanessa cardui*)

- coastal bar: white
- hind wing: row of four, small black eyespots
- usually the largest *Vanessa* out there

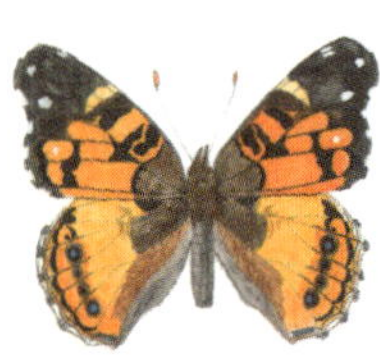

American Painted Lady (*Vanessa virginiensis*) female

- coastal bar: orange
- white dot on orange
- hind wing: blue eyespots heavily lined in black—all eyespots touch
- unique large eyespot ventrally
- only *Vanessa* being sexually dimorphic

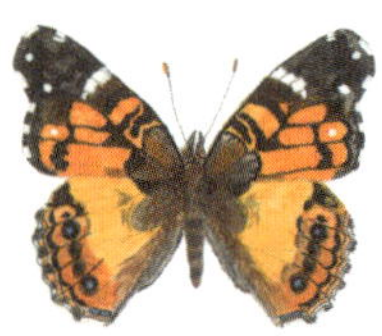

American Painted Lady (*Vanessa virginiensis*) male

- coastal bar: white
- forewing white dot on orange
- hind wing: blue eyespots heavily lined in black—all eyespots touch
- male somewhat smaller than female

TABLEAU 7

Telling Wood Nymphs Apart

The Common Wood Nymph is larger in overall size than the Great Basin Wood Nymph.

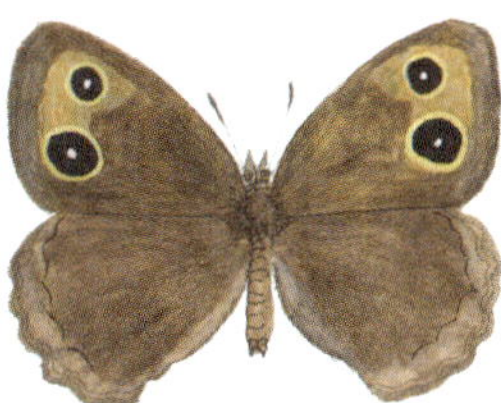

Common Wood Nymph
(*Cercyonis pegala boopis*)
male, dorsal

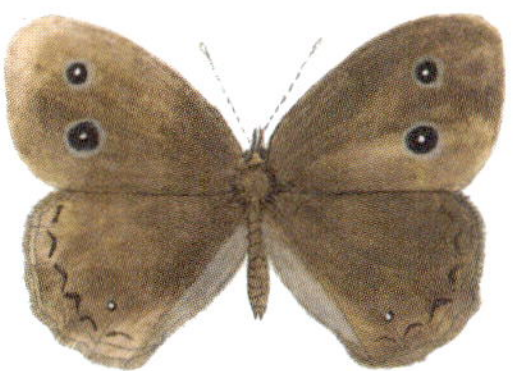

Common Wood Nymph
(*Cercyonis pegala boopis*)
male, ventral

second lower eyespot usually larger in the Common Wood Nymph than in the Great Basin Wood Nymph; can also be of same size

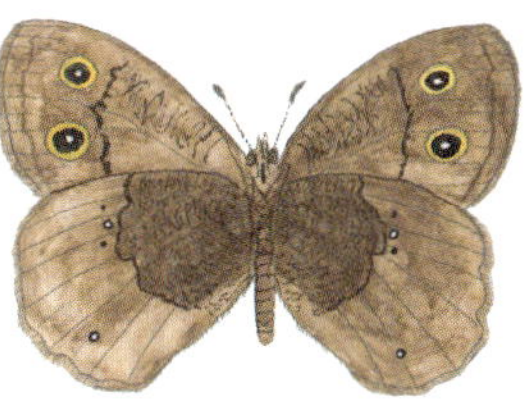

Common Wood Nymph
(*Cercyonis pegala boopis*)
female, ventral

The Great Basin Wood Nymph is smaller in overall size than the Common Wood Nymph.

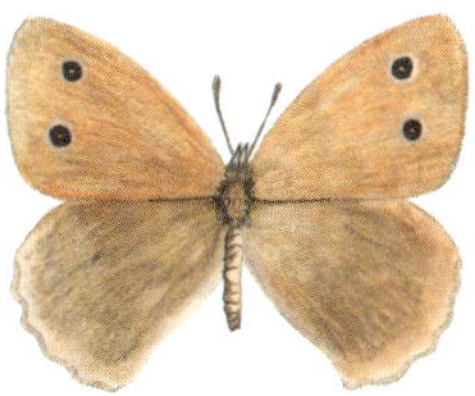

Great Basin Wood Nymph
(*Cercyonis sthenele*)
male, dorsal

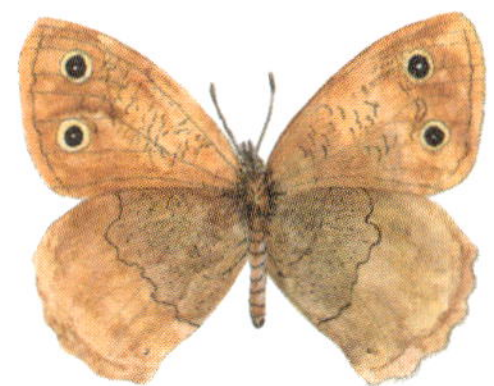

Great Basin Wood Nymph
(*Cercyonis sthenele*)
male, ventral

second lower eyespot usually smaller in Great Basin Wood Nymph than in Common Wood Nymph; can also be of same size

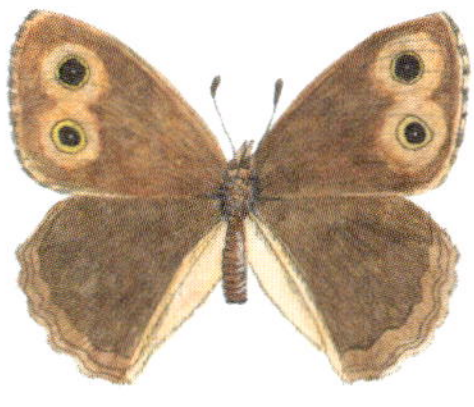

Great Basin Wood Nymph
(*Cercyonis sthenele*)
female, dorsal

I was taught to rely on the eyespot size to key these two apart, but that's more of an indicator of sexes within the species. I like to use time of year (Common Wood Nymph is on the wing first, and Great Basin is later in the summer) and overall size.

TABLEAU 8

Differences in the California, Sylvan, and Dryope Hairstreaks

All three with a field of muted gray below. It's the California and Sylvan Hairstreaks that look most alike—the Dryope should be relatively easy to key away from the others with the absence of a tail. They are in different habitats but easily could be nectaring together on a buckeye blossom. I usually see the two willow butterflies here just sitting on willow.

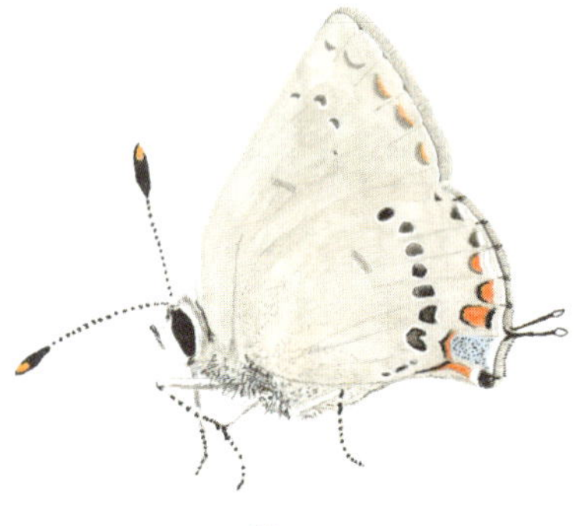

California Hairstreak (*Satyrium californica*)

- darker ground color (or field) than the other two
- black spots crisply delineated
- red and blue lunule spots are extremely defined
- creature of the oak forest (*Quercus* spp.)
- orange lunules extend up onto the forewing
- tailed

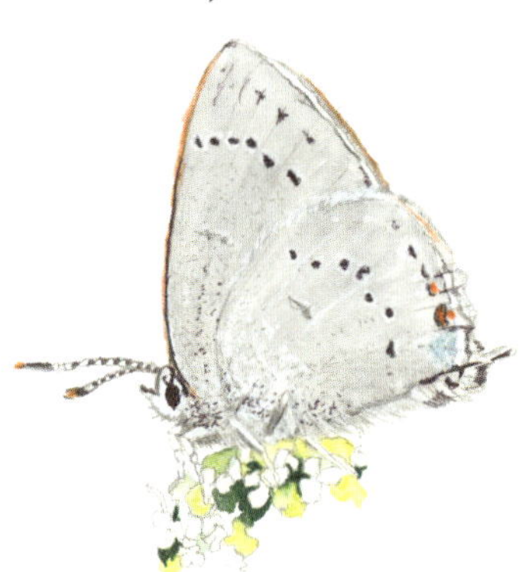

Sylvan Hairstreak (*Satyrium sylvinus*)

- chalky white/gray ground color, similar to Dryopes
- black spots are somewhat washed out
- red and blue lunule spots near the tail are less defined
- creature of the willows (*Salix* spp.)
- tailed

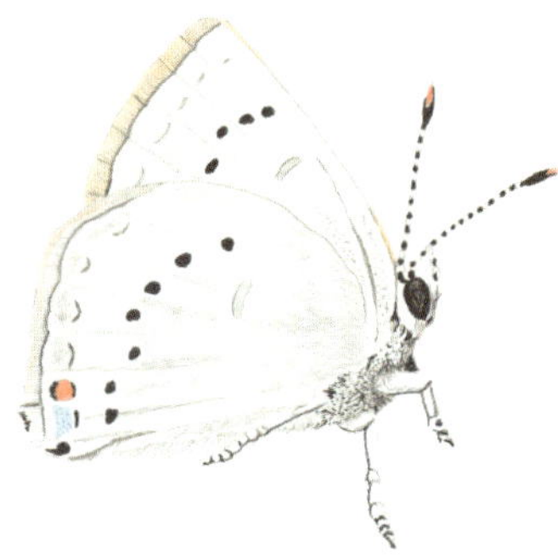

Dryope Hairstreak (*Satyrium dryope*)

- chalky white/gray ground color, similar to Sylvan
- black spots somewhat washed out
- red and blue lunules near where a tail would be are ill defined
- tailless

TABLEAU 9

Johnson's Hairstreak vs. Thicket Hairstreak

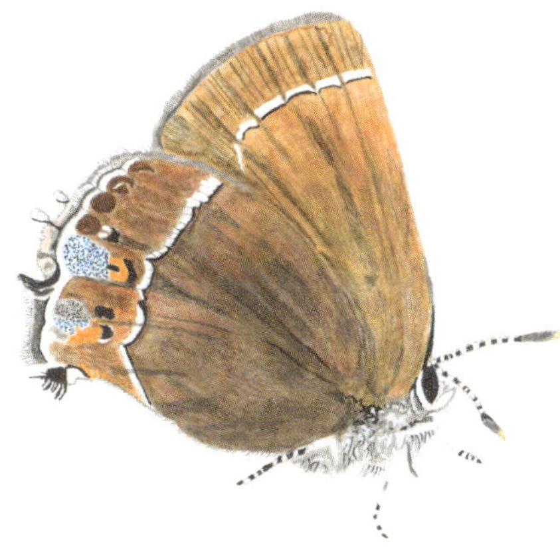

Johnson's Hairstreak (*Callophrys johnsoni*)

- chocolate brown ground (or field) ventrally
- white postmedian line
- less zigzag, more wavy
- small orange and blue spots reduced near the tail
- Johnson's is NOT teal blue above like Thicket
- both have two tails with the top one being smaller than the bottom one

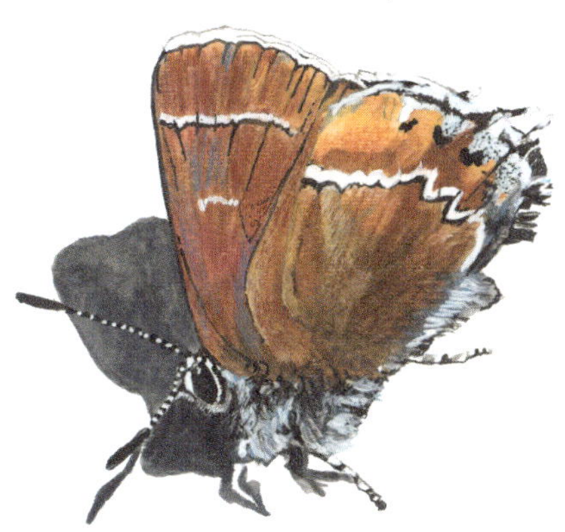

Thicket Hairstreak (*Callophrys spinetorum*)

- chocolate/mahogany brown field ventrally
- white postmedian line quite zigzag
- a white "W" is formed near the tails
- three small, vivid blue spots near the tail
- dorsal teal color is apparent in flight
- both have two tails with the top one being smaller than the bottom one

TABLEAU 10

Lotus Hairstreak vs. Coastal Green Hairstreak

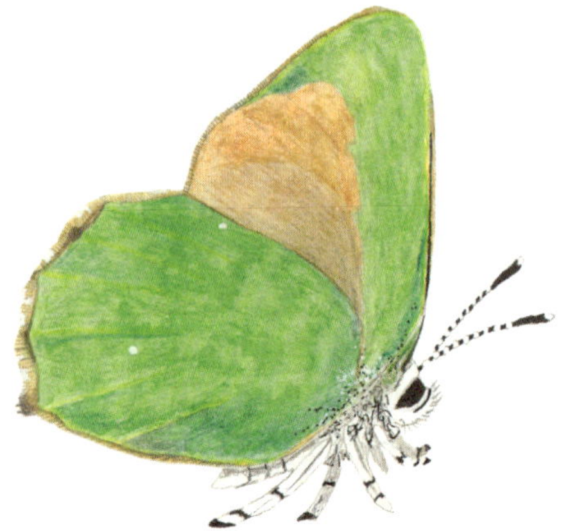

Lotus Hairstreak (*Callophrys dumetorum*)
ventrally (from below)

- brown notch on lower forewing
- few white spots on postmedian line
- no white fringe on wing borders

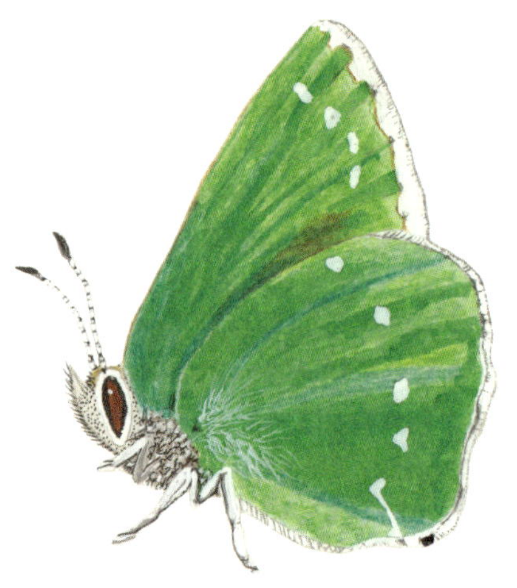

Coastal Green Hairstreak (*Callophrys viridis*)

- no brown notch on lower forewing
- many white spots along curved, postmedian line
- white fringe along both wing margins/borders

TABLEAU 11

Western Tailed Blue vs. Eastern Tailed Blue

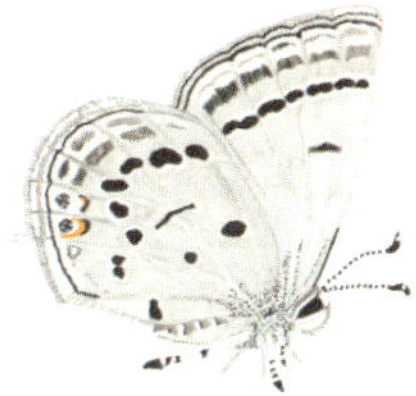

Western Tailed Blue (*Everes amyntula*)
ventral

- size of a nickel
- forewing border more straight
- black dots almost non-existent
- orange lunules spares and thin

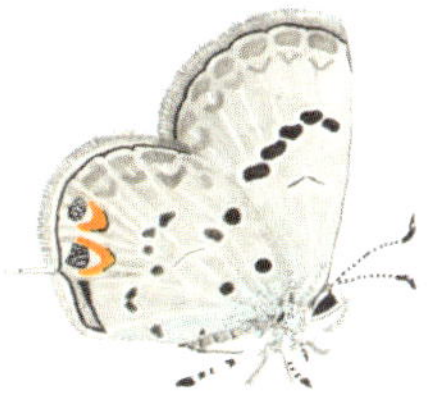

Eastern Tailed Blue (*Everes comyntus*)
ventral

- size of a dime
- forewing border more rounded
- black dots quite profound
- orange lunules substantial

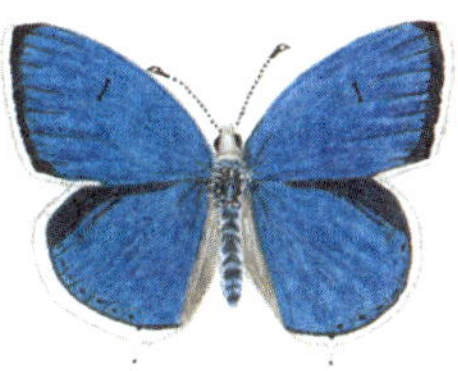

Western Tailed Blue (*Everes amyntula*)
male, dorsal

- from above, no orange dots present near tail

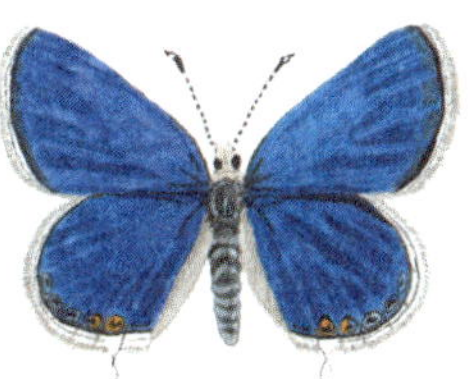

Eastern Tailed Blue (*Everes comyntus*)
male, dorsal

- from above, orange dots present near tail

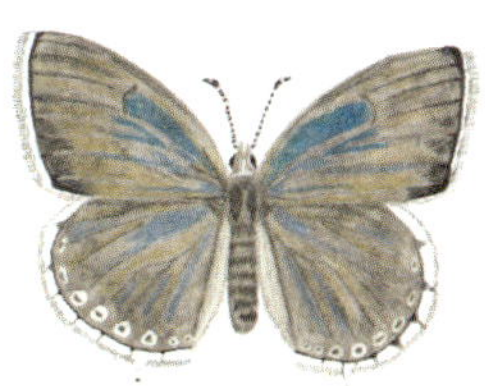

Western Tailed Blue (*Everes amyntula*)
female, dorsal

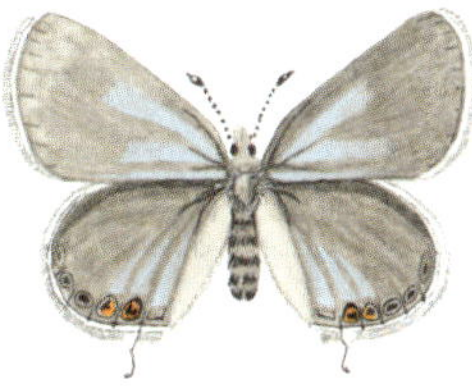

Eastern Tailed Blue (*Everes comyntus*)
female, dorsal

WARNING: Easy for novices to get tripped up on how much the **Summer Form** female Eastern Tailed Blue looks like another Lycaenid—the Gray Hairstreak. Gray Hairstreaks are larger and have orange-tipped antennae; the Tailed has black tips.

TABLEAU 12

The Almost Indistinguishable, Barely Discernible Differences between Acmon Blue and Clemence's Blue

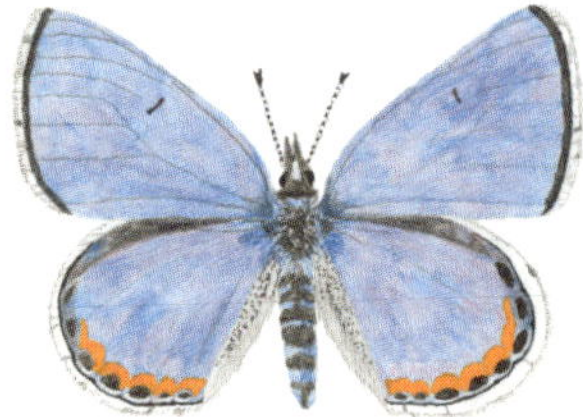

Acmon Blue (*Icaricia acmon*)

- smaller
- thinner orange lunule/markings
- the orange has a pink hue to its aurora
- females come in a dark blue form
- dorsal black borders thinner
- ventral spots defined poorly
- defined black line above aurora
- spot reduced in forewing discal cell

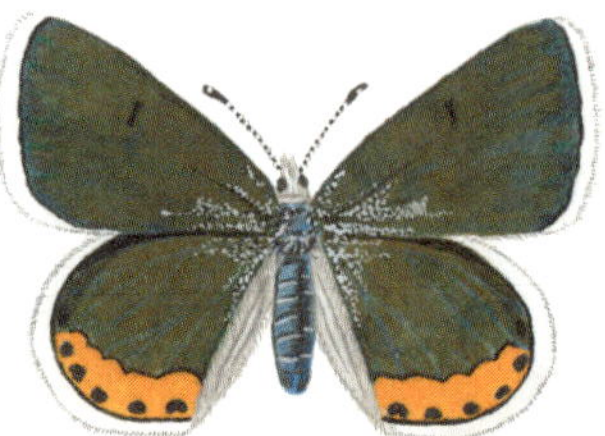

Clemence's Blue (*Icaricia monticola*)

- larger
- brighter orange lunules/markings
- vivid orange aurora
- females come in a dark blue/green form
- dorsal black borders are thicker
- ventral spots defined boldly
- ill-defined back line above aurora
- defined black spot in forewing disc cell

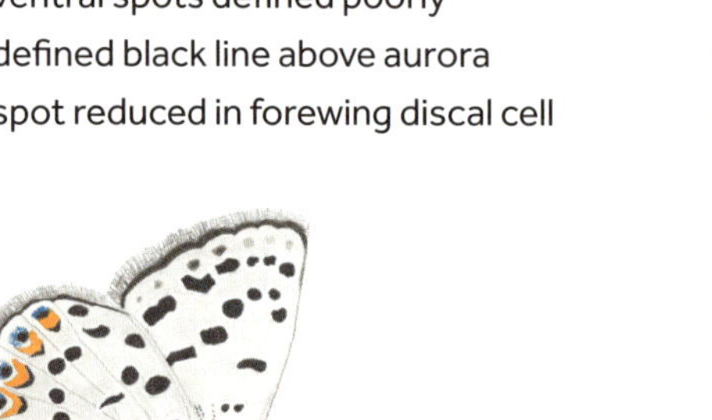

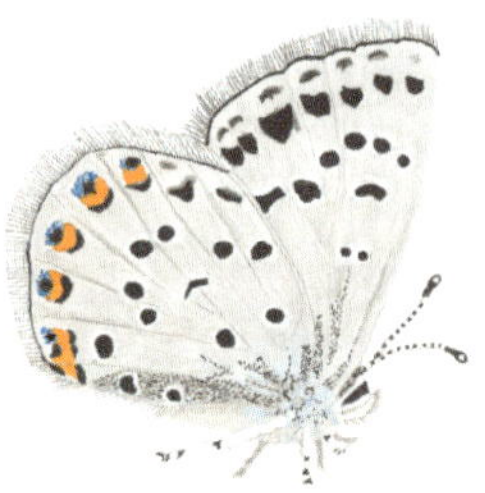

"It is not currently known how to distinguish female Clemence's Blues (Lupine Blues) from female Acmon Blues" —Jeffrey Glassberg

"Adults of Plebejus (Icaricia) lupini (monticola) and Plebejus (Icaricia) acmon have long been confused in Cascadia."
—James and Nunnallee (105)

"Have fun!" —Shapiro (106)

Organizations

The Lepidopterists' Society
lepsoc.org
The Lepidopterists' Society
c/o Chris Grinter
California Academy of Sciences
55 Music Concourse Drive
San Francisco, CA 94118

Based in the academic, but utterly embracing the amateur, this scientifically based group does not discourage collecting, and wielding a net is not looked down upon.

Monarch Watch
monarchwatch.org
Monarch Watch
University of Kansas
2021 Constant Avenue
Lawrence, KS 66047

All things Monarch.

North American Butterfly Association (NABA)
naba.org
North American Butterfly Association
4 Delaware Road
Morristown, NJ 07960

An international group for butterfly enthusiasts, with a focus on photography, gardening, conservation, and viewing. They run all the annual butterfly counts. If you join them, you'll receive the fantastic quarterly American Butterflies *magazine. This group has a strong stance against collecting or wielding a net.*

The Xerces Society for Invertebrate Conservation
xerces.org
The Xerces Society
1631 NE Broadway Street #821
Portland, OR 97232

The Xerces Society was created by Robert Michael Pyle. If you are interested in helping the Monarchs as they pass through your area on their annual migration, this is a group for you as well.

Glossary

aberration—a non-normal organism, created either genetically or environmentally
aedeagus—the male sex organ used for passing spermatophore packets into the female
allopatric—not occurring together; species flights do not overlap
apex—the tip of the forewing
aurora—the orange crescent-like lunules on the upper sides and undersides of the hind wings
basal—on the wing near its connection to the body
Batesian mimicry—an edible creature mimicking a poisonous one (see **Mullerian mimicry**)
bilateral—split down the middle
binomial system—developed by Linnaeus in the 1600s, the categorization of all living creatures. Genus + species + subspecies.
bivoltine—two flights or two generations per season
border—the edge of the wing
brindled—a coloration of patterned markings consisting of three colors
brood—a generation of butterflies all hatched from a previous season all flying during the same time
camouflage—an animal's natural coloration that enables it to blend in with its surroundings
caterpillar—the larva of a butterfly
chaparral—low, thick shrubs combined with low trees, common in California
chimera—a hybrid creature made up from different animal parts
clubbed antennae—antennae thickened at the end
coastal scrub—ecosystem that shows a preference for mild, maritime weather. Made up from a solid community of waist-high shrubs.
colony—a distinct group of butterflies defined by a boundary
conifer forest—an ecosystem made up of primarily cone-bearing trees
costa—the upper edge of both the forewing and the hind wing (*adj.* **costal**)
costal fold—in some butterflies, contains scent glands found along the top margin of the forewing. Almost all Duskywing skippers have this organ.
dimorphism—within a species, two distinct forms, i.e., a male and a female
discal cell—the area from the base to the middle of the wing
diurnal—active during the day
dorsal, -um—the upper side of the butterfly
dorsal basking—wings down and perpendicular to body; one is viewing the topside of the butterfly this way
eclose—to hatch
endemic—limited to a specific area
estivation—dormancy that a butterfly enters into during dry season
evergreen forest—a forest that retains its leaves or needles year-round
extinct—having no surviving individuals
extirpated—removed from a particular area
falcate—the tip of the forewing (the apex) coming to a curved hook
felt—interior of a stigma on males, mainly skippers
field—the overall background or area between all the markings, spots, and false eyes on a butterfly wing, aka the ground color

flight—a single generation from egg to adult, or a span of time in which they appear
foreleg—the front legs
forewing—the set of top two wings
form—a subgroup of species distinguished by looking different structurally
fulvous—tawny brown or dirty orange
genitalia—the sexual organs of a creature
genus—a basic grouping of formal taxonomy; a group of species more closely related to one other than to other species. The genus is always capitalized and italicized in the binomial system.
glade—an opening in a forest where sunlight penetrates the shade
grasslands—an ecosystem made up primarily of grasses, sedges, rushes, and clovers, all of which are very low growing
ground color—see **field**
gynandromorph—an individual displaying both female and male characteristics
head—the top of the butterfly, where eyes, proboscis, palpi, and antennae are located
hibernate—to go dormant during the cold of a season or overwinter; to diapause
hill-topping—a behavior found late in the day, during which butterflies congregate at a geographically high prominence in pursuit of the opposite sex
hind leg—the back legs
hind wing—the lower set of wings
horizontal corridor—a continuous ecosystem along the horizon; where butterflies are concerned, this usually means a canopy of trees along a line or street
host plant—the plant or the family of plants the female lays her eggs on
hyaline—glassy, see-through spots usually on the hind wing
imago, -oes, -ines—the adult, winged staged of a butterfly
instar—a stage of larval development between molts
lancelet—a single leaf of lupine
larva, -ae—the second phase of a butterfly's development, a lone caterpillar
lateral basking—wings are up over the back of the butterfly; one is viewing the underside of the butterfly this way
Lepidoptera—the order of insects containing moths and butterflies
lunule—a crescent or half-moon shape
margin—the top edge of the forewing from thorax to apex
metamorphosis—the development of a butterfly through many stages
mid leg—one of the set of middle legs
molts—the stages of each development of a larva or caterpillar
mud-puddling—taking nutrients from a moist area of bare soil
Mullerian mimicry—a poisonous creature mimicking another poisonous creature (see **Batesian mimicry**)
multivoltine—many flights or generations per season
nocturnal—active at night
nominate—the subspecies the entire species springs from, e.g. *Apodemia mormo mormo*
oak riparian—a stream course dominantly vegetated by oaks
ocellus, -i—an eyespot on the wing or larva
osmeterium—a fork-shaped organ that comes out of the head and emits a foul odor and fends off predators
ova, -um—the first stage of the butterfly's development, the egg(s)
oviposit—to lay eggs
palp, -us, -i—an appendage on the head used for smelling and cleaning
phenotype—the physical traits of a specific species
postmedian—just beyond the center of a wing, going toward the outer border
proboscis—the feeding tube of the adult butterfly
protandry—when males emerge before females, guaranteeing procreation for the next generation
pupa, -ae—the third phase of a butterfly's development after the larval stage and before metamorphosis; a hardened enclosed capsule
relict—a piece of land pushed by tectonic plants from its original location
riparian woodland—a habitat made up of willow, cottonwood, and other trees and bushes, usually along a creek
scintilla—the metallic silver or turquoise marking on the aurora
serpentine—an ecosystem made up of nutrient-poor soil due to the presence of the serpentine rock. Plant communities and creatures are highly endemic to this community. It can relate to a range of habitats, from a rocky, barren landscape to a seepy bog.
species—an animal or plant that is able to interbreed and create an offspring

spermatophore—the package of sperm transferred from the male to the female during copulation

sphragis—a hard substance left on the female by the male to ensure no remating with others occurs. The genera *Parnassian* and *Euphydryas* leave these "chastity belts."

stigma—a pheromone gland on the wings of male butterflies

subspecies—a subgroup of a species that occurs in a distinct range

sympatric—occurring together; species overlapping in flight

tail—an extended part of the hind wing that goes beyond the border

tarsus—the foot section of a butterfly leg

taxon—a group of species taxonomists decided to put together

thorax—the middle section of the body where the legs attach

tibia—the secondary section of a leg on a butterfly

toothed or **teething**—a pattern of sharp, toothlike indentations usually along the forewing of a skipper

tornus—the bottom corner of the wing

type locality—where the type specimen was from

type specimen—the original specimen that described a taxon

univoltine—one flight or generation per season

vagrant—a species not normally found in a particular area, a stray

ventral, -um—the underside of the butterfly

voltinism—the number of flights or generations or broods

weedy—an environment dominated by exotic vegetation

willow riparian—a stream course dominantly vegetated by willows

xeric—a dry, desert-like environment; arid

Bibliography

Behr, H. H.—1875, letter to Herman Strecker

Beidleman, Linda H., and Eugene N. Kozloff. *Plants of the San Francisco Bay Region: Mendocino to Monterey*. 2nd ed. Berkeley: University of California Press, 2003.

Bindman, Ariana. "Hundreds of Endangered Butterflies Slated to Be Released Along California Coast." *SFGATE*, January 17, 2024. https://www.sfgate.com/bayarea/article/butterfly-release-california-mendocino-coastline-18611671.php.

Broom, Alexander. "Is Antioch's Lange's Metalmark Butterfly Extinct?" *Save Mount Diablo* (blog), October 17, 2023. https://savemountdiablo.org/blog/is-antiochs-langes-metalmark-butterfly-extinct/.

Carter, David. *Butterflies and Moths*. Eyewitness Handbooks. London: Dorling Kindersley, 1992.

Coast Ridge Ecology. "Distribution Study for Lilian's Fritillary (*Speyeria calippe lilianna*)." Prepared for the U.S. Fish and Wildlife Service, 2009.

Coast Ridge Ecology. Speyeria Butterfly and Host Plant Surveys at Sears Point, CA, 2021.

Crabtree, Laurence L. *Discovering the Butterflies of Lassen Volcanic National Park*. Chester, CA: Hilltopping Publications, 1998.

Dornfeld, Ernst J. *The Butterflies of Oregon*. Forest Grove, OR: Timber Press, 1980.

Dupré, Bernard. *World Treasury of Butterflies in Color*. New York: Galahad Books, 1974.

Emmel, Thomas C., and John F. Emmel. *The Butterflies of Southern California*. Anderson, Ritchie & Simon. Los Angeles: Natural History Museum of Los Angeles County, 1973.

Ehrlich, Paul R., and Ilkka Hanski, eds. *On the Wings of Checkerspots: A Model System for Population Biology*. New York: Oxford University Press, 2004.

Feltwell, John. *The Encyclopedia of Butterflies*. London: Quarto Publishing, 1993.

Garth, John S., and J. W. Tilden. *California Butterflies*. Berkeley: University of California Press, 1986.

Glassberg, Jeffery. "There's No Need to Release Butterflies—They're Already Free." *American Butterflies*. 1999.

Glassberg, Jeffery. *A Swift Guide to Butterflies of North America*. Morristown, NJ: Sunstreak Books, 2012.

Holland, W. J. *The Butterfly Book*. New York: Doubleday & McClure, 1898.

Howe, William H. *The Butterflies of North America*. Garden City, NY: Doubleday, 1975.

James, David G., and David Nunnallee. *Life Histories of Cascadia Butterflies*. Corvallis: Oregon State University Press, 2003.

Kaufman, Kenn, and Jim P. Brock, *Field Guide to Butterflies of North America*. Boston: Houghton Mifflin, 2003.

Klots, Alexander B. *A Field Guide to the Butterflies of North America, East of the Great Plains*. Boston: Houghton Mifflin, 1951.

Kraybill-Voth, Kaitlin. "The Tiniest Landscapes Make the Brightest Colors." *Bay Nature*, Spring 2019.

Markoff, John. "The Butterfly Effect." *Alta Journal*, April 19, 2021.

Miller, Jeffrey C., and Paul C. Hammond. *Lepidoptera of the Pacific Northwest: Caterpillars and Adults*. Morgantown, WV: U.S. Department of Agriculture, Forest Service Division, 2003.

Monroe, Lynn, and Gene Monroe. *Butterflies and Their Favorite Flowering Plants: Anza-Borrego State Park & Environs*. Lyons, CO: Merry Leaf Press, 2004.

Nabokov, Vladimir. *Speak, Memory: An Autobiography Revisited*. New York: Vintage International, 1989.

O'Brien, Liam, and Matthew Zlatunich. *Butterflies of the Presidio*. San Francisco: Presidio Trust, 2012.

Opler, Paul A. and Amy Bartlett Wright. *Peterson Field Guides Western Butterflies*. Boston: Houghton Mifflin, 1999.

Powell, Jerry A., and Paul A. Opler. *Moths of Western North America*. Berkeley: University of California Press, 2009.

Pyle, Robert Michael. *Handbook for Butterfly Watchers*, Boston: Houghton Mifflin, 1984.

Pyle, Robert Michael. *Chasing Monarchs: Migrating with the Butterflies of Passage*. New York: Houghton Mifflin, 1999.

Pyle, Robert Michael. "Resurrection Ecology: Bring Back the Xerces Blue." *Wings: Essays on Invertebrate Conservation*, Fall 2001.

Pyle, Robert Michael. *The Butterflies of Cascadia: A Field Guide to All the Species of Washington, Oregon, and Surrounding Territories*. Seattle: Seattle Audubon Society, 2002.

Pyle, Robert Michael. *The Audubon Society Field Guide to North American Butterflies*. New York: Chanticleer Press, 1981.

Scott, James A. *The Butterflies of North America: A Natural History and Field Guide*. Palo Alto, CA: Stanford University Press, 1986.

Shapiro, Arthur M., and Timothy D. Manolis. *Field Guide to Butterflies of the San Francisco Bay and Sacramento Valley Regions*. Berkeley: University of California Press, 2007.

Sharnoff, Stephen. *A Field Guide to California Lichens*. New Haven, CT: Yale University Press, 2014.

Smart, Paul. *The Illustrated Encyclopedia of the Butterfly World*. London: Salamander Books, 1975.

Steiner, John. "Bay Area Butterflies: The Distribution and Natural History of San Francisco Rhopalocera." Master's thesis, Cal State Hayward, 1980.

Williams, Francis X. "The Butterflies of San Francisco." *Entomological News*. San Francisco, CA, 1910.

Xerces Society. *The Aurelian*, Seattle, WA, 1985.

Index of Butterflies

Index of Host Plants

V

W

Y

About the Author

Liam O'Brien is a self-taught lepidopterist and illustrator. He used to be a professional actor, having appeared in *Les Misérables* on Broadway, but shifted his powers of observation toward nature several decades back. He's fascinated not only by butterflies but also by our relationship to them. He surveyed the county of San Francisco, where he lives, for which butterfly species remained in 2007 and 2009. He is the creator of the Green Hairstreak Project for the organization Nature in the City, and he led efforts to restore Variable Checkerspots to the Presidio. Since 2015 he has helped monitor the endangered Mission Blue butterfly in the Marin Headlands. O'Brien was the recipient of *Bay Nature* magazine's Local Hero Award for Environmental Education in 2014.

A Note on Type

This book is set in the typefaces Effra and Playfair. Effra, the sans-serif typeface used for the main text, was designed by Fabio Luiz Haag and Jonas Schudel and was released through Dalton Maag in 2008. Inspired by Caslon Junior, one of the early sans-serif typefaces, Effra features a clean, open design that provides a modern but welcoming feel. Playfair, the serif typeface used for titles, was designed by Claus Eggers Sørenson and draws on the work of the eighteenth-century printer and type designer John Baskerville.

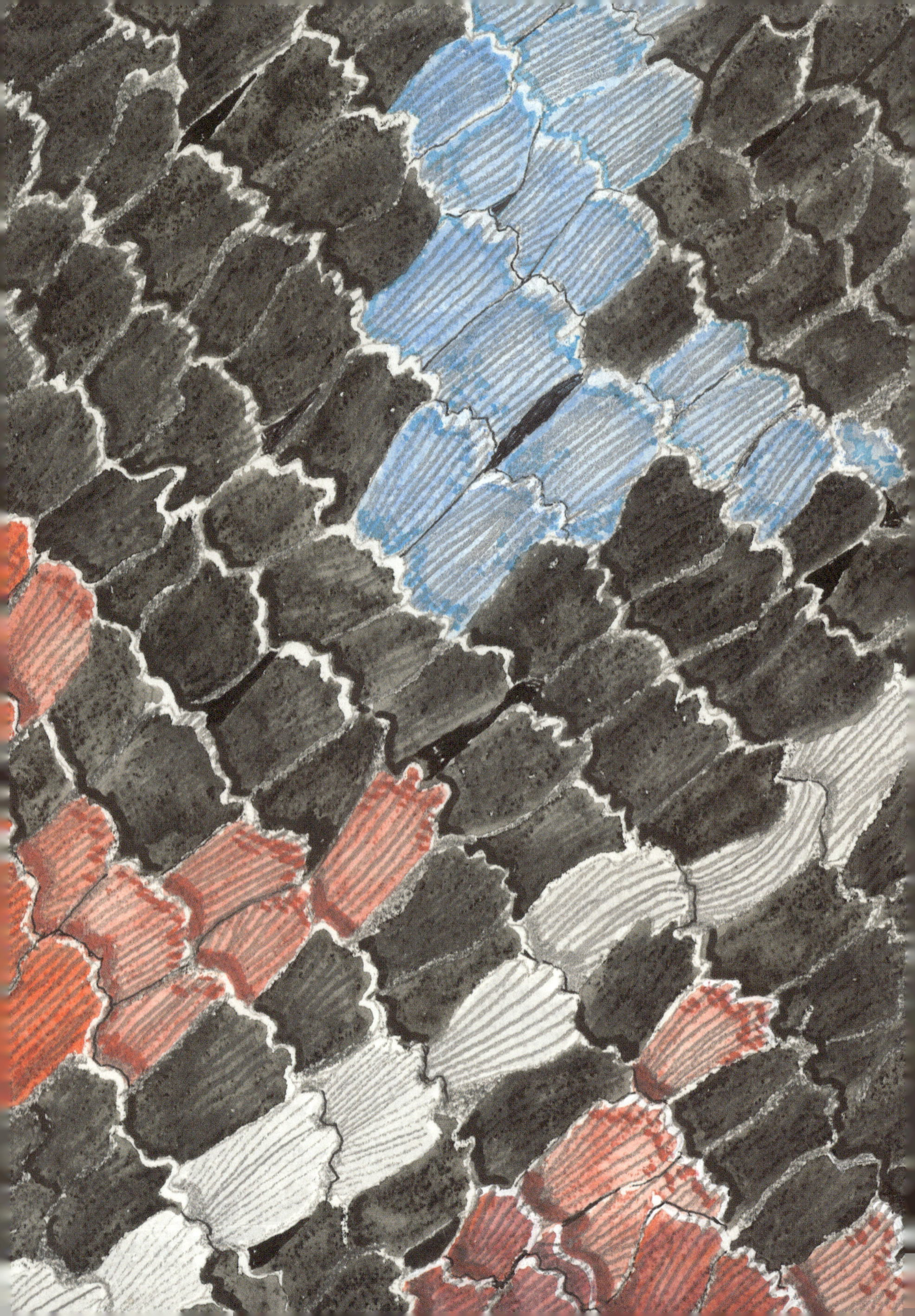